JOHN GLOAG ON INDUSTRIAL DESIGN

Volume 7

A HISTORY OF CAST IRON IN ARCHITECTURE

A HISTORY OF CAST IRON IN ARCHITECTURE

JOHN GLOAG
and
DEREK BRIDGWATER

LONDON AND NEW YORK

First published in 1948 by George Allen & Unwin Ltd.

This edition first published in 2023
by Routledge
4 Park Square, Milton Park, Abingdon, Oxon OX14 4RN

and by Routledge
605 Third Avenue, New York, NY 10158

Routledge is an imprint of the Taylor & Francis Group, an informa business

© 1948 John Gloag and Derek Bridgwater

All rights reserved. No part of this book may be reprinted or reproduced or utilised in any form or by any electronic, mechanical, or other means, now known or hereafter invented, including photocopying and recording, or in any information storage or retrieval system, without permission in writing from the publishers.

Trademark notice: Product or corporate names may be trademarks or registered trademarks, and are used only for identification and explanation without intent to infringe.

British Library Cataloguing in Publication Data
A catalogue record for this book is available from the British Library

ISBN: 978-1-032-36309-7 (Set)
ISBN: 978-1-032-36656-2 (Volume 7) (hbk)
ISBN: 978-1-032-36719-4 (Volume 7) (pbk)
ISBN: 978-1-003-33347-0 (Volume 7) (ebk)

DOI: 10.1201/9781003333470

Publisher's Note
The publisher has gone to great lengths to ensure the quality of this reprint but points out that some imperfections in the original copies may be apparent.

Disclaimer
The publisher has made every effort to trace copyright holders and would welcome correspondence from those they have been unable to trace.

PLATE I. Interior view of the Crystal Palace, the first large scale prefabricated building composed of units of cast iron and glass. The cast iron gates now divide Hyde Park and Kensington Gardens.

From *Dickinsons Comprehensive Pictures of the Great Exhibition of 1851*, painted by Messrs. Nash, Haghe and Roberts. Published by Dickinson Bros., 114 New Bond Street, 1854.

A HISTORY OF
CAST IRON IN ARCHITECTURE

JOHN GLOAG

AND

DEREK BRIDGWATER
B.ARCH., F.R.I.B.A

WITH A FOREWORD BY

PROFESSOR

SIR CHARLES REILLY
O.B.E., LL.D., F.R.I.B.A

LONDON

GEORGE ALLEN AND UNWIN LTD

First Published 1948

All rights reserved

PRINTED IN GREAT BRITAIN
in 12 point Baskerville type
By W. S. COWELL LTD, LONDON AND IPSWICH

DEDICATED TO THE MEMORY
OF
William Richard Lethaby
who, more than any other writer and architect
of his time, appreciated the
influence of industrial
materials upon architectural
design

CONTENTS

	PAGE
FOREWORD by Sir Charles Reilly	XIX
INTRODUCTION	1

SECTION ONE
The Cast Iron Industry Becomes Established:
1650 - 1750 37

SECTION TWO
The New Material in Architecture: the Rise of Cast Iron:
1750 - 1820 53

SECTION THREE
The Industrial Expansion of the Cast Iron Industry:
1820 - 1860 159

SECTION FOUR
Changes in the Industry following the Development of New Uses:
1860 - 1900 239

SECTION FIVE
Twentieth Century Developments:
1900 - 1945 313

APPENDIX I	375
INDEX	379

COLOURED PLATES

PLATE
I. *Interior of the Crystal Palace* — *frontispiece*

FACING PAGE

II. *Exterior view of the Crystal Palace in Hyde Park, 1851* — 160

III. *Interior view of the Crystal Palace showing cast iron columns* — 192

IV. *Hardware exhibits at the Great Exhibition of 1851* — 224

V. *Enamel and paint finishes for cast iron* — 320

VI. *Metal plating and spraying finishes, and "Parkerised" finishes for cast iron* — 336

LIST OF ILLUSTRATIONS IN THE TEXT

FIGURE		PAGE
1	*Cast iron lion, Chinese, sixth century, A.D.*	3
2	*Small standing figure of Kwan Yin, Chinese, sixth century, A.D.*	4
3	*Standing figure of Kwan Yin, Chinese, sixth century, A.D.*	4
4	*The great lion of Ts'anchou, Chinese, tenth century, A.D.*	5
5	*The Catalan Forge*	7
6	*Primitive hearth and bellows*	7
7	*The Osmund Furnace*	7
8	*The Joan Colins grave slab, Burwash Church, Sussex*	10
9	*The Anne Forster grave slab, Crowhurst Church, Surrey*	11
10	*Grave slab in Wadhurst Church, Sussex*	12
11	*Cast iron grave slabs in Wadhurst Church, Sussex*	13
12	*Elaborate example of a grave slab in Wadhurst Church, Sussex*	14
13 to 16	*Fifteenth century Sussex firebacks*	15, 16, 17 and 19
17 and 18	*Sixteenth century Sussex firebacks*	20 and 21
19	*Fireback with Royal Arms, sixteenth century*	22
20	*Fireback with Arms of Philip II of Spain*	23
21	*Part of a fireback with the Royal Arms and Supporters of Scotland*	24
22	*Heraldic fireback made in Sussex by John Harvo*	24
23	*Flemish stove plate in cast iron, sixteenth century*	25
24	*German stove plate, late sixteenth century*	25
25	*Flemish stove plate, seventeenth century*	26
26	*German stove plate, seventeenth century*	26
27	*Fireback with Royal Arms as used by the Stuart Sovereigns, seventeenth century*	27
28	*Seventeenth century heraldic fireback*	27
29	*Heraldic fireback, Sussex, seventeenth century*	28
30	*Heraldic fireback, Kent, late seventeenth century*	28
31	*Sussex fireback or stove plate, seventeenth century*	29
32	*The Lenard fireback, Sussex, seventeenth century*	30
33	*Firedog, probably cast to match the Lenard fireback*	30
34	*English firedogs, sixteenth and seventeenth centuries*	31

FIGURE		PAGE
35	*Eighteenth century fireback*	32
36	*Fireback with Arms of the Dauphin of France, eighteenth century*	32
37	*Nineteenth century Welsh fireback, with Arms and Supporters of Conroy of Llanbrynmair*	33
38	*Inscribed beams in a disused blast furnace at Coalbrookdale*	39
39	*Early water wheel at Coalbrookdale*	40
40	*Interior of old moulding shop at Coalbrookdale*	41
41	*View of the Coalbrookdale works in the sixteenth century*	45
42 and 43	*Early eighteenth century cast iron railings, designed by James Gibbs for the Senate House, Cambridge*	46 and 47
44	*The gates to Chirk Castle, Llangollen, eighteenth century*	49
45	*Thomas Farnolls Pritchard of Shrewsbury*	55
46	*Thomas Telford*	57
47 and 48	*Monument at Lindale, Lancashire, to the memory of John Wilkinson*	61
49	*Portrait plaque of John Wilkinson*	61
50	*Casting cannon balls*	62
51	*Monogram of George III*	65
52	*Part of an early cast iron steam cylinder made for James Watt*	67
53	*Cast iron stove from Compton Place, Eastbourne*	68
54	*German box stove in cast iron, late sixteenth century*	69
55	*Chamber stove designed by the brothers Haworth*	70
56	*Double Canada stove designed by the brothers Haworth*	70
57	*Back boiler range designed by the brothers Haworth*	71
58	*Portraits of George III and Queen Charlotte incorporated in a cast iron hob grate designed by Samuel Haworth*	71
59	*The hob grate in which the portraits of George III and Queen Charlotte appear*	71
60 to 63	*Late eighteenth and early nineteenth century hob grates in cast iron*	72 and 73
64 to 67	*Typical cast iron patterns for interiors and side jambs of late eighteenth century grates*	74
68 to 73	*Late eighteenth and early nineteenth century cast iron panels for hob grates*	75
74	*Early nineteenth century hob grate with unusual ornamental motifs*	76
75	*Fireplace and grate in the Commercial Rooms, Bristol*	77
76	*The oldest known cast iron water mains, Versailles, 1664*	79
77	*Early cast iron water main at Philadelphia, U.S.A.*	80
78	*Section of cast iron water main laid in London in 1810*	81
79	*Mid nineteenth century cast iron water main at Baltimore, U.S.A.*	81
80 to 82	*The cast iron bridge at Coalbrookdale*	83, 84 and 85
83 to 85	*The bridge over the River Wear, at Sunderland*	87, 88 and 89

FIGURE		PAGE
86	*Thomas Telford's proposed new London Bridge, 1801*	91
87	*Buildwas Bridge in Shropshire*	92
88 and 89	*Craigellachie Bridge near Banff*	93
90 and 91	*Galton Bridge at Smethwick, Birmingham*	94
92 and 93	*Waterloo Bridge at Bettws-y-Coed, Carnarvonshire*	95
94	*Ickneild Street Bridge at Birmingham*	96
95	*The Mythe Bridge at Tewkesbury*	96
96	*Stokesay Bridge, Shropshire*	97
97	*Chepstow Bridge, Monmouthshire*	97
98	*Boston Town Bridge, Lincolnshire*	98
99	*North Parade Bridge, Bath*	98
100	*Cleveland Bridge, Bath*	99
101	*Contemporary view of the New Iron Bridge at Bathwick, Bath, published in 1829*	100
102	*Abermule Bridge in Montgomeryshire*	101
103	*Design for a bridge over the Thames between London and Blackfriars Bridges*	102
104	*Two designs for cast iron bridges over the Thames, by Telford and Douglass*	102
105	*Vauxhall Bridge, London*	103
106	*Design for a cast iron bridge over the Thames at Kingston, Surrey*	103
107	*Cast iron bridge over the River Aire at Haddlesey, Yorkshire*	105
108	*Plan and elevation of the Chirk aqueduct*	107
109	*Section of the Chirk aqueduct*	107
110	*The Pont-Cysyllau aqueduct, carrying the canal across the River Dee at the bottom of the Vale of Llangollen*	108
111	*Elevations and plan of the Pont-Cysylltau aqueduct*	109
112	*Section through the Pont-Cysylltau aqueduct*	109
113	*Plan and sections of cast iron lock designed by Thomas Telford*	110
114	*Elevation, plan and section of cast iron swivel bridge at St. Katharine's Docks, London*	111
115	*Cross-section of cotton mill in Manchester, designed by Boulton & Watt, 1801, for Phillips, Wood & Lee*	112
116	*Plan and section of the Manchester cotton mill*	113
117	*Details of columns and bases in the Manchester cotton mill*	114
118	*Pair of gates from the garden entrance of Lansdowne House, Berkeley Square, London*	117
119	*Cast iron railings, Stone Buildings, Lincoln's Inn, London*	118
120	*Cast iron railings for Ely House, Dover Street, London*	119
121	*Cast iron and lead railings, Lincoln's Inn, London, eighteenth century*	121
122	*Cast iron balcony, Grand Parade, Brighton*	122

FIGURE		PAGE
123	*Nos. 17 - 19, Grand Parade, Brighton*	122
124	*Montpelier Road, Brighton*	123
125	*Edwardes Square, Kensington, London*	123
126	*Litfield Place, Clifton, Bristol*	123
127	*Balconies in Charlotte Street, Bristol*	124
128	*Savile Place, Clifton, Bristol*	125
129 to 136	*Examples of early nineteenth century balcony railings, London*	126 to 129
137	*Balcony railings, Earls Terrace, Kensington, London*	129
138	*Balcony railings at 43 Claremont Square, Pentonville Road, London*	130
139	*Balcony railings, Victoria Terrace, Weymouth*	131
140	*Balcony railings in Claremont Square, London*	131
141	*Cast iron balconies in Regent Terrace, Edinburgh*	132
142	*Balcony railings, Brunswick Place, Southampton*	133
143	*Balcony railings, Clarence Terrace, Regents Park, London*	133
144 and 145	*Balcony railings and railings in Chester Terrace, Regents Park, London*	134
146 and 147	*Railings in York Terrace, Regents Park, London*	135
148	*Balconies and railings in Dorset Square, London*	136
149	*Balcony railing in the Adelphi, London*	137
150	*Railings and lamp standards at Chandos House, Chandos Street, London*	137
151	*Railings in Portman Square, London*	138
152	*Railings at No. 5 Upper Harley Street, London*	138
153	*Railings and cast iron balcony at No. 4 Cleveland Square, London*	139
154 and 155	*Cast iron balcony at No. 5 Columbia Place, Winchcombe Street, Cheltenham*	140 and 141
156	*Cast iron railings and finial at Bryanston Square, Marylebone, London*	142
157	*Railings and finial in Manchester Square, London*	142
158	*Gate at Campden Hill Square, London, early nineteenth century*	143
159	*Railings at Park Crescent, Regents Park, London*	144
160	*Railings at Norfolk Square, Paddington, London*	144
161	*Railings at Gloucester Square, Paddington, London*	144
162	*Early nineteenth century cast iron gates in Portman Square, London*	145
163	*Railings and gate posts in Surrey Street, Norwich*	146
164	*Railings to Marlborough House, Brighton*	146
165	*Cast iron nameplate to Hall's warehouse, Worcester*	146
166	*Early nineteenth century cast iron bollards*	147
167	*Cast iron balusters in the Bedford Hotel, Brighton*	148

FIGURE		PAGE
168	*Cast iron balusters in the Octagon, Victoria Rooms, Clifton, Bristol*	148
169	*Cast iron stair balusters at the De Grey Rooms, York*	149
170 and 171	*Cast iron columns in the Royal Pavilion, Brighton*	150
172	*Cast iron staircase in the Royal Pavilion, Brighton*	151
173 to 175	*The head office of the Essex and Suffolk Equitable Insurance Society Limited, Colchester*	152 to 154
176	*Cast iron columns at Carlton House Terrace, London*	155
177	*Cast iron columns at North Lodge, Buckingham Palace, London*	155
178	*George Stephenson*	160
179	*Robert Stephenson*	161
180	*Cast iron railway footbridge*	163
181 to 183	*Cast iron bridge crossing the Regents Canal, near Chalk Farm, London*	164
184 and 185	*The Hampstead Bridge, London*	165
186 to 188	*Cast iron bridge over the Grand Junction Canal at Blisworth*	166 and 167
189 and 190	*The Park Street Bridge, London*	168
191	*Bridge over Spa Road, London*	168
192	*Cast iron bridge designed by Robert Stephenson and intended to cross the River Nene, near Wisbech*	168
193	*Nash Mill Bridge, near Kings Langley, Herts*	169
194	*Plan, section and details of circular engine house at Gorton*	170
195	*Circular engine house, Camden Town, London*	171
196 and 197	*The Nine Elms Goods Depot, London*	173
198	*Contemporary drawing of the interior of Paddington Station, London*	174
199	*Entrance to the offices at Paddington Station*	174
200	*Entrance to Euston Station, London*	175
201	*Statue of Robert Stephenson, outside Euston Station*	176
202	*Interior of Euston Station, from a contemporary drawing*	177
203 and 204	*Entrance and cast iron gates at Euston Station*	176
205	*Early print showing the roof of St. Pancras Station, London*	178
206 and 207	*Aldridge Road Bridge, Birmingham*	180
208 and 209	*The Town Bridge, Thetford*	181
210 to 213	*The Spey Bridge, Fochabers*	182
214 to 216	*Gingerbread Hall Bridge, Gt. Baddow, Essex*	183
217	*Stafford Bridge, Oakley, Beds.*	183
218	*Babraham Bridge, Cambs.*	184
219	*Some examples of cast iron beams, in section*	184

FIGURE		PAGE
220	*Gt. Barr Street Bridge, Birmingham*	185
221	*Islington Row Bridge, Birmingham*	185
222	*Tindal Bridge, King Edward's Road, Birmingham*	185
223	*Millington Hall Bridge, Retford, Notts.*	185
224	*Warrington Road Bridge, Culcheth, Lancs.*	185
225	*Eastleigh Bridge, Hants.*	185
226	*Cast iron arches intended for the Britannia Bridge, North Wales*	186
227	*Bridge over the River Trent, on the London, Midland & Scottish main railway line*	186
228 and 229	*Lambeth Suspension Bridge, London*	187
230	*Hammersmith Suspension Bridge, London*	188
231	*Chelsea Suspension Bridge, London*	189
232	*Chelsea Suspension Bridge, as rebuilt*	189
233 and 234	*Contemporary views of Westminster Bridge, London*	190 and 191
235	*Westminster Bridge, London*	191
236	*The Market Hall, Birmingham*	193
237	*Bridgewater House, London*	194
238	*Interior of the Riding School, Welbeck Abbey*	195
239 and 240	*The City Temple, London*	195 and 196
241 and 242	*The Metropolitan Tabernacle, Newington Butts, London*	197
243	*Nos. 219 - 221 Chestnut Street, St. Louis, U.S.A.*	198
244	*St. George's Church, Birmingham: cast iron gallery front and arches*	199
245 to 252	*The Crystal Palace, London*	200 to 205
253 and 254	*The Palm House, Kew Gardens, London*	206 and 207
255	*The New York Crystal Palace*	208
256	*An exhibit at the Great Exhibition of 1851*	211
257 to 262	*Cast iron exhibits at the Great Exhibition of 1851*	212 and 213
263	*Cast iron plaque*	214
264	*German filigree plate in cast iron*	214
265	*Cast iron lamp standard at Northampton*	215
266	*Cast iron plaque of the first Duke of Wellington*	215
267	*Railings in the Museum grounds, Leicester*	216
268 and 271	*The cast iron lion designed by Alfred Stevens*	216 and 217
269 and 270	*Railings and entrance gates to the British Museum, London*	217
272	*Gates to Hyde Park, Marble Arch, London*	218
273 and 274	*Cast iron gates and railings designed for the Great Exhibition of 1851*	218 and 219

FIGURE		PAGE
275	*Cast iron grille at the Albert Hall, London*	220
276 to 292	*Designs for railings, gates, grilles, lamp standards, etc., in cast iron, taken from L. N. Cottingham's* Smith and Founders Director, *published in 1840*	221 to 235
293	*Blackfriars Bridge, London*	241
294 and 295	*Railway Bridge at Blackfriars, London*	242 and 243
296 and 297	*The Maquis Viaduct on the Santiago and Valparaiso Railway, Chile*	244
298	*The Albert Suspension Bridge, London*	245
299	*Railway bridge over the River Bremer, Queensland, Australia*	245
300 to 302	*The Trent Bridge, Nottingham*	246 and 247
303	*Conservatory in Glasgow*	248
304	*The Arcade, Johannesburg, South Africa*	249
305	*Winter Garden, designed for the gardens of the Royal Horticultural Society, London*	250
306	*Kiosk designed for erection in India*	251
307	*Design for a cast iron pier*	253
308	*Winter Palace, Dublin*	254
309	*Winter Garden, the Infirmary, Leeds*	255
310 to 312	*Structural cast iron columns*	256 and 257
313 to 316	*Amsterdam Station, Dutch Rhenish Railway*	258 to 261
317 to 319	*Woodside Station, Birkenhead*	262 and 263
320 to 325	*Victoria Station, London*	264 to 268
326	*Cast iron bandstand, Southend-on-Sea*	269
327 to 332	*Late nineteenth century brackets, lamps and lamp standards*	270 to 273
333	*The Royal Coat of Arms, in cast iron*	274
334 and 335	*Dogs in cast iron, London*	275
336	*Cast iron gates, Gothic Revival*	276
337 and 338	*Cast iron railings and terminals*	277
339 and 340	*Cast iron stair balusters*	278 and 279
341	*Cast iron balcony railings, supports, verandah and area railings at Lansdowne Place, Bristol*	280
342	*Cast iron balcony railings*	281
343	*Mid nineteenth century balcony and area railings at Craven Hill Gardens, Paddington, London*	282
344	*Balcony and area railings in Munster Square, Regents Park, London*	283
345	*Railings at Silwood Place, Brighton*	283
346	*Mid nineteenth century balcony railings and brackets in Clifton, Bristol*	284
347	*Balcony railings at Alexander Place, Kensington, London*	285
348	*Railings outside the Prison, Leicester*	286

FIGURE		PAGE
349	Balcony railings at No. 14 Albert Road, Regents Park, London	287
350	Garden gate and piers in New Walk, Leicester	287
351	Cast iron railings and gates at the High School, Edinburgh	288
352	Gateway to the Sailors' Home, Liverpool	289
353	The Lecture Theatre in the Medical Institution, Liverpool	290
354	Cast iron capitals, Gothic Revival	291
355	Cast iron tracery panel in Holy Trinity Church, Coalbrookdale	291
356 to 366	Late nineteenth century firegrates	292 and 293
367 to 370	Late nineteenth century firegrates and mantelpieces	294 and 295
371 to 373	Late nineteenth century heating stoves	296
374 and 375	Late nineteenth century ranges and boilers	297
376	Cast iron gratings	298
377	Cast iron edging for garden paths	298
378	Cast iron gratings	299
379	Cast iron coal cellar plates	300
380 to 382	Cast iron garden chairs and seat	301
383 to 386	Cast iron hall and umbrella stands	302 and 303
387 and 388	Cast iron garden frames	304
389	Cast iron plant stand	304
390	Cast iron mud scrapers	304
391 and 392	Cast iron fountains	305
393	Cast iron drinking fountains	305
394 and 395	Cast iron window frames	306
396 to 398	Cast iron windows, frame, fanlights and door lintel	307
399	Cast iron used for framing shop windows and entrance doorways	308
400	Cast iron shop facade, and bridge	314
401	Example of cast iron seaside architecture	315
402 and 403	Cast iron shelters	316 and 317
404	Cast iron window surrounds, panels and mullions at Unilever House, London	318
405	Cast iron window mullions and apron panels at the Scottish Legal Building, Glasgow	319
406	Cast iron window breast panels at Lothian House, Edinburgh	320
407	Cast iron window bays in the Adelphi Building, London	320
408	The G.P.O. telephone kiosk	321
409 and 410	Cast iron lamp standards	322
411 to 413	Victorian and modern pillar and posting boxes	323

FIGURE		PAGE
414 to 417	Road traffic signs in cast iron	324
418	Railings to the Embankment Gardens, London	325
419 and 420	Railings, lamp standard and bollards, London	326 and 327
421 and 422	Hayes Bridge, Cardiff	328 and 329
423	Croxdale Bridge, Sunderland	330
424 and 425	Cast iron bridge parapet at Cardiff	331
426 and 427	Cast iron window and doors at Sassoon House, Shanghai	332 and 333
428	Cast iron flower box for house at Northolt, Middlesex	334
429	Cast iron staircase, the Students Union, Liverpool University	334
430	Cast iron gates, Clive Buildings, Calcutta	335
431 to 434	The Mersey Tunnel, Liverpool	336 to 338
435 to 439	Examples of cast iron paving tiles and road setts	339
440 and 441	Cast iron water storage tanks	340
442	Victorian kitchen range	341
443	The same range, re-designed between the two wars	341
444	Contemporary version of the same design	341
445 to 453	Cast iron solid fuel cookers	342 to 344
454 and 455	Cast iron hot water heaters	345
456	Cast iron solid fuel cooker, wall panels and table top	345
457 and 458	Cast iron cookers, boilers and hot closets for institutional cooking	346
459 to 462	Gas cookers and fire in cast iron	347
463	Nineteenth century design for an independent stove in cast iron	348
464	Solid fuel space heater, in cast iron	348
465	Domestic hot water boiler in cast iron	348
466 to 470	Slow burning solid fuel space heaters and stoves, in cast iron	348 and 349
471 to 475	Gas fires, in cast iron	350
476 to 478	Cast iron radiators	351
479	Gas heater in cast iron	351
480	Electric heater in cast iron	351
481	Range of typical cast iron rainwater heads	352
482	Cast iron sink, with draining board	352
483	Cast iron wash basin	353
484	Kitchen installation with cast iron sink, draining boards and splashback tiles	353
485	Cast iron shower bath tray	353
486 and 487	Cast iron fuse boxes and switch gear	354

FIGURE		PAGE
488	*Diagrammatic section through a blast furnace*	356
489	*Diagrammatic section through a cupola*	357
490 to 497	*Stages in the casting of a slow combustion stove*	359 to 361
498 to 502	*Plant in a mechanised foundry*	363 to 367
503	*Sand casting pipes vertically in a pipe pit*	368
504	*Centrifugal casting of pipes*	368
505	*Cooking table ware in cast iron*	371
506	*Cast iron letters*	372
507	*Cast iron panel at the Oratory Central Schools, Chelsea, London*	373

FOREWORD

By Sir CHARLES REILLY

IT sounds a very ordinary thing to say of any book that it fills a distinct want, but it is nevertheless true of John Gloag and Derek Bridgwater's *History of Cast Iron in Architecture*. Those of us who are architects have been using cast iron all our days for a great variety of purposes. It is indeed one of the most ordinary of building materials, finding its way in half-a-dozen forms into every building, yet it is safe to say very few architects have given to it or to its manufacture more than a few moments of thought. We have taken it for granted, often assuming that it is a very humble material always the same and always to be relied on if used in certain directions. It has never occurred to most of us that it can be had in many varieties for each of which some uses are better and some are not. When we read this book, however, we quickly realise we are in a new country and that cast iron has not only had a romantic history of its own, playing a great part in many famous buildings, but that it has indeed many qualities, finishes and uses we have not dreamt of. Clearly, in these days of functionalism, this is all wrong and it is up to us to know and understand all there is to know about this, as of all our materials. It may have been legitimate for the Victorians, with their limited outlook, to seize upon the capacity of cast iron for the cheap reproduction of ornament and to smother it with that, but much more is expected of us. We must know all there is to know about it, the forms it likes and the forms it dislikes, its various kinds and their qualities and finishes. It is just because this book tells us so much about all this that it is so welcome. With its help architects no doubt will discover many new uses for cast iron and laymen new standards of judgment.

ACKNOWLEDGMENTS

WE wish to acknowledge with gratitude our indebtedness for advice and assistance to Professor A. E. Richardson, A.R.A., F.R.I.B.A., Mr. H. S. Goodhart-Rendel, PP.R.I.B.A., Professor Sir Charles Reilly, O.B.E., LL.D., F.R.I.B.A., Mr. H. M. Fletcher, F.R.I.B.A., and Mr. J. G. Pearce, M.SC., F.INST.P., M.I.E.E., M.I.MECH.E., Director and Secretary of the British Cast Iron Research Association, and to the staff of that Association. For suggestions and practical help in the task of collecting illustrations, we must thank Mr. John Summerson, F.R.I.B.A., F.S.A., Mr. Cecil Farthing, and the staff of the National Buildings Record, Mr. F. R. Yerbury, Hon. A.R.I.B.A., the Librarian and staff of the Library of the Royal Institute of British Architects, the Birmingham Central Reference Library, the Birkenhead Public Library, and the Librarian of the Iron and Steel Institute; also Mr. C. S. Chettoe, B.SC., M.INST.C.E., the Sussex Archæological Society, the Building Research Station at Watford, the London, Midland & Scottish Railway Company, the American Cast Iron Pipe Association, the firms and individuals in the foundry industry, and particularly the members of the British Ironfounders Association, who have allowed so many of their photographs and original drawings to be reproduced in this book. Finally our thanks are due to Miss Dora Ware for tackling the formidable task of collating and arranging the material.

JOHN GLOAG
DEREK BRIDGWATER

INTRODUCTION

IRON has been the most influential metal in forming civilisation, and the vast mechanical and architectural facilities we now enjoy would have been impossible had not the earth possessed an abundant supply of iron ore. It has been the basic material for industrial development, both in the days of handicrafts and mechanised production.

Something must be known of the use of iron in historic and pre-historic times before we can study the much later discovery of the art of casting the metal, and of the particular significance the history of cast iron has had on the rise of British industry and the character of British architecture since the first industrial revolution.

1. The Stone Ages and the Early Use of Metals

Until about 4000 B.C., nearly all tools and implements were made of stone. The Neolithic age ended when metals came into use; but towards the end of that age—no one can say exactly when—some knowledge of metals undoubtedly existed. These would be found in the raw state, and early finds of iron would almost certainly be of meteoric origin. This iron would be treated as a super hard stone and probably considered a gift from the gods. G. F. Zimmer gives ample evidence to prove that meteorites all over the world have been chipped and the fragments used by primitive peoples.[1]

2. Early Type of Furnace

It is impossible to say who first discovered that metal could be produced from ores by heat. It has been suggested that following early forest fires primitive man would discover the pasty or even molten metal and learn from this to use it and perhaps even use the same method to procure further supplies. It is more probable, however, as Professor William Gowland points out that the discovery of metals "was brought about in a more commonplace and more humble way. It had its origin in the domestic fires of the Neolithic Age.

"The extraction of the common metals from their ores does not require the elaborate furnaces and complicated processes of our own days, as pieces of ore, either copper carbonate or oxide, cassiterite, cerussite, or mixtures of these, and even iron oxides which by chance formed part of the ring of stones enclosing the domestic fire, and which became accidentally embedded in its embers, would become reduced to metal. The camp fire was, in fact, the first metallurgical furnace, and from it, by successive modifications, the huge furnaces of the present day have been gradually evolved".[2]

From this a crude form of furnace would be developed. At first, it probably had just a shallow cavity scooped out in the hearth of the fire to receive the pasty or molten metal, and then gradually the sides would be built up. This kind of technique would evolve and spread throughout the world as culture was diffused. Improvements would be made periodically; it would be found that furnaces were more successful on certain hill sides where the wind helped the draught, and so some primitive method would be devised for providing extra draught, thus leading to the use of bellows.

3. EARLY EVIDENCE OF IRON IN EGYPT, ASSYRIA, INDIA, CHINA, EUROPE AND BRITAIN

The Stone, Bronze and Iron Ages overlapped in various parts of the world, for tribes and nations in different stages of culture were only separated geographically. Discoveries in Central America indicate that the Mayan civilization was still Neolithic as late as A.D. 400, when nearly all Europe had passed from the Bronze to the Iron Age. The Britons were using stone implements at a time when Assyria, Egypt, Greece and China were long familiar with iron.

EGYPT

4000 *B.C.* Oxidised remains of iron beads were found in an Egyptian cemetery at El Gerseh during excavations organised by the British School of Archæology. These were examined by Professor Gowland who stated: "They do not consist of iron ore, but of hydrated ferric oxide, which is the result of the rusting of the wrought iron of which they were originally made".

3100 *B.C.* In 1837 in the Great Pyramid of Gizeh a piece of iron was discovered which was presumed to be a tool and was dated by the archæologists as 3100 B.C. It is now in the British Museum.

2800 *B.C.* Sir William Flinders Petrie states "an absolutely dated case is that of the mass of rust, apparently from a wedge of iron, found stuck together with copper adzes of the Sixth Dynasty type, at the level of floors of that age in the early temple of Abydos" *circa* 2800 B.C.[3]

ASSYRIA

2000-1500 *B.C.* Professor Gowland tells us in describing the ruins of the Palace of Sargon, King of Assyria, 722 B.C., that Victor Place "found a storehouse containing, according to his estimation, not less than 160,000 kilogrammes of iron. The greater part consisted of iron bars from 12 to 19 inches in length, and $2\frac{3}{4}$ to $5\frac{1}{2}$ inches in thickness. They were roughly drawn out at each end and pierced with a hole and weighed from about 8 to 44 lbs. Place supposed them to be work tools of some kind, but they are really bars of iron forged at the furnace of the mines into this shape for convenience of transport by men, horses, or camels. It is worthy of note here that similar forms survived for iron for transport and trading in Roman times, and even up to thirty or forty years ago in Finland and Sweden. The collection was chiefly a store of unworked iron held in readiness by the King for the instruments of war and for building construction. It contained also, however, many kinds of finished iron articles, such as chains, horse-bits, etc., all arranged in regular order. This vast accumulation of iron indicates incontestably that the metal had been in use for many centuries previous to the time of Sargon, so that it will not be unreasonable to assume that the Assyrians were acquainted with iron certainly earlier than 1500 or even 2000 B.C."[4]

Fig. 1. The earliest known dated example of cast iron is Chinese, sixth century A.D. One of a pair of recumbent lions. Base 7¼ ins. by 1 ft. 2⅜ ins., height 5½ ins., weight 26 lbs. The translation of the inscription on the base is: "Made on the twenty-fourth day of the seventh month of the third year of Ching Ming of Great Wei." (September 11th, A.D. 502)

Reproduced by courtesy of Professor T. E. Read, The School of Mines, Columbia University, New York.

805 *B.C.* An extremely interesting small bronze casting was found at the Nimrod excavations which had a core of iron over which the bronze had been cast. This suggests that the Assyrians had some considerable metallurgical knowledge.

INDIA AND CHINA

It is from writings rather than excavations that we learn most about the early use of iron in India and China.

It was probably known and used in 2000 B.C. Professor Panchaman Neogi concludes from certain passages in the *Black Yajurveda* that some form of iron cannon or engine of war was in use between 2000 B.C. and 1000 B.C.[5] The famous pillar at Delhi, 23 feet high and 16½ inches in diameter at the base is a remarkable example of early wrought iron about A.D. 300.

Dr. Friedrich Hirth describes a federal state of China that levied a tax on iron about 675 B.C.; but it was certainly known and used much earlier.[6]

Some of the oldest examples of cast iron known are Chinese; and recent research work by some American scientists seems definitely to establish the fact that iron castings had their origin in China, and not, as was formerly believed, in eastern Europe. A paper given before the American Institute of Mining and Metallurgical Engineers in February, 1938, by Maurice L. Pinel, Thomas T. Read and Thomas A. Wright, was devoted to the results of this research work, and was entitled the "Composition and Microstructure of Ancient Iron Castings". There is direct evidence that iron castings were produced in China as early as A.D. 502, and enough evidence to suggest that the technique of casting iron was known a thousand years earlier. Of the nine specimens collected by the authors of the paper in a tour of China, five had actual dates cast on them, and two could be accurately dated. The specimens

Chinese figures in cast iron: sixth century A.D..

FIG. 2 (*Above*). A small standing figure of Kwan Yin on a rectangular base, the overall height being 1 ft. 8 ins., and the weight 16 lbs. On the reverse of the halo about the head is the inscription: "This iron image was made on the twenty-fifth day of the ninth month of the sixteenth year of Wu Ting above (alone?) for the Emperor and after him for the multitude of lives." (October 22nd, A.D. 558)

FIG. 3 (*Right*). A standing figure of Kwan Yin, on a 7¼ ins. by 3¼ ins. round (lotus) pedestal, the overall height being 2 ft. 7 ins., and the weight 50 lbs. The inscription on the base is: "An image respectfully made by Chang Wen for his parents at Cloud Light Temple, Bell Rock Mountain, on the twenty-eighth day of the third month of the first year of Tien Pao." (April 30th, A.D. 550)

Reproduced by courtesy of Professor T. E. Read, The School of Mines, Columbia University, New York.

Fig. 4. Chinese, tenth century, A.D. The Great Lion of Ts'angchou, fifty miles south of Tientsin. 18 ft. long, 20 ft. high, A.D. 953.
Reproduced by courtesy of Professor T. E. Read, The School of Mines, Columbia University, New York.

varied in date from A.D. 502 to A.D. 1093; and one of the undated specimens—a sample of metal from an early cast iron stove, showing that the metal was used for practical as well as decorative purposes—is considered almost certainly older than A.D. 200. Each specimen was subjected to chemical, spectrographic and metallurgical study. According to Professor Read—one of the joint authors of the paper—the method of moulding was by carving rammed sand, or by using the modern method of building up the moulds with cores.

EUROPE

It seems probable that the Greeks were the first people in Europe to use iron. Montelius puts the Grecian Iron Age as early as 1400 B.C.[7] Homer (880 B.C.) often writes of the metal, and Aristotle (380 B.C.) describes in some detail the different qualities of various irons and how the metal was obtained from the ore. Virgil in about 40 B.C. shows that the Romans were familiar with steel, and Pliny's *Natural History* indicates that by A.D. 70 the Romans had a quite remarkable metallurgical knowledge of iron and its ores. He gives a complete and interesting account of the chief iron ore localities, the character of the ores, the method of extraction of the metal, and the methods of hardening by quenching with water and oil to form an early type of steel.

Between 1847 and 1864 important excavations were carried out in the old burial grounds of pre-historic men at Hallstatt in Austria and at La Tène in Switzerland. Some thousands of iron objects were found dating back as far as 750 B.C. The discoveries of Professor K. Absolon, made in 1946, in the Býči Skála cave, north-east of Brno in Czechoslovakia, suggest that the casting of iron was practised as early as 600 B.C.[8]

BRITAIN

Iron was certainly known to the Britons in 200 B.C., probably earlier, though it is uncertain which part of the country was the original home of the iron trade. Pre-Roman iron remains have been found at various centres. In 1907 work was completed on the excavations of a lake dwelling near Glastonbury which dated back to about 100 B.C. A large number of iron objects were found, and the presence of pieces of slag among these showed that the villagers could obtain the metal from the ore. One bill hook was so well preserved in the peat that the ash wood handle was still intact and kept in position by an iron rivet. Similar finds were made at Hod Hill (Dorset), Ham Hill (Somerset), Hunsbury Camp (Northants), and Wookey Hole (Somerset), and include reaping hooks, sickles, saws, gouges, adzes, files, keys, daggers, spearheads and knives. There are also many instances of currency bars of iron, and these were mentioned by Cæsar in describing his invasion of Britain.[9] J. Newton Friend says: "The Britons at the time of the Roman conquest were famous for their chariots of war, drawn by two horses abreast, with a pole between them along which the charioteer would on occasion run, even with the horses at full gallop. It is stated that the Britons who opposed Julius Cæsar had some 4,000 of these chariots. When a British chieftain died he was frequently buried in his chariot together with his horses and their equipment. Remains of these chariot burials, as they are called, have been found in different parts of the country, notably in the East Riding of Yorkshire. The wooden parts of the chariots have of course long since mouldered away, but fragments of the iron rims of the wheels have remained, sufficient to show that the wheels were originally nearly 3 feet in diameter. In some cases, too, iron mirrors have been discovered! The modern practice of fitting motors with mirrors to enable the driver to see what is going on behind would thus appear to be merely the resuscitation of a device used 2,000 years ago by the British charioteer. Well may one ask if there is anything new under the sun!"[10]

This authority doubts whether British chariots were fitted with scythed axles, pointing out that Julius Cæsar, "who saw the scythed chariots in Pontus in B.C. 47, would hardly have failed to mention the fact had he encountered them or even heard of them in Britain".

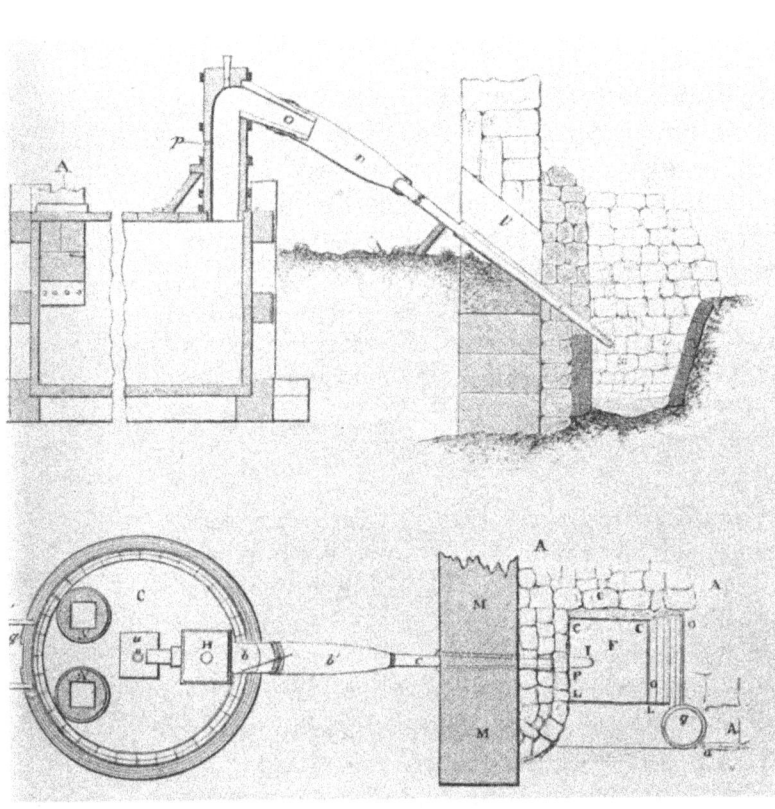

Fig. 6. A primitive hearth and bellows.

Fig. 5 (*Above*). The Catalan Forge. The furnace consisted of a four-sided hearth containing the ore and charcoal. (See page 8.) The blast was produced by a trompe, an apparatus invented in Italy in mediæval times, which provided a pressure of air by means of falling water.

Fig. 7 (*Right*). The Osmund Furnace. Similar to the Catalan Forge, but built with a higher hearth. The old print shows the method of charging the bellows for production of the blast and the tapping hole.

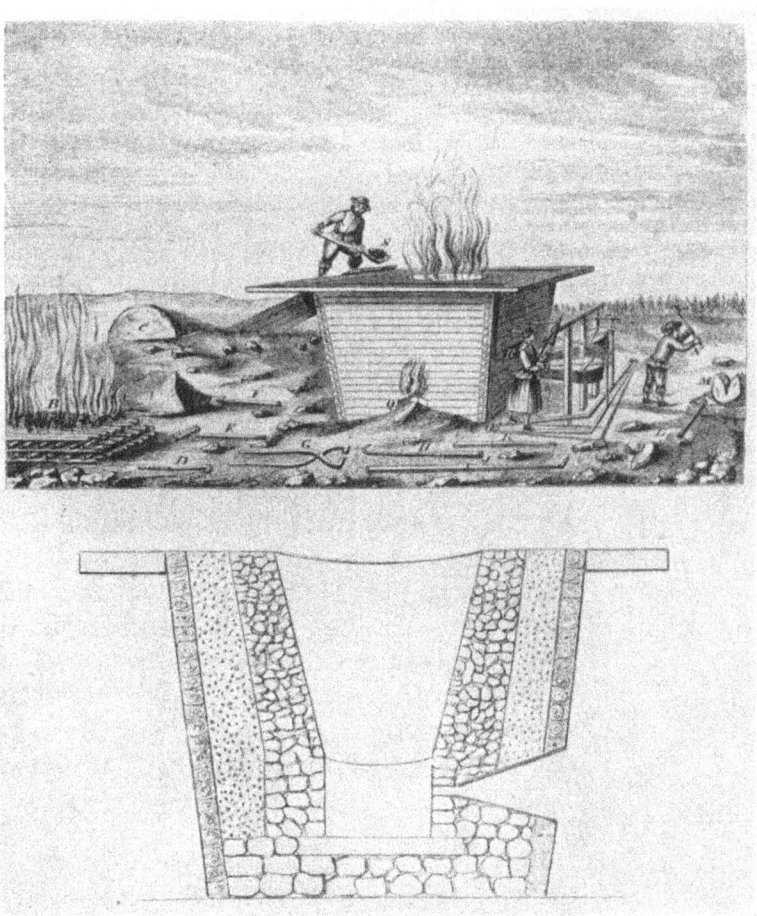

On the method employed for extracting the metal, R. A. Smith concludes that as no furnaces of the earliest period have been found, the ancient Britons probably used the simple low hearth, similar to the Catalan furnace of the Pyrenees, which has been employed there from ancient times to our own day. He suggests that "the source from which Britain derived the furnace and art of extracting iron from its ores, seems to have been the Mediterranean region, either the Eastern Pyrenees or North-West Italy; but it may also be reasonably held that the first iron furnace of the Britons was derived from that used so successfully in the extraction of tin. It is not, however, probable that our islands were the earliest centre for the metal".[11]

Iron in Roman Britain

After Britain became a Roman Province, the production of iron increased considerably. In Sussex and Gloucestershire, Romano-British ironworks were so extensive that large deposits of slag have been found. In Gloucestershire the slag was used in the sixteenth century and later; being worked again for iron to save digging the ore.

In A.D. 120 the Emperor Hadrian worked iron in the Forest of Dean. M. P. Charlesworth, in *Trade-Routes and Commerce of the Roman Empire*, refers to "a third-century Birmingham" which has been excavated near Ross. He states that iron in the Roman province of Britain was plentiful, and was worked in several places. He tells us that "carpenters' tools, hammers, axes, saws, chisels, files, tools for farriers and shoemakers, shears, scythes, knives, padlocks and various other implements made from iron have been found all over the country, and we have remains that shew that iron was worked in Sussex and by the Forest of Dean. In that region within recent years excavations have been made on a site which must have been of some importance; near Ross, at the ancient site of Ariconium, there have been uncovered remains that indicate that a great industry was pursued there; iron was smelted and forged and worked into various implements and an industrial town had grown up round the ironworks; it has indeed been called a third-century Birmingham by the excavator".[12]

An account, published in 1912, of the excavations carried out at Corstopitum near Corbridge on the North bank of the Tyne, shows that an actual bloom of Roman iron was found.[13] Important Roman iron remains have also been found at Richborough Castle near Sandwich, Kent; at East Cliff, Folkestone, the site of a Roman villa occupied from about A.D. 90 to A.D. 388 which was excavated by S. E. Winbolt in 1924;[14] and at Viroconium in Shropshire. It has been suggested by J. Storrie, who carried out considerable excavations near Cardiff in 1894, that the Roman iron remains in that part of Wales indicated that manganiferous ores were imported from Spain for the manufacture of steel.

Some of the developments, industrial and commercial, which were to take place fourteen hundred years later, were foreshadowed in the Roman province of Britain. The stimulating effect of Roman civilisation upon British industry is described by Collingwood and Myres in their comprehensive work on *Roman Britain and the English Settlements*.[15] The part played by iron was significant then; its significance was prophetic, for Romano-British industrial development disclosed the immense reserves of skill that existed in the brains and hands of the inhabitants of this remote province of the Western Empire. That skill was driven underground by the barbarian invasions and settlements of the fifth and sixth centuries; it emerged sporadically during the four hundred years that preceded the Norman Conquest and flourished vigorously in mediæval times when English smiths became the most famous and expert of craftsmen.

The Ironworkers of Mediæval England

The iron industry was not re-established until the end of that period of intermittent civilisation England enjoyed under the independent Saxon states. By that time Gloucestershire had a considerable trade and the Forest of Dean again became an important centre. But under the Normans the industry declined, and to such an extent that by the fourteenth century the metal was considered a rare and costly material and Harry Scrivenor relates how the Scots invading England in 1317 "met with no iron worth their notice until they came to Furness, in Lancashire, where they seized all the manufactured iron they could find, and carried it off with the greatest joy, though so heavy of carriage, and preferred it to any other plunder".[16]

The chief centres of the iron industry in mediæval England were Kent, Sussex, the Forest of Dean, and Rockingham Forest in Northamptonshire; but there were many other centres of less importance where small amounts were produced, some only meeting local demands. For example some of the great Abbeys had their own iron producing plant for their agricultural implements, and for trading in the metal. The Abbey of Flaxley, founded in 1140, organised iron production in many districts of the Forest of Dean until well into the seventeenth century. The Bishops of Durham owned the minerals within the county and in 1408 began to work iron at a forge in Weardale. Ironworks also existed at the monastic centres of Wenlock, Kirkstead, Fountains, Rievaulx, and Conishead. G. T. Lapsley gives an account of this early Weardale ironworks where nearly two tons of metal was produced each week.[17]

Though there were undoubtedly many local variations, the usual method of extracting the metal from the ore is thus described by Professor T. S. Ashton: "The ore was first crushed into fragments and mixed with a small quantity of marl and lime which served to bind it together, and the mass was then divided into lumps which were placed on a forge and surrounded with charcoal. By means of a blast produced with leathern bellows, worked either by manual labour or by water power, the fire was maintained at a moderate temperature; and the metal was brought to a pasty, rather than a liquid form, the impurities being removed by repeated hammering, 'even,' says Sturtevant, 'as the whey is wrung out by the violence of the Presse, and so the curds are made into a cheese.' Several heatings and hammerings were necessary before the ore was finally transformed into a bloom of wrought iron, ready to be worked into implements by the smith. Though the quality of the finished metal was high, the quantity produced at any one forge was necessarily small, and a considerable weight of metallic iron was left behind in the cinders or slag".[18]

In the more advanced districts improvements would be made and eventually a crude type of furnace was evolved, which has been described by J. Dearden as built up of clay, to form a vertical shaft about 4 ft. high and 18 in. diameter. "At the base were holes through which a blast of air was blown by crude bellows, or even by the wind itself on an exposed hill-side. A charcoal fire was lighted at the bottom of the furnace, and the shaft filled up with layers of iron-stone and charcoal. After many hours the furnace was allowed to cool, and was then demolished. If the process had been successful, a cake of iron, 30-60 lb. in weight, was found at the bottom. This had to be reheated and hammered to expel pieces of charcoal and slag and to forge it into some useful shape such as a tool or a sword".[19] This led naturally to the building of furnaces in masonry or brickwork and to the improvement of methods for producing draught.

CAST IRON IN BRITAIN

Until the thirteenth or fourteenth centuries the object of any furnace had been to produce a "bloom" of malleable or wrought iron. If the furnace was too efficient or the draught too great the ore became too hot; as a result the iron would dissolve carbon from the charcoal, lowering the melting point, and the metal would run out at the bottom of the furnace, to the dismay of the ironmaster. This undoubtedly happened quite often, and the resulting brittle metal would be destroyed as useless.

FIG. 8. The Joan Colins grave slab, originally in the floor, but since removed for preservation, to the wall of Burwash Church, Sussex. The inscription is: "ORATE P. ANNEMA JHONE COLINS". ("Pray for the soul of Joan Colins".) In the opinion of some antiquaries it may be dated as mid-fourteenth century, and is the oldest piece of cast iron known in England.

Reproduced by courtesy of Mrs. E. H. Straker, from Wealden Iron *by Ernest Straker. (G. Bell & Sons Ltd., 1931.)*

Fig. 9. The Anne Forster grave slab in Crowhurst Church, Surrey, dated 1591.
Reproduced by courtesy of Mrs. H. E. Straker, from Wealden Iron *by Ernest Straker. (G. Bell & Sons Ltd., 1931.)*

It is impossible to say when it was discovered that this cast iron or pig iron was a useful product. In the Middle Ages, the iron trade in Germany and Spain was in advance of England, both in quantity and quality of production, and the Germans were using in the fifteenth century a larger and more efficient type of furnace—the Stuckofen. This was a small blast furnace constructed of masonry and consisting of two truncated cones placed base to base. The front of the hearth was a thin, temporary wall, which was broken down at the end of the operation to allow the bloom of wrought iron to be removed. Again as the efficiency and size of furnaces grew the ore would be longer in contact with the fuel and the iron would become carburized and run out in a fluid condition.

This knowledge is often supposed to have come to England from Germany during the fifteenth century; but Mr. Starkie Gardner has suggested that a grave slab of cast iron at Burwash in Sussex must have been produced at a much earlier time, probably the middle of the fourteenth century, and that though the early cannon of the fifteenth century were formed by binding together bars of wrought iron to construct a barrel, these bars sometimes encircled an inner chamber of cast iron.

The chief products of the mediæval Sussex foundries were grave slabs, firebacks, andirons, cannon, anvils, mortars, and cooking utensils, and some fine examples of grave slabs and firebacks still exist though there are many modern copies of the latter. In Wadhurst Church there are some thirty grave slabs dating from 1617-1771. In Burwash Church is the famous Joan Colins cast iron grave slab to which the date of 1350 has been attributed, and which is probably the oldest-known piece of cast iron in Britain. In Crowhurst Church in Surrey is an example known as the Anne Forster grave slab dated 1591, and there are many others distributed throughout Sussex, Surrey and Kent, and others in the Midlands, particularly at Wellington and Bridgnorth in Shropshire. The design is often crude, and obviously the ironfounders were experimenting with the new material, much in the same way as they did in the design of firebacks.

When fireplaces were inset in walls, with flues to carry the smoke to the open air, instead of escape vents for the smoke of a central fire in the roof of

FIG. 10. Cast iron grave slab in Wadhurst Church, Sussex.

Fig. 11. Some of the thirty grave slabs in Wadhurst Church, Sussex, which date from 1617 - 1771.

Fig. 12. The most elaborate of the many grave slabs in Wadhurst Church, Sussex. The grave slab of the Barham family, 1701.

the great hall, the problem of preserving the brick or stonework from heat became important and the cast iron fireback was the solution. Mediæval firebacks may be roughly classified in four groups. The earliest firebacks were cast in open sand moulds into which small objects were pressed, such as pieces of rope, fragments of wood carving from panelling, and heraldic emblems or shields. These were rather crude and simple, and date from the latter part of the fifteenth century. The second type, though there is no rigid division, was a little more complicated and generally formed some composition, such as a large coat-of-arms. These devices were probably carved in wood and then impressed upon the sand mould; this type appeared

FIG. 13. Fifteenth century Sussex fireback. Three crosses of cable twist within a rope moulding. Dimensions: 2 ft. ¼ in. by 3 ft. 9½ ins.

From the Victoria and Albert Museum, by whose courtesy this is reproduced. Crown copyright reserved.

in the latter part of the sixteenth century. The third type was more ambitious in design, having emblematic, allegorical and pictorial subjects, and dates from the latter part of the seventeenth century. The fourth type was influenced strongly by German design, with pictorial subjects, often scriptural, executed in a naturalistic and much freer style.

In old open fireplaces, the firebacks were made low and wide, but as the openings grew smaller the shape of the fireback became squarer, and the late seventeenth- and eighteenth-century firebacks had considerably greater height than width. Many bear coats-of-arms, scriptural subjects, classical stories and allegories. Professor Lethaby says "These were pleasant enough conceits for the fireside, and seem to suggest something of the kind of life which demanded such things. The technical achievement of the poorest of these firebacks is very far from high, but they all seem to have a real motive or reference to something pleasant and suggestive. All are treated with great simplicity and that instinct for balance and spontaneous freedom which is the mark of traditional crafts . . . They were decorated plates of iron rather than cast sculptures".[20]

Fig. 14. Fireback with roughly formed M in cable twist, and three medallions. Sussex, fifteenth century. Dimensions: 2 ft. 1½ ins. by 3 ft. 4½ ins.
From the Victoria and Albert Museum, by whose courtesy this is reproduced. Crown copyright reserved.

The famous Richard Lenard fireback is of particular interest as one of the few contemporary illustrations of the foundry process of the times. It is dated 1636, and shows the founder in his forge surrounded by the tools of his trade and some of the products of the foundry (Fig. 32, page 30).

The andiron actually precedes chronologically the mediæval fireback. Though used to support the logs in a built-in fireplace, andirons formerly fulfilled the same purpose in the open fires in the centre of the great hall. Some had figures, others were of vase or obelisk form, but all showed the characteristic lumpiness of the material. (The example, illustrated on Fig. 33, page 30, is interesting, as being obviously designed to match the Richard Lenard fireback.)

By the fourteenth century the ironmaster had enough metallurgical knowledge to realise that molten metal, run off into rough moulds in the ground, need not be discarded if not used for direct casting. It could be again refined at the forge and converted into malleable iron. Thus by an indirect process cast iron came to be made at small fineries into wrought iron, and though in Edward III's reign (1327-77) most if not all of the iron used was the product of the bloomery or forge, we know that by Henry VIII's reign (1509-47) most of it was smelted in a furnace and either cast into pigs for subsequent refining as wrought iron, or cast direct into finished articles. Previously iron had been produced only in a pasty form which for a large or intricate piece of work demanded careful and laborious forging. But when

Fig. 15. Fireback with rope moulding, and three repetitions of a dagger. Sussex, fifteenth century. Dimensions: 1 ft. 6 ins. by 2 ft. 4½ ins.
From the Victoria and Albert Museum, by whose courtesy this is reproduced. Crown copyright reserved.

once the ironmasters had apprehended the properties of cast iron they had virtually a new material which could be cast into any shape or form desired.

Whether the art of casting iron was discovered in England or was brought from the continent, it is certain that its early history was greatly affected by foreign influence. Henry VIII employed many foreign specialists at his gun foundries, including the famous Peter Baude, Aroanus de Cesera and Van Cullen. At the Tower of London there are still two cast iron cannon brought over from Ireland about 1500, and in 1516 a very large gun called the Basiliscus weighing nearly 5 tons was cast in London. Until recently the first authentic documentary evidence of the early cast iron cannon had been that referred to in Holinshed, recording the cannon cast at Buxted in Sussex in 1543 by Ralph Hogge and Peter Baude. Now Dr. H. Schubert, shows that Symart made "three caste gonnes of irron with seven chamberys of yrron" in 1509-10.[21]

Sussex, through its abundant supplies of fuel and its proximity to London, was the natural centre in the sixteenth century of the chief foundry work of the country—the casting of cannon and shot. In 1546 Lord Seymour of Sudley was producing cannon and shot at Worth and Sheffield in Sussex. In 1573 Ralph Hogge took action over the infringement of his patent for casting cannon, and a list was drawn up of the chief ironmakers of Sussex. This included: Her Majesty the Queen; the Earls of Derby, Surrey, and Northumberland; the Lords Abergavenny, Montague, Buckhurst, and Dacre;

Sir Thomas Gresham, Sir John Baker; Sir Richard Baker; Sir Alexander Culpepper; Sir John Pelham; Sir Robert Tirwett, and Sir Henry Sydneye. As a result of this action the Privy Council ruled that no one could make or sell cannon without a licence from the Queen, and to ensure that this armament industry was controlled the Master of Ordnance had to be provided with information regarding each piece cast and to whom it was sold. In 1576 all casting of ordnance in the Weald was stopped on the grounds that the country's supply was adequate. But this does not seem to have been an effective deterrent to the export trade; gun running continued; and similar measures were taken in 1579, 1588, 1589 and 1602. That this cannon casting was not confined to the Weald, is disclosed by documents referring to guns made by Edward Matthews of Radyr near Cardiff in 1602.

The greatly increased production of iron in the sixteenth century led to various difficulties in the supply of fuel. The reduction of the forests that had seemed of benefit to agriculture and transport in earlier times now became a serious menace, and in 1543 an Act was passed regulating the cutting of woods and coppices; but this Act did not include Surrey, Sussex and Kent. The shortage of timber continued, and threatened the ship building industry so that in 1558 an Act was passed prohibiting the use as fuel of any timber growing within fourteen miles of the coast or navigable rivers. But again the Weald was exempted. This fuel shortage and the actual or threatened control stimulated ironmasters to search for an alternative to charcoal for smelting and refining.

Several patents were granted later in the sixteenth century dealing with the use of stone coal, pit coal and sea coal; but it is not clear whether these refer only to refining. In 1611 the first patent concerning coal used for smelting was granted to Simon Sturtevant. Nothing practical came of this and the patent was withdrawn and another issued to John Rovenzon, Sturtevant's assistant. Again nothing seemed to mature, and in 1621 Lord Dudley was granted a patent for a process worked out by his son, Dud Dudley.

Dud was the son of Edward, Lord Dudley and Elizabeth Tomlinson "a base collier's daughter", and was brought down from Balliol to supervise his father's furnaces at Pensnett in Worcestershire. It was here that he claimed to have made the revolutionary discovery that ore could be smelted with pit coal, and to have fined many tons at the Cradley Forges. He claimed that ill fortune and antagonism alone stopped him from completing his work and in his famous *Metallum Martis*, 1665, he sets forth his claims and also his misfortunes. Writers in the past have usually taken him at his own valuation as an inventor, but Professor Ashton has made a careful investigation of the man and his claim to have made with coal, iron that was superior in quality to that smelted and fined with charcoal. "This claim at least was exaggerated —unless indeed Dud Dudley had succeeded beyond the powers of all other ironmasters down to the present day. But there are reasons for doubting whether iron of even moderate quality was ever produced by him with mineral fuel. No mention is made in his treatise of any attempt to coke the

Fig. 16. Part of an heraldic fireback, with a lion rampant reguardant, two roses and two apes. Sussex, fifteenth century. Dimensions: 2 ft. ½ in. by 1 ft. 11 ins. *From the Victoria and Albert Museum, by whose courtesy this is reproduced. Crown copyright reserved.*

coal, and with the blowing apparatus of the seventeenth century it would appear to have been impossible to produce sound iron with raw fuel. If, as he would have us believe, he was a high-minded patriot actuated only by a desire to save the timber vital to England's security, it is strange that he allowed his knowledge to die with him. But the few details of his life gathered from sources other than his own writings do not by any means indicate high-mindedness. That the memorial he erected to his dead wife should record his own rather than her virtues, is indicative of a boastful and assertive nature."

After referring to his litigiousness, and giving various examples of his singularly cantankerous character, Ashton points out that though such details of personal conduct do not annul Dud Dudley's claim to be an inventive genius, "they do decidedly modify the picture of the unrequited patriot that has so often been exhibited. And they discipline one from accepting without reservation a story the evidence for which is his own word unsupported by that of any less partial witness. That Dudley did produce some sort of iron with mineral fuel is probable enough, but that this was of sound merchantable quality is very unlikely; and there is no valid reason why this Balliol under-

FIG. 17. Fireback with two star-shaped panels, each enclosing within a wreath an heraldic crest of a greyhound, or talbot, a dog once used for hunting, that resembled a hound and a beagle. Sussex, sixteenth century. Dimensions: 2 ft. 4¼ ins. by 4 ft. 10¾ ins. *From the Victoria and Albert Museum, by whose courtesy this is reproduced. Crown copyright reserved.*

FIG. 18. Fireback with close, repeating, decorative pattern. Sussex, sixteenth century. Dimensions: 2 ft. 1 in. high. From the Victoria and Albert Museum, by whose courtesy this is reproduced. Crown copyright reserved.

Fig. 19. The Royal Arms, as used by the Tudor Sovereigns, except Henry VII. Sixteenth century. Dimensions: 1 ft. 11½ ins. by 2 ft. 8½ ins. *From the Victoria and Albert Museum, by whose courtesy this is reproduced. Crown copyright reserved.*

Fig. 20. The Arms of Philip II of Spain. Late sixteenth century Spanish fireback. Dimensions: 2 ft. ½ in. high. *From the Victoria and Albert Museum, by whose courtesy this is reproduced. Crown copyright reserved.*

FIG. 21 (*Above*). Part of a fireback bearing the Royal Arms and Supporters of Scotland, with motto: "In defens". Usually attributed to Sussex, but possibly made by English craftsmen working in Scotland. Late sixteenth century. Dimensions: 1 ft. 4¼ ins. by 1 ft. 10¼ ins.

FIG. 22 (*Below*). Fireback inscribed: "Made in Sussex by John Harvo". The heraldic panel is flanked by strips of the cable twist device that appears on much earlier examples. (See Figs. 13, 14, 15 and 16.) Sussex, sixteenth century. Dimensions: 2 ft. 6 ins. by 3 ft.

Both from the Victoria and Albert Museum, by whose courtesy they are reproduced. Crown copyright reserved.

FIG. 23 (*Above*). Stove-plate. Allegorical subjects with inscriptions. Flemish, late sixteenth century. Dimensions: 2 ft. 6¾ ins. by 2 ft. 2½ ins.

FIG. 24 (*Left*). Stove-plate. The marriage feast in Cana of Galilee. (St. John I, 2, 1.) German, late sixteenth century. Dimensions: 1 ft. 9 ins. by 2 ft. 3⅝ ins.

Both from the Victoria and Albert Museum, by whose courtesy they are reproduced. Crown copyright reserved.

FIG. 25 (*Left*). Stove-plate. A vase of flowers. Flemish, seventeenth century. Dimensions: 2 ft. 11⅜ ins. by 2 ft. 2⅛ ins.

FIG. 26 (*Right*). Stove-plate. Moses and the Serpent in the wilderness. German, seventeenth century. Dimensions: 2 ft. 11¾ ins. by 2 ft. 1¾ ins.

Both from the Victoria and Albert Museum, by whose courtesy they are reproduced. Crown copyright reserved.

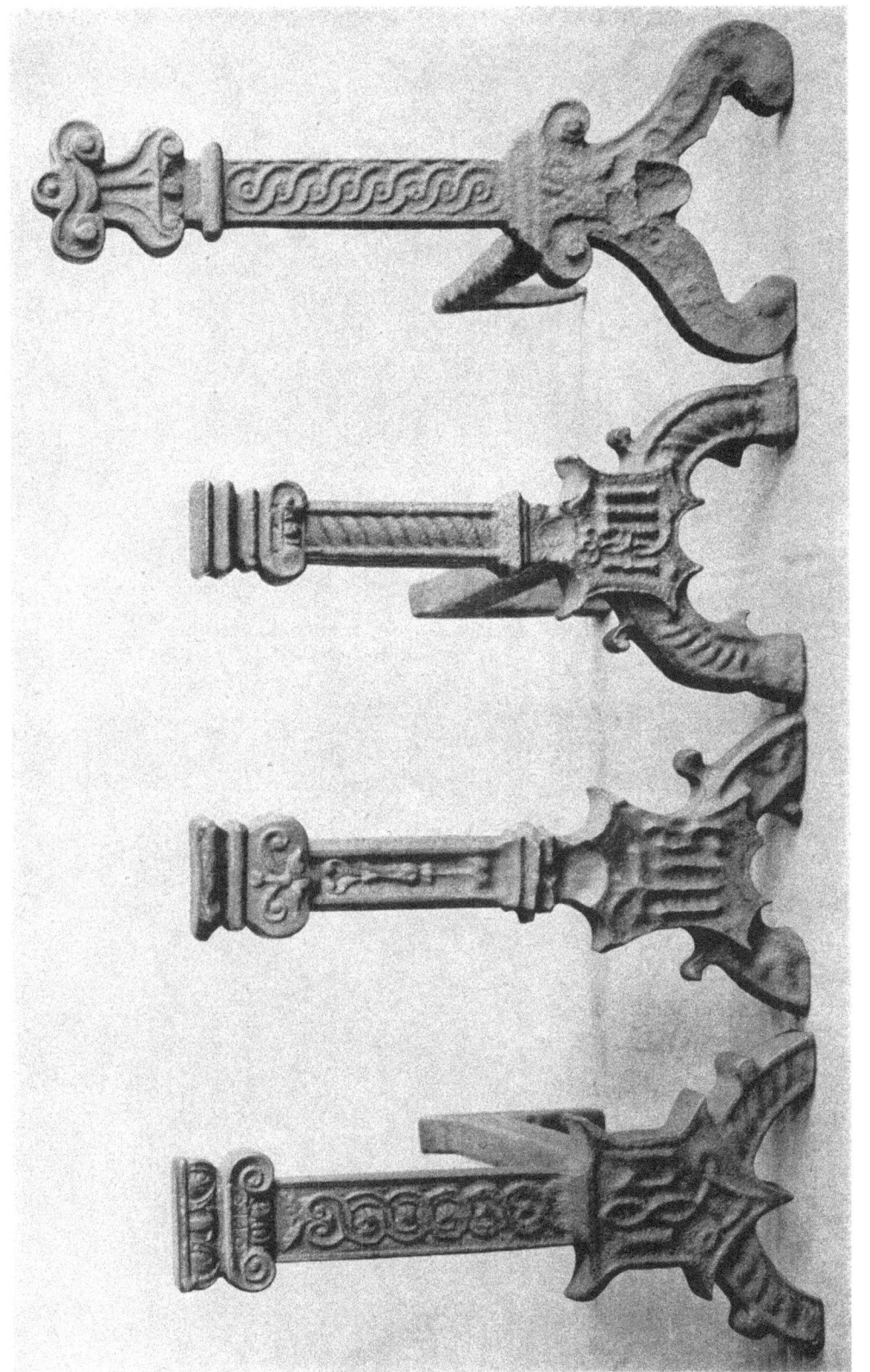

Fig. 34. English firedogs, sixteenth and seventeenth centuries. The Ionic capital has inspired three of them, and as usual in this period of fumbling with classic motives, an assortment of ornamental oddments, drawn from antique sources, are casually assembled in all four examples. They lack the independent vigour of the Lenard firedog shown opposite in Fig. 33.

From the Victoria and Albert Museum, by whose courtesy this is reproduced. Crown copyright reserved.

FIG. 35 (*Above*). Fireback. A cock with an oak wreath. Eighteenth century. A design that exemplifies firmness and restraint. Dimensions: 1 ft. 3⅞ ins. by 1 ft. 4 ins.

FIG. 36 (*Below*). Fireback. Arms of the Dauphin of France. French, eighteenth century. Dimensions: 1 ft. 8½ ins. high.

Both from the Victoria and Albert Museum, by whose courtesy they are reproduced. Crown copyright reserved.

Fig. 37. Fireback bearing the Arms and Supporters of Conroy of Llanbrynmair, Montgomeryshire. Welsh, nineteenth century. Dimensions: 1 ft. 10¼ ins. by 1 ft. 8⅜ ins. The improvement in foundry technique was not accompanied by a corresponding improvement in design: the old, fluent ease of composition, even when disciplined by the requirements of an heraldic subject, has been replaced by rigid and uninspired realism. Compare this with the vigorous design of the sixteenth century fireback by John Harvo (Fig. 22) or the seventeenth century Lenard fireback (Fig. 32). The earlier examples illustrate an imaginative mastery of techniques—this Welsh fireback discloses how technique can intimidate imagination.

From the Victoria and Albert Museum, by whose courtesy this is reproduced. Crown copyright reserved.

graduate, rather than any one of a dozen other projectors of the seventeenth century, should have been singled out for fame".[22]

Many other patents were granted right up to 1692, but none seems to have had any practical result and the fuel shortage and controls continued to curtail the activities of the industry. But although the Civil War in the middle years of the seventeenth century interrupted the industry, and many Royalist ironworks in the Weald, the Forest of Dean and elsewhere were demolished entirely, it was in this century that the cast iron industry really began to be established; from 1650 it was destined to grow with gradually increasing vigour for a hundred years, and to play a leading part in the industrial revolution.

SOURCES OF REFERENCES IN THE INTRODUCTION

[1] "The use of meteoric iron by primitive man", by George Frederick Zimmer, A.M.Inst.C.E. *The Journal of the Iron and Steel Institute*. Vol. XCIV, 1916, p. 306.

[2] "Copper and its alloys in early times," by William Gowland, Assoc. R.S.M., F.R.S., Emeritus Professor of Metallurgy at the Royal School of Mines. *The Journal of the Institute of Metals*. Vol. VII, No. 1, 1912, p. 24.

[3] *Ancient Egypt*, edited by Sir William M. Flinders Petrie, F.R.S. (Macmillan & Co., London and New York, 1915). Part I, p. 20.

[4] "The Metals in Antiquity", by William Gowland, Assoc. R.S.M., F.R.S., F.S.A., Emeritus Professor of Metallurgy at the Royal School of Mines. *Journal of the Royal Anthropological Institute of Great Britain and Ireland*. Vol. XLII, 1912, p. 281.

[5] "Iron in Ancient India." Notes on pamphlet by Mr. Panchaman Neogi, Professor of Chemistry at the Rajshahi College. *Journal of the Royal Society of Arts*. Vol. LXIII, Nov. 27th, 1914, p. 43.

[6] *The Ancient History of China*, by Friedrich Hirth, Ph.D. Columbia University Press, 1923, p. 203.

[7] Extract from paper by celebrated Swedish archeologist Montelius, at 31st General Meeting of the German Anthropological Society at Halle. *The Journal of the Iron and Steel Institute*. Vol. LVIII, No. 2, 1900, p. 514.

[8] "A Cast-Iron Ring 2,500 years old: Further discoveries in the Býčí Skála Cave", by Professor K. Absolon. (Founder of the Anthropos Institute of Czechoslovakia.) *Illustrated London News*, Vol. 209, No. 5614, November 23rd, 1946.

[9] Caesar. *De Bello Gallico*. Book V. Ch. 12. Translated by MacDevitt.

[10] *Iron in Antiquity*, by J. Newton Friend, D.Sc., Ph.D., F.I.C. Charles Griffin & Co. Ltd. 1926. p. 116.

[11] *A guide to the antiquities of the early iron age of central and western Europe*, by R. A. Smith. The British Museum, 1905, p. 4.

[12] *Trade-Routes and Commerce of the Roman Empire*, by M. P. Charlesworth. (Cambridge University Press, 1924). Chapter 12, pages 216-217.

[13] "Notes on a bloom of Roman iron found at Corstopitum (Corbridge)", by Sir Hugh Bell, Bt. *The Journal of the Iron and Steel Institute*. Vol. LXXXV, No. 1, 1912, p. 118.

[14] *Roman Folkestone*, by S. E. Winbolt. (Methuen, 1925).

[15] *Roman Britain and the English Settlements*, by R. G. Collingwood and J. N. L. Myres. (Oxford University Press, 1936). Pages 226 and 232-234.

[16] *History of the Iron Trade*, by Harry Scrivenor. Longman, Brown, Green and Longmans. 1854, p. 33.

[17] "The account roll of a fifteenth-century ironmaster", by Gaillard Thomas Lapsley. *The English Historical Review*. Longmans Green & Co. Vol. XIV, No. 53, 1899, p. 509.

[18] *Iron and Steel in the Industrial Revolution*, by T. S. Ashton, M.A., University Press, Manchester and Longmans, Green & Co. 1924, p. 2.

[19] *Iron and Steel Today*, by John Dearden, B.Sc., A.M.I.Mech.E. Oxford University Press. 1943, p. 16.

[20] "English Cast-iron II", by Professor W. R. Lethaby. *The Builder*, Nov. 5th, 1926, p. 741.

[21] "The first cast-iron cannon made in England", by Dr. H. Schubert, M.A., D.Phil. *The Journal of the Iron and Steel Institute*. Vol. CXLVI, No. 2, 1942.

[22] *Iron and Steel in the Industrial Revolution*, by T. S. Ashton, M.A. University Press, Manchester and Longmans, Green and Co., London, 1924, p. 12.

SECTION ONE

SECTION ONE

THE CAST IRON INDUSTRY BECOMES ESTABLISHED: 1650 - 1750

BY the mid-seventeenth century the fuel situation in Sussex had become serious. There was a threatened shortage of domestic fuel and timber for shipbuilding which alarmed the government: the Civil War had restricted trade; and after the destruction of Royalist ironworks, many Sussex ironmasters removed their works to other centres, chiefly Merthyr Tydfil and Aberdare, where there were natural deposits of the ore and plenty of woodland for fuel. The ironworks of the Weald were gradually disappearing, though there were still some isolated works and foundries in Sussex. In a petition of the Sussex Ironmasters, reference is made to twenty-seven furnaces and forty-two forges being in operation in 1653; but by 1664 there were only eleven furnaces and eighteen forges. Even as late as 1770 there were foundries working at Heathfield and Fernhurst, and in 1724 the Lamberhurst furnace, at which the railings for St. Paul's Cathedral were cast, was still operating.[1]

There was a general tendency at that time to concentrate on the smelting side of the industry rather than the refining side, as this required less charcoal fuel per ton of metal handled. This naturally helped the foundry which still worked with the molten metal direct from the furnace.

Robert Plot, in his *Natural History of Staffordshire*, 1686, states:

"For the backs of chimneys, garden-rolls and such like, they use a sort of cast iron which they take out of the Receivers of the Furnaces as soon as it is melted, in great Ladles, and pour it into moulds of fine sand in like manner as they cast other softer Metals." In this contemporary account, a detailed description is given of the method of producing iron at the end of the seventeenth century, as follows:

"When they have gotten their Ore before 'tis fit for the furnace, they burn or calcine it upon the open ground, with small charcoal, wood, or sea-cole, to make it break into small pieces, which will be done in three days, and this they call annealing it, or fiting it for the furnace. In the meanwhile they also heat their furnace for a week's time with charcoal without blowing it, which they call seasoning it, and then they bring the Ore to the furnace thus prepared, and throw it in with the charcole in baskets vicissim, *i.e.* a basket of Ore, and then a basket of Coal S.SS. where by two vast pair of bellows placed behind the furnace, and compress'd alternatly by a large wheel turned by water, the fire is made so intense, that after three days' time the metall will begin to run, still after increasing, till at length in fourteenights' time they can run a Sow and piggs once in twelve hours, which they do in a bed of sand before the mouth of the furnace, wherein they make one larger furrow than the rest, next the Timp (where the metall comes forth) which is for the Sow, from whence they draw two or three and twenty others (like the labells of a file in Heraldry) for the piggs, all which too they make greater or lesser according to the quantity of their Metall: into these when their Receivers are full they let it forth, which is made so very fluid by the violence of the fire, that it not only runs to the utmost distance of the furrows but stands boiling in them for a considerable time: before it is cold, that is when it begins to blacken at top, and the red to goe off, they break the Sow and the pigs off from one another, and the sow into the same length with the pigs, though in the running it is longer and bigger much, which is now done with ease; whereas if let alone till they are quite cold, they will either not break at all, or not without difficulty.

"From the Furnaces, they bring their Sows and pigs of Iron when broken asunder, and into lengths, to the Forges; which are of two sorts, but commonly (as at Cunsall) standing together under the same roof; one whereof they call the Finery, the other the Chafery: they are both of them open hearths, upon which they place great heaps of coal (*i.e.* charcoal), which are blown by bellows like to those of the Furnaces, and compressed the same way, but nothing near so large. In these two forges they give the Sow and piggs five several heats, before they are perfectly wrought into barrs. First in the Finery they are melted down as thin as lead, where the Metall in an hour thickens by degrees into a lump or mass, which they call a loop, this they bring to the great Hammer raised by the motion of a waterwheel, and first beat it into a thick square, which they call a half bloom. Then 2ly they put it into the Finery again for an hour, and then bring it again to the same Hammer, where they work it into a bloom, which is a square barr in the middle, and two square knobs at the ends, one much less than the other, the smaller being call'd the Ancony end, and the greater the Mocket head. And this is all they doe at the Finery. Then 3 the Ancony end is brought to the Chafery, where after it has been heated for a quarter of an hour it is also brought to the Hammer, and there beat quite out to a bar, first at that end: and after that, the Mocket head is brought also 4 to the chafery, which being thick, requires two heats, before it can be wrought under the Hammer into bars of such shapes and sizes as they think fittest for Sale.

"Whereof, those they intend to be cut into rodds, are carryed to the slitting Mills, where they first break or cut them cold with the force of one of the Wheels into short lengths; then they are put into a furnace to be heated red hot to a good height, and then brought singly to the Rollers, by which they are drawn even, and to a greater length: after this another Workman takes them whilst hot and puts them through the Cutters, which are of divers sizes, and may be put on and off, according to pleasure: then another lays them straight also whilst hot, and when cold binds them into faggots, and then they are fitting for sale.

Fig. 38. Inscribed beams in the breast of a disused blast furnace at Coalbrookdale, Shropshire. *Reproduced by courtesy of the Coalbrookdale Company.*

"Thus I say the Iron-works are exercised in their perfection, and all their principal Iron undergoes all the foremention'd preparations; not but that for several purposes, as for the backs of Chimneys, Garden-rolls, and such like; they use a sort of cast-Iron which they take out of the Receivers of the Furnaces as soon as it is melted, in great Ladles, and pour it into moulds of fine sand, in like manner as they cast the other softer Metalls. Thus the ingenious Will. Chetwynd of Rugeley, Esq., at Madeley furnace, cast Iron-Rolls for gardens, hollow like the Mills for Sugar Canes, of 5, 6, 7 or 800 weight a piece. For such purposes as these, this serves well enough, but for others it will not, for it is so brittle, that being heated, with one blow of a hammer it will break all to pieces."[2]

At this period there is evidence of considerable imports into the country, not only of ore and bar iron but of actual finished cast articles, which suggests that the casting art had advanced further in Europe than in England, and perhaps had not been so affected by fuel shortage. Professor Ashton quotes two references in the *Records of Cardiff*. One in 1686 refers, among other items brought from Rotterdam, to "Three hundred and forty Iron pots and

Fig. 39. An early water wheel at Coalbrookdale.
Reproduced by courtesy of the Coalbrookdale Company.

Kettles"; and another, in 1698, to "Seaven hundred and fifty Iron potts and Keetles."[3]

In 1676, at Sedgeley, near Dudley, on the borders of Worcestershire and Staffordshire, Abraham Darby was born—the man who was to be known throughout the world for his pioneering work in the iron industry and particularly on the foundry side. His forbears were yeomen, but his father and grandfather had combined agricultural work with the making of nails and locks. After being apprenticed to a maker of Malt Mills he went to

Fig. 40. The interior of the old moulding shop at Coalbrookdale, showing the foundry cranes. *Reproduced by courtesy of the Coalbrookdale Company.*

Bristol, and in 1699 started an iron and brass works at Baptist Mills.

Abraham Darby experimented with the casting of hollow ware in iron, and in 1704 visited Holland to discover all he could about Continental practice. He returned to Bristol with skilled Dutch workmen and continued his experiments with John Thomas until, in 1707, he took out a patent on a "new Invention of Casting Iron-bellied Pots and other Iron-bellied Ware in Sand only without Loam or Clay". Cast iron pots had been known in England for a long time and had also been imported. Rhys Jenkins quotes

The Postman of December 24-26, 1700, where an advertisement of a Mr. Stringer's Iron Foundry refers to "Potts and Kettles".[4] But it seems that Darby's work had produced *bellied* pots, and that his chief secret was in the making of his moulds or casting boxes.

In 1708 Abraham Darby moved to the district that is always associated with his name, Coalbrookdale in Shropshire, where iron ore and coal were both to be found near the surface, where there was limestone in the vicinity, a good supply of timber for charcoal, and a fast running tributary of the Severn available for power. There were already two furnaces at Coalbrookdale in 1696. These were sub-let to Darby, and he appears to have spent some time and money on preparing the works before they came into operation in 1709.

There have been many references in documents, manuscripts and patents, to various attempts to overcome the difficulty of fuel shortage and to replace charcoal with coal or coke for the fuel used in smelting. This is what Dud Dudley claimed to have done when he took out a patent for the process in 1621. Other patents relating to the use of mineral fuel had been taken out in 1589, 1595 and 1607, though they do not specifically mention the *smelting* of the ore as contrasted with the refining of the metal at a later stage; but those of Simon Sturtevant of 1611 and Rovenzon, like Dudley's, do refer to the production from the ore.

All these early attempts were abortive; either they came to nothing, or, in some way, failed to appeal to the industry, and it is generally agreed that Coalbrookdale was the locality where the actual discovery of the successful, commercial use of mineral fuel in smelting iron ore was made, though whether it was the first Abraham Darby or his son, born in 1711, also called Abraham, who did the pioneer work, is a highly controversial subject. Scrivenor and Beck put the date of the discovery as 1713; then Dr. Percy suggested a much later date, 1730-35, basing his view on an account given to him in 1864 by the wife of a descendant of Abraham Darby.[5] This account originated the now well-known story of the second Abraham Darby who, having produced a suitable coke, spent six days and nights on the bridge of his furnace superintending the charging until he finally saw the molten metal flow out, and was then removed, unconscious, to his home. Toynbee gave the date as 1740-50; Dr. Clapham as 1740; Cunningham as 1735; Mantoux as 1735; and Meade as 1730-35. But Mr. W. G. Norris proved that coal was coked and used in the blast furnaces at Coalbrookdale as early as 1718. Then in 1924 Professor Ashton in a brilliant analysis of the Journal of Abraham Darby and his accounts, completes this proof, only to find, as the final proofs of his book were being read, an original letter in the possession of Alfred Darby, Esq., of Shrewsbury, written in about 1775 by Abiah Darby, the wife of the second Abraham, which seems finally to settle the matter, wholly supporting his analysis by naming Abraham Darby the First as having smelted iron ore with coke at Coalbrookdale.[6] (This letter is reproduced in Appendix I, page 375.) Though no date is given, it would seem that Scrivenor's first date of 1713 was very near the truth. By that date anything from five to ten tons of iron was being made each week at Coalbrookdale. Some of this was run into pigs and the remainder cast, direct from the furnace, as pots, kettles and other hollow ware. Later, in the Journal of Abraham Darby, grates, smoothing irons, pestles and mortars are mentioned, and a reference which shows that as many as 150 pots and kettles were cast in a week.

By 1700, owing to the restrictions which had been imposed over a long period on the number and location of furnaces, and the exhaustion of fuel supplies, a general dispersal of the various centres of smelting had occurred. New centres were set up where timber was available as charcoal, and this movement continued until the end of the eighteenth century. From 1680 up to 1775 the old centres in Sussex, the Forest of Dean and Shropshire were gradually declining in importance as others were created or enlarged.

The accompanying table shows the location and number of furnaces and forges and is based on the figures given by David Mushet about 1720. In *Iron and Steel in the Industrial Revolution*, Professor Ashton suggests that these figures are under-estimated (see Section Two, page 56); but they give a general indication of the location and number of furnaces and forges of the period:

Location	Furnaces	Forges
Kent	4	1
Sussex	10	8
Hants	1	1
Berks	–	1
Gloucester	6	7
Hereford	3	5
Monmouth	2	8
Shropshire	6	10
Worcester	2	10
Warwick	2	4
Stafford	2	12
Chester	3	3
Flint	–	1
Denbigh	2	1
Montgomery	–	1
Cardigan	–	1
Brecon	2	1
Glamorgan	2	1
Caermarthen	1	4
Pembroke	–	1
York	6	9
Derby	4	4
Notts	1	4
Lancs	–	1
Cumberland	–	1

This dispersal is an eighteenth-century characteristic of all sections of the industry. Furnaces, forges, smithies, rolling mills and foundries were often remote from each other, though in general the foundries were inclined to stay near the furnaces from which the molten metal was still more often than not, used direct. But the melting of pig iron for foundry uses was already practised, and isolated foundries were to be found, for example, at Bristol and Liverpool, where American as well as British pig iron could be delivered by ship. The mills were set up where water power was available, and the forges and smithies near coal that was easily accessible. The shortage of timber fuel also meant a shortage of British pig iron, and bar metal was imported on a large scale from Norway, Sweden, Spain and Russia.

The industry in the early eighteenth century was very clearly divided into two branches—one dealing with wrought iron, the other with cast iron, the former being far larger and having little interest in the foundry and its experiments with coke pig iron. It was this fact, perhaps, together with the natural dispersal and the difficulties of transport and travel, that accounted for the long period between Abraham Darby's first use of coke for smelting and its eventual use on a large scale throughout the country. While there seems no reasonable doubt that Abraham Darby I did use coke in this way very early in the eighteenth century, the fluctuating success of his experiments, the occasional failures, and his modest reticence as a Quaker, would account for the lack of publicity given to the discovery.[7]

By no means all coal was satisfactory when coked, and there is evidence that in 1718 there was an accumulation of defective and waste iron at Coalbrookdale indicating that the smelting was not always successful.[8]

Charcoal, owing to its liability to be crushed by the weight of iron ore in the furnace, limited the size of the furnace, and from the ironfounders' point of view was not as satisfactory as the early coke, for the latter produced a metal so molten that it could easily be poured into fine and intricate moulds. The fused metal from the charcoal furnace was much less easily poured, and resulted in the heavier castings of the earlier period, typified by fire-backs and grave slabs. In the forges and refineries, the iron produced by smelting with charcoal was still preferred by the forgemasters for its finer quality and easier working, and it was not until about 1750 that coke smelting became general. Shortly after 1750 new coke furnaces were erected at Coalbrookdale, Horsehay, Ketley, Madeley Wood, Lightmoor, Willey, and in South Wales. Until the end of the century there were still a number of charcoal furnaces producing iron for special high-quality work; but by the middle of the century enough experience had been acquired and enough knowledge of the type of coal to use and the control methods to be adopted for ironmasters generally to realise that this cheaper fuel was a commercial proposition.

It is not certain when pig iron was first remelted on any large scale for foundry purposes. Casting was generally done direct from the furnace, the molten metal being run straight into moulds, or carried to them in ladles. It is recorded how Isaac Wilkinson, about 1730, when working at the Backbarrow furnace in Lancashire, carried molten iron in ladles from the blast

Fig. 41. "View of the upper works at Coalbrookdale, in the county of Salop." A print designed and published by G. Perry and T. Smith in 1758. (T. Vivares, Sculp.) The foundry buildings are in the foreground, and on the right are four heaps of burning timber, forming charcoal for the furnaces. A large cast iron cylinder is being dragged by a team of horses.
Reproduced by courtesy of R. Darby Esq.

FIG. 42. Some very early examples of cast iron railings designed by James Gibbs for the east front of the Senate House at Cambridge, 1722 - 30. These have a wrought iron bar between each cast iron baluster, and are similar to the railings that surround the statue of Henry VI at Eton. The first recorded use of cast iron railings in England is of those fixed round St. Paul's Cathedral in 1714. (See page 37.) A close-up detailed view of the Senate House railings is shown opposite in Fig. 43.

Copyright: Country Life.

furnace and took it across the road to the foundry where he cast it into pots.[9] But a reference found by Rhys Jenkins shows that pig iron must have been remelted for foundry use at a much earlier date.[10]

In *The Postman* of December 24-26, 1700, there is the advertisement, referred to on page 42 which says, "At Mr. Stringer's Iron Foundry and Refinery in Blackfryars near Ludgate, are cast without Wood, Charcole or Bellows, Cannons, Bombs, Shot Sheels, etc., Bells of any size or tone, Potts and Kettles, hollow Rolls, Stoves, Cockles, and Bars for Sugar-Works, solid large Rolls for flatting of Iron, Brass, Copper or Lead, Rolls for Mints, Stoves, Backs and Hearths for Chymneys, Flower Pots, and Balconies, and

FIG. 43. See Fig. 42 opposite.
Copyright: Country Life.

Hatter Basons, Plates for Packers and Hotpresses, very large plates for looking-glass Grinders, Cylinders for Water-works, various things for Millwork, Boxes for Coaches, Carts and Drays, Anvils for Smiths and Forges. All sorts of Chymical Vessels that can be made in Iron or in Stone Glass. Iron is there made of any temper desired, either so hard that no file cannot (can?) touch it, or so soft as to Bore or Turn as Wood. Those that have any quantities of broken Guns, or other old cast iron, may have money for them. As also money for course Copper, old Brass or Bell-Metal, Copper ores, Lead ores, or any sort of Metallick Bodies, whether Foreign or Domestick. This Iron Foundery meeting with such encouragements, requires the best Founders in Loom, and those that can mould in Sand, will find suitable encouragement, with all necessary provisions of Life".

As Rhys Jenkins remarks, "The articles are cast without wood, charcoal or bellows, so that the blast furnace is ruled out, and we are left with the reverberatory furnace using pit coal". This type of furnace had been known for metallurgical work for some time prior to this date. Samuel Smiles quotes a letter from Mr. Reynolds of Coalbrookdale, dated 1766, in which he mentions "Thos. Tilleys air-furnace", which can refer only to the reverberatory furnace. A little later there were similar furnaces for remelting pig iron in use at Newcastle and Carron, which are referred to and described by M. Jars.[11] In the reverberatory furnace the metal was not allowed to come in contact with the fuel, but only with the flame which was forced down, or reverberated, from the roof of the furnace. Thus it was possible to use coal as a fuel, and this method was used in many foundries until a later date when the cupola for melting pig iron for foundry work was adopted after W. Wilkinson's invention in 1795.

During the eighteenth century, cast iron was introduced for machinery; and the advance in foundry technique, which produced better and more reliable castings, made possible many of the engineering achievements of the industrial revolution. The earliest use made of cast iron in machinery was probably by Sorocold in the London Bridge Works for pumping Thames water to supply the City of London. Peter Morris in 1582 had erected the first Works, but these were obsolete and Sorocold began to rebuild the scheme about 1704. From Henry Beighton's description, the spindles, four-throw crankshafts, pump barrels and pipes were of cast iron.[12]

Abraham Darby I died in 1717 when his son was only six years old, but his widow, with Thomas Goldney and Darby's son-in-law, Richard Ford, formed a new partnership, and under the energetic managership of the latter the firm continued to prosper. Mortars, boilers, firebacks, garden rollers and pipes, were the chief products of the firm, which, though having its own furnaces and forges, concentrated on the foundry branch of the industry; and Ford was at this time producing some of the first sound machinery castings.

In 1705, Newcomen, a blacksmith who had been employed by Captain Savery to work on his invention of the fire-engine, patented in 1698, produced many improvements of his master's design. These engines had previously had cylinders of brass, pipes of lead, and a beam of wood, but after Newcomen had visited the Midlands in 1715, he apparently heard of Darby's work, and in 1718 the Darby foundry was turning out iron castings for these engines. By 1724 the firm had a reputation for this kind of work and were producing cast iron cylinders, pipes, and other parts for the Newcomen engine regularly and in some quantity.

By 1743 Abraham Darby II was in charge at Coalbrookdale. This, like

FIG. 44. The gates to Chirk Castle, near Llangollen, Wales, *circa* 1721. Cast iron is used for the bases, capitals and moulded balusters of the piers. This represents a very early use of the material in this particular form.
Copyright: *Country Life*.

other districts relying on water power, suffered serious loss during a summer drought, as the water was often insufficient to work the furnace bellows, necessitating the shutting down or "blowing slow" of the furnaces. Abraham Darby II conceived the idea of installing a fire engine to pump the water back from the lower river and return it for use again over the wheels, and in this year, 1743, an engine was erected. It was such a success that soon most furnace owners were using the same device. (See Appendix I, page 376). The successful casting of engine parts in the Dale was helping and cheapening the production of engines which, in turn, were helping to step up the production of iron. Darby's reputation for this kind of work spread, and in 1756 and 1763 James Brindley ordered from this firm the parts for engines he was erecting at Fenton Vivian and the Walker Colliery near Newcastle.

Cast iron was beginning to affect the character and efficiency of mechanical engineering; it was soon to influence the technique of structural engineering and architecture.

SOURCES OF REFERENCE IN SECTION ONE

[1] "Ironworks of the County of Sussex", by Mrs. M. A. Lower. *The Sussex Archæological Collections.* Vol. II, 1849, p. 203.
[2] *The Natural History of Staffordshire*, by Robert Plot, LL.D., Keeper of the Ashmolean Museum. Printed at the Theatre, 1686.
[3] *Iron and Steel in the Industrial Revolution*, by T. S. Ashton, M.A., University Press, Manchester and Longmans, Green & Co. 1924, p. 27, Note 1.
[4] "The beginnings of iron founding in England", by Rhys Jenkins. *The collected papers of Rhys Jenkins.* Printed for the Newcomen Society at the University Press, Cambridge. 1936, p. 120
[5] *Metallurgy*, by Dr. John Percy, M.D., F.R.S., p. 888. Published by John Murray, Albemarle Street, London, 1864.
[6] *Iron and Steel in the Industrial Revolution*, by T. S. Ashton, M.A., University Press, Manchester and Longmans, Green & Co. 1924. Appendix E, p. 249.
[7] *Ibid*, pp. 32, 33, 34.
[8] "Ironworks", by John Randall, from *The Victoria History of Shropshire*, edited by William Page, F.S.A., Vol. 1, p. 462. Published by Archibald Constable & Co. Ltd., London, 1908.
[9] *John Wilkinson, Ironmaster*, by H. W. Dickinson, A.M.I.Mech.E. Published by Hume Kitchin, 7 Market Street, Ulverston, Lancs. 1914. p. 11.
[10] "The beginnings of iron founding in England", by Rhys Jenkins. *The collected papers of Rhys Jenkins.* Printed for the Newcomen Society at the University Press, Cambridge. 1936, p. 120.
[11] *Voyages Metallurgiques*, by M. Jars. Published in 1774.
[12] "A description of the water-works at London Bridge", by H. Beighton, F.R.S. *Philosophical Transactions.* Vol. XXXVII. 1731-2. Printed for the Royal Society, 1733. p. 5.

SECTION TWO

SECTION TWO

THE NEW MATERIAL IN ARCHITECTURE: THE RISE OF CAST IRON: 1750-1820

WE have seen that the change from charcoal smelting to coke smelting was a slow process, hindered by the suspicions of the ironmasters, the difficulties of transport if coal were not available locally, and the fact that all coal was not satisfactory when coked. But by 1750 the new method of production by mineral fuel had sensibly affected the charcoal production of iron in the country. Towards the end of the eighteenth century various factors were to influence production still further. Cast iron was helping the inventors and engineers working on the steam engine. That engine was used to produce blast for the furnace, thus effecting a more thorough reduction of the ore. Transport was improved by the construction of new roads and bridges, and particularly by the new network of canals. The canal—and later the steam engine—diminished the importance of rivers as the means of transport and power, and mineral fuel released the ironmaster from the necessity of siting his works near woods and forests. Instead of the various branches of the industry being scattered, the tendency was for them to be concentrated round the coal fields. The canal from Birmingham *via* Smethwick and Bilston to Wolverhampton, built in 1767, resulted in coal in Birmingham dropping from 13/- a ton to 8/4 a ton.[1]

The old charcoal pig iron was apt to "run thick", but the coke-smelted iron was free from this objection and was more manageable, and this gave a great stimulus to the foundry side of the industry. Before the end of the century many foundries were started in various parts of the United Kingdom which not only still exist but still lead in the industry, some remaining in the hands of the same family of ironfounders until the present day.

Prior to 1750 cast iron had been used very little in engineering and building, and was employed chiefly for tools, utensils, firebacks and andirons, grave slabs, and, particularly, cannon and implements of war. For other uses it was a new material and was regarded with the same cautious expectation that until recently was aroused by reinforced concrete, and is now aroused by such substances as plastics. But the seventy-year period from 1750 to 1820 is crowded with the names of adventurous, innovating, mechanical engineers, civil engineers and architects who appreciated the possibilities of the new material. Men like the Darbys, John Wilkinson and his brother, William Wilkinson, John Smeaton, John Roebuck, Matthew Boulton, James Watt, Brindley, Cranage, George Stephenson, John Rennie, Tom Paine, Andrew Meikle, Ransom, Thomas Telford, Murdoch, Richard Trevithick, John

Rastrick, David Mushet, William Wilkins, John Nash, Sir Robert Taylor, George Dance, Sir John Soane, James Gibbs, the brothers Adam and Sir Robert Smirke. These men used cast iron for engineering, for building and for decoration. Apart from the purely engineering uses in machine and engine construction, the material was employed for the first time for bridges, canal banks, locks and lock gates, aqueducts, stanchions, beams, lintels, sills, windows, tram and railway lines, columns, and often took beautiful and highly decorative forms as railings.

English and Scottish ironfounders were progressive and energetic men, with profound faith in the attributes and future of the material they had mastered. The great growth of the foundry at this time is indicated by examining the work of famous ironmasters, civil engineers and architects of the period.

THE IRONMASTERS

The Darbys: A wide expanse of the Midlands came under the influence of the iron industry during this 1750-1820 period; but in 1750 Shropshire was still all powerful and the "Dale" was the centre where experiments and improvements were conducted. Abraham Darby I died in 1717 when his son, Abraham Darby II, was only six years old; and the elder Darby's son-in-law, Richard Ford, carried on the business in partnership with Thomas Goldney and the widow of Abraham Darby I. Ford's enthusiasm and spirit were responsible for the great name this centre acquired as ironfounders for the first important parts of engines. By 1724 cast iron cylinders, pipes and barrels were ordered on a large scale from the Dale Company for use in Newcomen engines, and there appears to be no evidence of cylinders being cast elsewhere until 1760.

Abraham Darby II, when he grew up, showed much of his father's initiative. In 1743 he erected the first engine at Coalbrookdale for pumping back the water to the reservoir for re-use over the wheels driving the bellows for the blast. It is claimed, and substantiated by his wife, Abiah Darby (see Appendix I, page 376), that he was the first to fine coke-smelted pig iron into bars. He continued to improve on the casting of engine parts and on his father's invention of coke smelting, with which Coalbrookdale is enduringly associated.

Coke smelting, apart from accounting for a considerable improvement in the quality of the iron, provided important by-products from the manufacture of the coke. In *The World*, a London newspaper, of October 23rd, 1790, the following item appeared:—

"Lord Dundonald's Mineral Tar, extracted from Coals, at Colebrook Dale, seems to bid fair to be of national utility; it has been tried by several Bristol merchants, who find it answer the purpose, both of preserving the bottoms of ships from being injured by the worms, and also proving more durable than the Vegetable Tar, extracted from the Pine. It answers another useful purpose. The cannon cast by means of the coals thus charred, and freed from the sulphureous bitumen, are not liable to burst—a matter of great consequence.

FIG. 45. Thomas Farnolls Pritchard of Shrewsbury, the first architect responsible for the large-scale use of cast iron. He designed the first cast iron bridge which was erected in 1779 over the River Severn. This portrait is reproduced from a drawing made in 1824 by C. F. Thatcher, from an original by Worlidge. The lettering and date were probably added by Thatcher, as Thomas Worlidge died in 1766.

Pritchard was invited to produce the design by the bridge company, which had been influenced by John Wilkinson, and the designs were cast by the local foundries of the Darbys. (See pages 62 and 82, and Figs. 80, 81 and 82.)

The drawing from which this illustration is reproduced is in the possession of the Royal Institute of British Architects.

55

"This is an old discovery which has lain dormant above a hundred years, and is to be found in the Philosophical transactions."

It was natural that the production of coke pig iron was most conspicuous in this neighbourhood and spread only gradually to the rest of the country. But the effect on iron production before the end of the century was vast. In 1720 the total tonnage of pig iron for England and Wales was 17,350 tons, but by 1788 the production was 61,300 tons, and of this 48,200 tons had been smelted by coke.[2] Professor Ashton shows in his *Iron and Steel in the Industrial Revolution*[3] that these figures were almost certainly under-estimated, and have other inaccuracies, but as an approximate guide to the proportions and location of the main centres of the industry, they are of some value. (See table on page 43.) The 1788 list shows that Shropshire had twenty-one coke furnaces and produced 21,300 tons of coke pig iron, nearly one-half of the total quantity for England and Wales. The production of 1,100 tons per furnace was greater than that of any other district in England and was only equalled by Glamorganshire. At this date there were still three charcoal furnaces in Shropshire, each averaging 600 tons per year. By the end of this great period of expansion, 1820, the number of coke furnaces in Shropshire had doubled, and the annual production tonnage of coke pig iron had risen to 55,000. But by this time the ironmasters had been converted, and the increase in the rest of the country exceeded even that in Shropshire; Staffordshire had as many furnaces, and one of the Welsh works, Cyfarthfa, exceeded all others in production.[4]

Coke smelting, the successful casting of bellied iron pots, and the making of engine parts was to be followed in Coalbrookdale by yet another invention which was to influence the iron industry. The niece of Abraham Darby I married Thomas Cranage who was in charge of the Company's forge at Bridgnorth and whose brother George was a founder at the Dale Works. These brothers, in 1766, took out a patent for "making pig iron or cast iron malleable in a reverberatory furnace with pit coal only." The pig iron was separated from the coal fire by a low wall of bricks over which the flame was carried by the draught to play on the metal, without the latter being in direct contact with the fuel. In spite of difficulties experienced with the burning out of the furnace bottom some success was certainly attained in achieving an iron suitable for all purposes and of a uniform quality. Earlier attempts had been made to use the reverberatory furnace before fining at the forge. but the successful work of the brothers Cranage marked the first real gain of the mineral fuel industry over charcoal in the wrought iron field, and influenced the industry considerably before Henry Cort perfected the puddling process in 1783-4. Though not directly affecting the foundry, the use of mineral coal in this way greatly increased the demand for cast iron in the form of pig iron.

There are several stories about the first use of cast iron rails for tramways. One describes the use of this material by Abraham Darby II on the tramway between Horsehay and Coalbrookdale between 1750 and 1763. Others give

Fig. 46. Thomas Telford, 1757 - 1834. The first British civil engineer to appreciate the possibilities of cast iron for large-scale engineering works.
From an engraving from W. Holl after the portrait by Samuel Lane.

the credit to Richard Reynolds, Manager from 1763-1768. It is claimed that Reynolds was responsible for the laying of a considerable run of rails between Ketley, Horsehay and the Dale, and that from 1768 to 1771 some 800 tons of cast iron rails were laid. As Rhys Jenkins says, "It is possible that a short experimental length of line had been laid by Darby, and this proving satisfactory, upon his death in 1763, it fell to Reynolds to carry out the idea on a large scale." [5] We do know that in 1767, during a trade slump that affected the iron industry, the Coalbrookdale Company used its surplus metal to make rails of cast iron to replace the wooden ones on its waggon ways. In the course of a tour to North Wales in 1784, the Hon. John Byng—who became the fifth Viscount Torrington—visited the neighbourhood of

Coalbrookdale, and recorded the flourishing state of the iron foundries in the locality. He observed that "every cart belonging to this trade is made of iron, and even the ruts of the road are shod with iron!" [6]

The influence of the Coalbrookdale group of firms started by Abraham Darby I continued to spread to the iron industry over nearly the whole country. Though there is no Darby in the Dale today, there are still men of that name associated with the industry—direct descendants of Abraham Darby I.

The Darbys created a powerful family tradition. From the first, their Quaker background precluded them from many of the ordinary activities of the locality, and they seem to have concentrated wholly on the business and the welfare of the neighbourhood and their employees, thereby fostering an affectionate respect, that clearly emerges from contemporary accounts of functions in the district. The scale on which these functions were held can be understood from the following contemporary newspaper account—though later than the period under consideration—of the festivities provided for the workpeople on the birth of a son to Mr. and Mrs. Alfred Darby in 1850.

"Grand Procession at Horsehay.

"Yesterday between 3,000 and 4,000 of the men employed in the Horsehay Ironworks and Dawley Collieries (the property of the Coalbrookdale Company) walked in procession, each wearing a pink and white scarf, and a band round their hats upon which was printed the words 'Long life to Edmund William Darby.' The procession was headed by two men on horseback, one named William Ball, a giant about twenty-two score weight, and the other Benjamin Poole, a dwarf, scarcely as many pounds. The giant was lifted upon his horse by means of two large cranes erected purposely for the occasion. The route taken was through Coalbrookdale, Ironbridge, Madeley, and through Dawley. On their return to Horsehay, upwards of 15,140 lbs. of meat (19 beasts and 42 sheep having been cut up), 1,700 loaves, and 1,000 gallons of ale were distributed among the workmen. Three excellent bands of music were in attendance, and some first-rate flags. On the banners were various mottoes, amongst which the following were particularly conspicuous—'He who renders Education more accessible confers a lasting boon upon the community,' 'Hail, child of our hopes, may happy years be thine,' 'Long live the name of Darby'.

"On Wednesday next about 1,000 children are to be regaled at Horsehay with tea and cake; and on Friday next the agents, clerks and chartermasters are to dine together in the large school room recently erected by the Messrs. Darby.

"There is also to be a grand ball at the Tontine Inn, Ironbridge, on Wednesday evening, as a wind-up to the rejoicings on this memorable occasion."

The Wilkinsons: Isaac Wilkinson, a workman of the Little Clifton Furnace near Workington, moved in the seventeen thirties to North Lancashire where he was engaged as a pot-founder by the Backbarrow Company before starting on his own account at Wilson House, near Lindale, in 1744. It is recorded how he brought iron in a molten condition from the blast-furnace, and carried it in ladles across the road to his foundry where he cast it into pots. He took out a patent in 1738 for a new type of cast iron box smoothing iron, which proved very successful, and being a versatile and talented man he gradually extended his operations.

The Backbarrow furnace, erected in 1711, is of historic interest. Some years ago a cast iron lintel was found buried in the stone walls of the furnace, bearing the date 1711 and the initials "I.M.W. R.S.C.", presumably those of the landowners, about whom little is known. This furnace continued to produce charcoal iron of the famous "Lorn" brand, probably the oldest brand of pig iron in the country, until the old furnace ceased to operate in 1920.

In 1753 Isaac Wilkinson moved south and took over the Bersham furnace near Wrexham, where he was helped by his two sons, John and William. Here he engaged in casting on a large scale; his wares including guns, cannon, fire engine parts, cylinders, pipes and sugar rolls; and it was here that he took out his fourth and last patent. This covered a new method of casting guns in dried sand in iron boxes made for the purpose. The method had probably been used before, but great secrecy had always been observed in all branches of the moulders' art, and the account of the process in the patent is, therefore, of particular interest as one of the earliest written descriptions, and is given here in full:

"A.D. 1758, April 21st (No. 723) Isaac Wilkinson, of Barsham Furnace, in the Parish of Wrexham, in the County of Denbigh, Gentleman.

"A New Method or Invention for Casting of Guns or Cannon, Fire Engines Cylinders, Pipes and Sugar Rolls, and other such like Instruments in Dried Sand, in Iron Boxes made for that Purpose; whereby the said Guns or Cannon Fire Engines Cylinders, Pipes and Sugar Rolls, and other such like Instruments will be made and Cast in a much more Neat, Compeat Exact and Usefull, as well as Cheap and Expeditious manner, than any Method hitherto known and made use of"!

The description is:

"The outside or cope of the mould or moulds in which the guns . . . are intended to be cast must be made of sand, mixt with a little horse or cow dung or any other thing to make it porous. This sand is made wett and then rammed up, and the patern being first put in iron boxes made for that purpose of two . . . or any number of parts or peices as the nature of the instrument to be cast requires; then the boxes are to be taken asunder into peices and the patern taken out; then the sand in the boxes is dried in a stove; and when dry it must be blacked or faced with some wett charcoal dust or black lead or any other mixture or thing to make the same come off or part from the metal when cast. The insides or cores of all the different instruments above mentioned are made with iron bars either hollow or full of holes; or sollid and traced or fluted, and if the core is large it may be made of bricks walled and the barrs of iron or bricks are to be wraped round with ropes made of straw or hay to take the air of, and must then be covered with a proper thickness of the said sand, and then dried or blacked as before directed; and then the moulds are put together and the instruments cast . . .".

John Wilkinson was later employed as Manager of the New Willey Company at Broseley, near Coalbrookdale; the concern eventually coming under his sole control in 1763. The Bersham furnace at Wrexham failed financially, but was reconstituted as the New Bersham Company, controlled by John and William. In 1770 John Wilkinson set up a furnace at Bradley, near Bilston, in Staffordshire, and in 1774 he invented a new method of boring cast iron cannon, which put the Bradley works in the forefront of ordnance manufacture. Orders for ordnance came from the Government of France; some of which were alleged to have been filled while that country was at war with Britain—the sort of tale most wars have produced and perhaps exaggerated. But the invention for boring cast iron cannon was to have a peace-time application which accelerated the development of the machine age, and really made the steam engine a practical possibility.

In 1774 James Watt had moved to Birmingham and came into close contact with Boulton at the famous Soho Works, where Birmingham "toys" and various artistic objects were made. Boulton, probably more than the inventor, saw the possibility of Watt's new steam engine and encouraged Watt, finally setting up the firm of Boulton & Watt alongside the Soho factory. Watt had experienced difficulty in getting cylinders from Scotland of sufficiently accurate workmanship, and he subsequently had cylinders and parts cast at Coalbrookdale. But though this was the recognized centre for such work, they were still unsatisfactory and the bore not sufficiently accurate to prevent the leakage of steam. Wilkinson was approached, and with his cannon lathe at Bersham (a lathe which caused the cast cylinder to revolve around a fixed rod on which were the cutters, instead of a rotating rod working inside the casting as in all previous work), he produced the first satisfactory cast iron steam engine cylinders. For many years Boulton & Watt insisted that all cylinders for engines designed by them were to be cast by John Wilkinson at the Bersham or Bradley foundries. Other engine parts were cast by local foundries, particularly at Coalbrookdale and at the famous Eagle Foundry in Birmingham, then controlled by Richard Dearman; but the majority were cast at Coalbrookdale or at the Wilkinsons' foundries, and Abraham Darby II, in 1762, entered into an agreement with the brothers Wilkinson whereby the price for certain cylinders, pipes and other articles was controlled and made uniform. In spite of Wilkinson's patents, other ironmasters in various parts of the country were certainly casting cylinders by the new method, and this led to considerable litigation.

In 1776 Wilkinson erected one of Boulton & Watts' engines to supply blast to his New Willey blast furnaces, and this, with an engine set up for pumping at Bloomfield Colliery, near Tipton in Staffordshire, shares the distinction of being the first of Watt's engines to be erected outside the Soho

FIG. 47 (*Right*). A monument erected at Lindale, Lancashire, to the memory of John Wilkinson, the great ironfounder. This illustration shows the detail of the base.

FIG. 48 (*Below, left*). General view of the Wilkinson monument, which was made in cast iron.

FIG. 49 (*Below, right*). The portrait plaque of John Wilkinson, which was removed for safety, during the second world war.

These photographs are reproduced by courtesy of M. C. Oldham, Esq.

Works, and it was certainly the first to be applied to purposes other than pumping. It was John Wilkinson's influence, perhaps more than any other factor, which made possible the production of the first cast iron bridge in the world, in 1779—the famous bridge over the Severn near Broseley, at a spot since known as Ironbridge. When the project was contemplated Wilkinson's advice was sought, although many people thought he was "iron mad". He strongly urged the use of cast iron for the work of spanning the river to replace the ferry. He was by this time a powerful man, and his opinion being upheld by the architect, the work proceeded. (See Figs. 45, 80, 81, 82, and pages 82 and 86).

FIG. 50. Casting cannon balls.
From Pyne's 'Microcosm', 1803 - 1806.

The brothers Wilkinson, through their work for Boulton & Watt, became famous not only in this country but throughout Europe, and innumerable foreign engineers came to see the work going on in the Midlands and to ask advice. William handled much of this consultative and export work, and in 1770 he visited France where he put down a cannon foundry, furnaces, and a green sand foundry; again in 1777 he was in Paris, advising the French engineer Perrier about a scheme for supplying Paris with water from the Seine, for which the firm supplied some forty miles of cast iron pipes before the scheme was completed in 1786. In that year he was giving advice at Le Creusot, then an insignificant iron centre, and he remodelled the whole place, putting down four blast furnaces, five kilns, a foundry and boring mill, staying there three years and ultimately acting as consultant for another ten. William was deeply impressed with the possibilities of the area and its resources, and in a letter to his brother said: "Whenever Frenchmen relinquish their fiddling and dancing and cultivate the art of ironmaking, etc., England will tremble." While at Le Creusot he introduced cast iron railway lines, and is sometimes wrongly credited by the French with their invention. He also erected here one of the first Watt engines to be used out of England.

In 1789 William went to Prussia in an advisory capacity, and found, as he had at Le Creusot, that there were endless resources. He was chiefly concerned there with the possibility of coke smelting, and though the weather stopped his work on this visit, he subsequently got out designs and details for coke blast furnaces which were successfully erected and initiated the great Upper Silesian industry. In all this work abroad the Wilkinsons supplied much of the material from the Bradley and Bersham foundries, the whole arrangement proving most lucrative.

Beck says of the brothers' influence abroad: "John Wilkinson's activity was not limited to his native country for he introduced new discoveries on the continent and was the prophet and founder of modern ironwork in France and in Germany".[7]

To William Wilkinson the credit is usually given for an invention that affected the foundry perhaps even more than that of coke smelting, which influenced the whole iron industry. Previously iron for casting had been taken directly from the blast furnace or had been re-melted in an air or reverberatory furnace. William produced the foundry cupola for the re-melting of pig iron before casting. He took out no patent and it may be that he was influenced by his brother's patent of 1794 which covered a furnace not more than ten feet high with an outer casing of iron, and which was to be used for converting pig iron into malleable iron. There is evidence that many people had been experimenting over a long period with a similar method of dealing with the cast pig iron in order to be able to control the re-melted metal for accurate casting. Réaumur as early as 1722 pointed out that small blast furnaces would be very suitable for re-melting pig iron, and illustrated in his *L'Art d'Adoucir le Fer Fondu* small portable furnaces of the cupola type.

John Farey made surveys of Derbyshire in 1807, 1808 and 1809, and in "Derbyshire" (1811) he says:

"The small Cupolas, or Hells as they are called, which are used in the Foundries here, were introduced about thirty years ago, for heating Pig-iron instead of Air or Reverberatory Furnaces, which, as I am told, though they answer for Cannon balls and some other purposes, making very Solid castings, yet the iron becomes whiter, and nearer to the quality of bar iron in infusibility, every time it is melted in such furnaces, losing 30s. per ton of its value at each melting; by adding small quantities of oyster-shells or limestone to the cokes in the hells, the quality of the iron can be preserved, in two or three successive meltings, but not more, I understand."

Perhaps the earliest sketch of an English cupola is to be found in Pyne's *Microcosm*, 1803-6, which shows a small cupola blown by hand bellows and is entitled "Casting Cannon Balls". (See Fig. 50, opposite.)

John Wilkinson's hard and obstinate character, by the end of the century, had produced so many sordid quarrels, not only with his brothers but with Boulton & Watt, that the team work of the inventor, engineer and ironmaster and founder was wrecked, and in 1796 Boulton & Watt opened their own Smethwick Foundry "which, like Soho, was destined to become a nursery of men of invention and industry".[8]

He was certainly the great ironmaster of the Midlands, but in a different way from the quiet Quaker masters of Coalbrookdale. His workmen sang songs extolling his great career:

> "But before I proceed any more with my lingo
> You shall all drink my toast in a bumper of stingo;
> Fill up, and without any further parade,
> JOHN WILKINSON, boys, that supporter of trade.
>
> "May all his endeavours be crown'd with success,
> And his Works, ever growing, posterity bless;
> May his comforts increase with the length of his days,
> And his fame shine as bright as his furnaces blaze."

But they respected, rather than loved him.

He gathered round him a group of very able men such as Thomas Pearce and Gilbert Gilpin. In a letter, Gilpin has recorded the consistent enthusiasm for iron John Wilkinson displayed throughout his life:

"J. W. is expected at Rowley's daily . . . He has two coffins ready in his hot-house at Bradley, the first being a blank with spanners, etc., to screw him up. He sent the order from London, and was very pressing for its execution, which made his people conceive the devil had at length sent him his route and passport."

Samuel Smiles writes:

"John Wilkinson had, as some thought, an extravagant, but, as results have proved, a truly prophetic, appreciation of the extensive uses to which iron might be applied . . . Everybody knew of Mr. Wilkinson's hobby, and of his prognostication that the time would come when we should live in houses of iron, and even navigate the seas in ships of iron".[9]

The great ironmaster did actually build barges of cast iron. His first experiment, in 1787, was with a barge made to bring peat moss to his iron furnace at Wilson House, near Castle Head in Cartmel, Furness, and was followed by larger cast iron vessels made to carry iron down the Severn. He erected in Bilston a chapel in which the doors, window frames and pulpit were made of cast iron, all cast in the Bradley foundry. The pulpit still exists in a new chapel on the same site.

He was determined that when he died his body should be encased in his favourite metal, and directed in his Will that he should be buried in his garden in a cast iron coffin with an iron monument over him of twenty tons weight. He was three times buried, twice being disinterred, and found a final resting place at Lindale where the great cast iron monument still stands and which was saved from destruction by public subscription some twenty years ago.

The Carron Works: The Carron ironworks differed from the other great eighteenth-century iron businesses in that from the first it was planned on a relatively large scale and did not grow gradually from a very small beginning. John Roebuck, son of a cutlery manufacturer of Sheffield, was trained as a doctor and practised as a physician in Birmingham in the seventeen forties.

Fig. 51. Monogram of George III in cast iron, which illustrates the great designing ability of the eighteenth century pattern makers.

Reproduced by courtesy of the Carron Company.

He spent much time on chemical research and with such success that he established a business for the manufacture of sulphuric acid and opened a branch works at Prestonpans, near Edinburgh. In this work he was associated with Samuel Garbett, a business man of some standing in Birmingham. It was in 1759 that these two, with their friend William Cadell, ironmaster, shipowner and merchant, set up ironworks on the banks of the river Carron, near Falkirk, where ironmaking began on the first day of the year 1760. Great care was taken over the choice of site. Ironstone, limestone and coal could all be found locally. It was a short journey to Glasgow, and sea transport for long distance trade with Europe and England was directly available.

Dr. Roebuck, an experienced mechanic as well as a man of science, was responsible for much of the planning, siting and lay-out of the works, and also for the introduction of some important improvements in iron production. He became too interested in other enterprises, including coal mines and soda works, and these activities, together with the unfortunate flooding of his mines, resulted in a financial disaster, involving his friends and a certain amount of the capital of the Carron Company. In 1773 the Company was reconstituted and a Royal Charter was granted.

The early Carron works were modelled very much on the lines of the Darbys' in Shropshire, and workmen from this centre were employed, with Robert Hawkins, a relative of the Darbys, as superintendent. Dr. Roebuck consulted the eminent engineers, John Smeaton and James Watt, on many of his problems, and Smeaton, apart from his valuable work in harnessing the water power of the River Carron to control the works' machinery, produced, about 1768, his device of blowing cylinders, made of cast iron, to supply the powerful blast required for smelting the iron ore, and later a mill for boring guns and cylinders. The secrecy that cloaked the pioneer work of the industry and continued into the nineteenth century is well illustrated by Samuel Smiles in his *Lives of the Engineers*, where he relates how in 1858 he tried to see Smeaton's long disused blowing apparatus, but that the reply of the Manager was "Na, na, it canna be allooed—we canna be fashed wi' straingers here".

Much of James Watt's early experimental work on the steam engine was done at Carron with the help and encouragement of Dr. Roebuck. This early work did not meet with much success, and the engine did not become a commercial proposition until Watt moved to Birmingham and had the advantage of the Wilkinsons' improved boring machine. But experimental engines were set up. Incorporated in the stone-work of the great entrance gates of the present Carron Works is a part of an original steam cylinder cast for James Watt and dated 1766. Perhaps the most interesting point of this relic is the evidence it gives of the natural designing ability of the eighteenth-century pattern makers. (See Fig. 52 opposite.)

An interesting contemporary reference appears in one of James Anderson's essays:

"He now reached the extensive ironworks erected at Carron about thirty years ago, which, being upon a large scale, and embracing a variety of objects, afforded full scope to his active mind for the best part of a day.

"He was so lucky as to see a larger cylinder for a steam-engine cast that day; and he describes the process in a manner that may be expected from one who never saw an operation of that sort before, but which as it would be nothing new to most of our readers, we here omit; as also his account of the *tremendous* bellows, as he styled them, whose stunning roar almost deafened him . . . Near the centre of this vale, the Carron Works, from which issues by day an amazing pillar of smoke, and by night immense volumes of fire, is at all times an object of powerful effect.[10]

The principal products of the works were those of the foundry and the most important of these were concerned with ordnance. Macpherson says the great guns being made there in 1777 were "cast solid, and bored by a

Fig. 52. Part of an early cast iron steam cylinder made at the Carron Works, for James Watt, and now incorporated in the stonework at the main entrance of those Works.

drill worked by the whole force of the River Carron, and were exported to Russia, Denmark, Spain, etc."[11] These exports were so considerable that the Government felt it unwise for such goods to be carried in ordinary trading vessels for fear of attack by American cruisers during the War of Independence, and the company, therefore, fitted out stout, well-armed and manned ships of their own, thus founding a line run by Carron between Scotland and London to carry their own products as well as passengers and general merchandise.

The company produced castings of all kinds for machinery and engines, and acquired a great reputation for its domestic stoves and grates—a speciality of the firm to this day.

It has been seen how the foundry work of the great ironmasters of the eighteenth century affected building—the work of the Darbys and the

Fig. 53. A cast iron stove from Compton Place, Eastbourne, Sussex. The height is 5 ft. 8 ins. This late eighteenth century design is attributed to Robert Adam, and it certainly has the characteristics of that architect's work. The mastery of classical detail; the restraint and delicacy with which ornament is used; the excellence of the proportions, and the urbane disregard of any sense of obligation to acknowledge the particular character of the material, all suggest that Robert Adam or one of his many imitators or pupils was responsible for its conception. For this purpose, cast iron was a convenient material; but so far as the outward form is concerned, it might, apart from the functional need to employ a fireproof material, be equally well constructed of wood, stone or plaster. Still, elegance is achieved; not the ornate rococo elegance of France, but a sober, almost Puritanical elegance. Compare this with the German box stove of the same period on the opposite page.

From the Victoria and Albert Museum, by whose courtesy this is reproduced. Crown copyright reserved.

Fig. 54. A German box stove in cast iron, *circa* 1770. This is typical of many stoves of this period, which still survive in country houses in Central Europe.

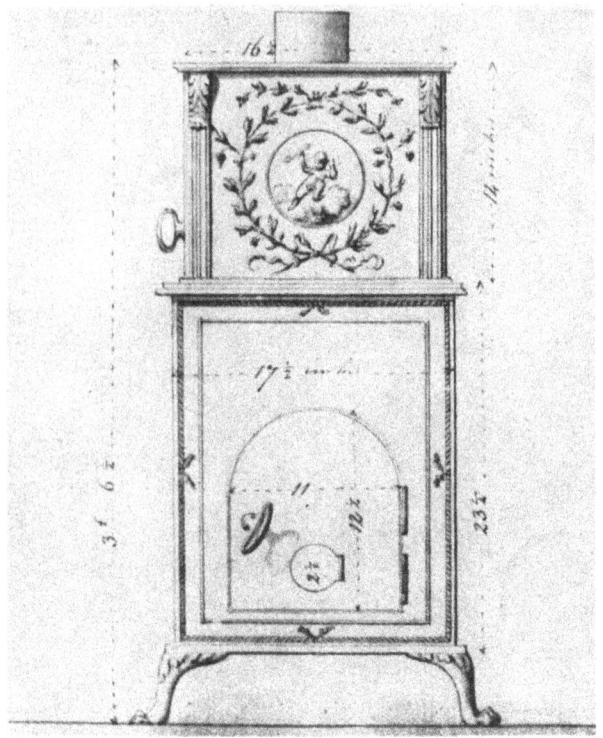 

Fig. 55. A chamber stove designed by the brothers Haworth.

Fig. 56. A double Canada stove by the brothers Haworth.

Wilkinsons. But more than any other firm, Carron left their mark on architecture, and this was perhaps due to their foresight in finding and employing the most competent designers. The firm derived great advantage from their connection with the Adam family. As early as 1764 John Adam was a partner in the Carron Company while carrying on his father's architectural practice; but it was his younger brothers, Robert and James, who were to produce a new and elegant interpretation of the Greek and Roman orders, thus evolving a style that had far-reaching effects on English architecture. Much of Robert Adam's delicate ornamental detail was especially applicable to cast iron, and where previously railings, gates, verandahs, fireplaces, vases and urns had been of wrought iron, Adam used either a mixture of wrought and cast, or cast iron alone, or a mixture of cast iron and steel. It was natural that he should turn to the Carron Company for much of his work, and the influence of the brothers Adam may be traced in the designs of many of the Company's early castings, particularly in panels, grates and fireplaces.

During this period Carron employed as designers the brothers William and Henry Haworth, who had studied at the Royal Academy School during the presidency of Sir Joshua Reynolds. Henry started as designer and carver at Carron in 1779, and did much good work, but unfortunately died two years later. He was succeeded by his brother William who worked as designer and carver at Carron for the next fifty-six years. The brothers came of an artistic family, their father and grandfather being noted carvers. The father,

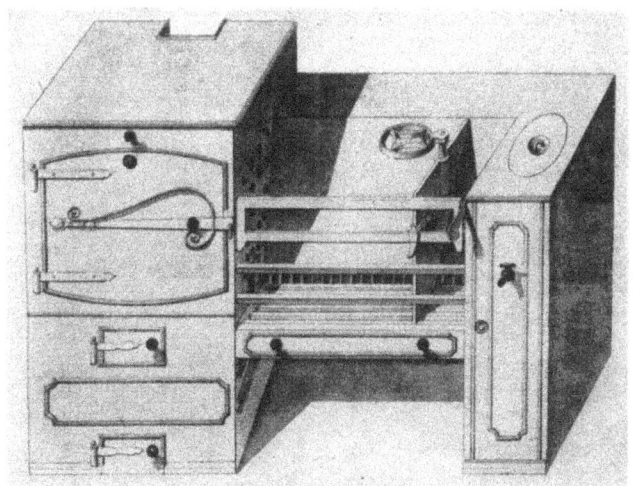

Fig. 57. A back boiler range designed by the brothers Haworth. Typical work of the late eighteenth and early nineteenth centuries.

Samuel Haworth. to commemorate the granting of the Royal Charter in 1773, carved for Carron the famous portraits of King George III and his Queen, which were to adorn so many of the fashionable hob grates of the period. The brothers led a simple life and can hardly be said to have become famous in their day. But, with their father, they created a distinctive style of ornament which has characterised much of Carron's work. They have left much carving in wood of the most delicate order: whether classical figure work, ornamental panels, mouldings, decoration, or balcony and balustrade work, their craftsmanship is faultless, and shows a complete mastery of the casting processes. Of the merits and antecedents of the Haworths, Grey Wornum has written as follows:—

"Both brothers were masters of classical figure work, as well as being quite accomplished painters. Understanding of foundry technique ensured for their products beautiful quality and definition, when finished with light black lead. They were of a London family and represented the third generation of craftsmen. Their father, who died in 1779, had a flourishing carving business in Denmark Street, St. Giles-in-the-Fields, employing with his two sons

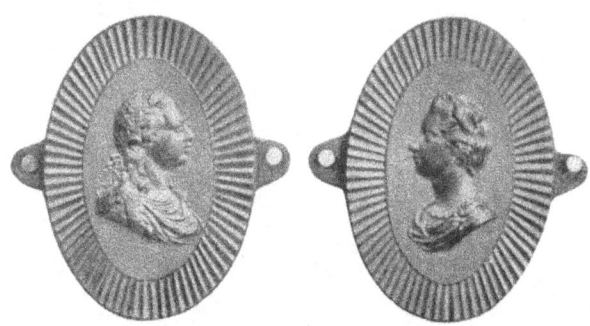

Fig. 58 (*Above*). Portraits of King George III and Queen Charlotte, designed by Samuel Haworth to commemorate the granting of a Royal Charter to the Carron Company in 1773.

Fig. 59 (*Left*). The hob grate in which the portraits of King George III and Queen Charlotte are incorporated.

Reproduced by courtesy of the Carron Company.

Fig. 60. Hob grate for small fireplace opening. This, like the other examples on this page, was probably designed by the brothers Haworth, and is typical of the late eighteenth and early nineteenth centuries.

Fig. 61. Hob grate of the same period as that shown in Fig. 60, but with a more lavish use of ornament to accentuate its lines. Skilled designers could be lavish with ornament without masking good proportion by rococo extravagance. Fig. 62 (*left*) exemplifies this capacity for ornamental treatment.

Fig. 62 (*Left*). Hob grate with semi-circular space for fire, and inward curving hobs. The decorative reeding, the fan devices in the spandrels, and the boldness of the decoration generally, illustrate another aspect of the skill and capacity of late eighteenth and early nineteenth century ironfounders.

Reproduced by courtesy of the Carron Company.

Fig. 63. Another example of a late eighteenth century hob grate, with outward curving fire bars and decorative reeding in the interior. The scene from one of Æsop's Fables, depicted in the panel, resembles the free, naturalistic treatment apparent in Fig. 73 on page 75.

Henry and William, some thirty carvers, many Dutch and Flemish craftsmen among them. His establishment received occasional visits from George III himself".[12]

The Walkers: In 1741 three brothers, Samuel Walker, schoolmaster, Aaron Walker, mechanic, and Jonathan Walker, farmer, started a small foundry at Grenoside, near Sheffield. By 1746 the business had grown sufficiently to admit another partner, John Crawshaw, and a furnace and smithy were started at Masborough, near Rotherham. Here water for power and transport were available and local iron ore and coal, and the concern flourished, growing into the great iron-producing and foundry works of the late eighteenth century. Though the Walkers specialised in steel production by the crucible process invented by Benjamin Huntsman (and said to have been copied by Samuel Walker), much ordinary foundry work was executed. Again the background was ordnance, and during the American War of Independence and the Napoleonic wars, a large amount of cannon was produced. In 1781 Wilkinson's new boring apparatus was installed for this type of work, and in 1782 an engine, to Watt's plan, was installed to blow the furnaces at Rotherham. In 1788 arrangements were made to set aside a room as a workshop for Tom Paine, and it was here that so much of his experimental work in bridge design was conducted, the foundry eventually casting the work for the bridge over the Wear at Sunderland in 1796. After the end of the French wars in 1815, the orders for ordnance dropped and Walkers again

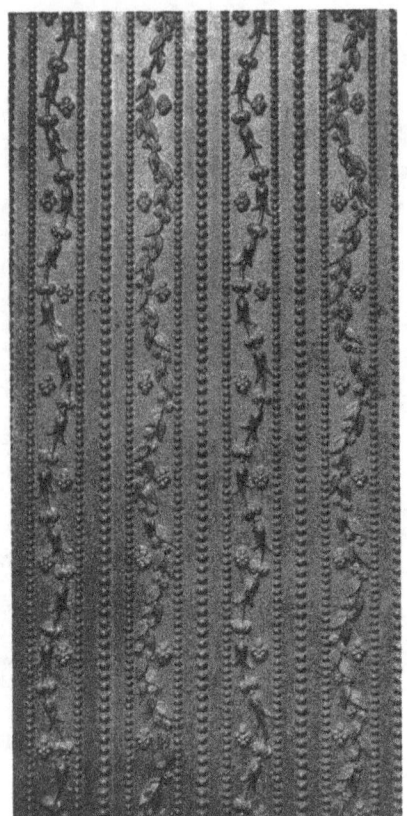

FIG. 64.

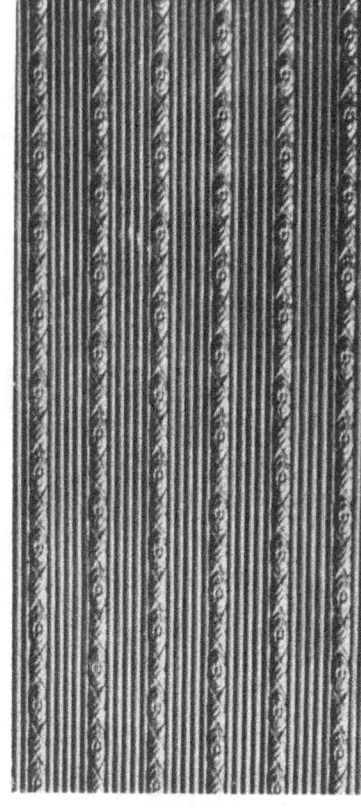

FIG. 65.

FIGS. 68 to 73 (*Opposite*). With the exception of Fig. 73, all the original wood and plaster carvings for the cast iron panels shown on the opposite page are based on classical models. They were made between 1782 and 1838 by Henry Haworth.

FIG. 73, like the panel of the hob grate in Fig. 63 on page 73, has mediæval affinities: here is the work of the craftsman-designer rather than the work of the designer directing the craftsman.

Reproduced by courtesy of the Carron Company.

FIGS. 64 to 67. These are typical cast iron patterns for the interiors and side jambs of late eighteenth century grates.

FIGS. 64, 65 and 66 have an unmistakable classical origin reflecting the prevailing taste of the closing decades of the eighteenth century, and reproducing patterns in iron which would have been equally suitable for reproduction in plaster. But Fig. 67 has an almost mediæval flavour. There is a vivid boldness about it which is not wholly attributable to the fleur-de-lys device, but which recalls the vigour of some of the sixteenth century Sussex firebacks. Compare this with Fig. 18 on page 21.

FIG. 66.

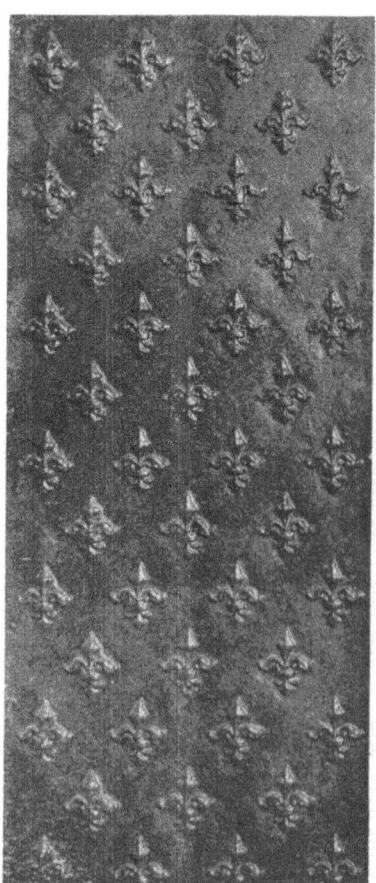

FIG. 67.

Fig. 68.

Fig. 71.

Fig. 69.   Fig. 70.   Fig. 73.

turned their attention to bridge work, obtaining the order for Rennie's great cast iron bridge over the Thames at Southwark. The work of casting these bridge members was looked upon as a great wonder of the day, and the central arch was erected at the Holmes works where vast crowds went to see it.

The four large concerns which have been briefly described, were selected because they are typical of the period and because they exerted an extensive influence on architecture and building.

The Darbys, through their pioneer work in iron production, particularly on the foundry side, made cast iron a commercial material, capable of being used for structural work. Coke smelting stepped up production, and better moulding and pattern work made delicate work possible as well as the large structural units, such as those used for the first cast iron bridge. The Wilkinsons' foundry produced the cylinders which made the steam engine a practical power unit, and thus affected not only iron production but mechanised industry generally, which demanded new and specially planned buildings to house steam driven machinery. The Walkers built up a foundry organisation capable of making big castings for bridges; and the Carron works were, perhaps more than any other foundry, directly connected with building and decoration.

There were many other foundries in the eighteenth century engaged on work of direct architectural importance, or indirectly through the production of household goods and fittings. Izons & Company, now in West Bromwich, and who claim to be the oldest makers of cast hollow ware in the world, established their works in Birmingham in 1763. An old catalogue of their goods includes, amongst other items, door knockers, kitchen stoves, various household utensils and cast iron hinges.

Thomas Chambers, who received his early training with the Walkers,

FIG. 74. An early nineteenth century hob grate with an unusual association of ornamental motifs. The panel below the fire bars with the vine leaves, grapes and tendrils intertwined, has great delicacy of composition, contrasting strangely with the vigorous, rather clumsy application of two stunted legs, terminating in claws, powerful supports which support nothing, but accord with the decorative conventions of the English Empire period. The fireback with the central thistle motif has a delicacy comparable with the vine leaf panel. All these diverse ornamental forms show the increasing mastery of casting technique.

Reproduced by courtesy of the Carron Company.

FIG. 75. Fireplace and grate in the Commercial Rooms at Bristol. This is the main room fire grate, and is a typical example of early nineteenth century design. The urns, placed somewhat unnecessarily on the upper part of the hobs, have lost the light and elegant slimness they would have had in the closing decades of the eighteenth century: they are bold and simple in form, like the reeded surface of the hobs.

Reproduced by courtesy of the National Buildings Record.

joined with George Newton in 1792 to set up the Phoenix foundry. Here stoves, grates, wheels and machinery were cast, and the success of the scheme, together with the difficulties experienced in obtaining supplies of pig iron, led to the firm starting its own blast furnaces and foundry, and moving from Sheffield to found the great Thorncliffe Ironworks at Chapeltown in 1795. Later, James Malom, who had been connected with Boulton & Watt, joined the firm and introduced the manufacture of plant for gasworks, a branch of the industry in which the firm has continued to play an important part, supplying plant for the early gas lighting of London, Brighton, Yarmouth, Hamburg and other cities and towns. The Sheffield district acquired a reputation for producing artistic and ornamental castings—a section of the industry which demands highly skilled moulding and pattern making technique. H. G. Hoole & Company were later famous for this kind of work, especially in grate and fireplace designs in which, like Carron, they employed the most competent designers, among whom were Alfred Stevens and Godfrey Sykes. This firm were later taken over by the famous Falkirk Ironworks, which continued to produce many of the fireplaces originally designed by Stevens.

So efficient did the Yorkshire foundrymen become that they were able, at this period, to cast knives, razors, scissors and forks in active competition with forged steel, and in 1780 attempts were made to control this activity by law. In 1819 an Act was passed regulating the cutlery trade in England, and these facts are a significant indication of the advances made in the malleable cast iron industry.

Réne Antoine Ferchault de Réaumur in 1722 in his treatise *L'Art d'Adoucir le Fer Fondu* described a process of softening iron castings by heating them while they were embedded in red oxide of iron. It is clear from early patents that Samuel Lucas, of Dronfield, was working, perhaps independently, on similar lines. In 1804 he patented his process and thus gave the impetus to that large section of the ironfounding industry which to-day specialises in malleable cast iron castings, many of which—hinges, locks, door furniture, pipes, pipe joints and fittings—are of importance in building.

Though South Wales had produced iron for many years, the industry began its great period of growth there when, in 1760, John Guest of Broseley in Shropshire, started the Dowlais Ironworks, and a few years later Anthony Bacon, of Whitehaven, set up ironworks at Cyfarthfa. In 1782 the three brothers Humfray (also from Broseley) joined Bacon. Later again, Richard Crawshay took over the works at Cyfarthfa where, by 1803, he was employing over 2,000 men.

Meanwhile iron works were springing up all over Scotland, often started by enterprising employees of the Carron Works. In 1786 the great iron industry of the West of Scotland was founded by William Cadell who, after leaving Carron, set up with Thomas Edington the Clyde Ironworks near Glasgow, and "as pig iron became more plentiful a crop of foundries began to spring up in Stirlingshire under the shadows of the Carron furnaces".[13] In 1819 other Carron workers established the Falkirk Ironworks, later to

Fig. 76. The oldest known cast iron water mains. These were installed in 1664 to supply the town and parks of Versailles. The illustration shows the present day condition of these pipes, which are still in use.

Reproduced by courtesy of the Cast Iron Pipe Research Association, of America.

become the second largest foundry in Scotland and, as Section 3 shows, to exert a profound influence on architectural castings.

The Newland Iron Company, founded in 1735, spread over the Furness district, eventually absorbing the whole of the furnaces of the locality, and in 1818 acquiring the famous old Backbarrow Furnace which had existed since 1711.

In Nottinghamshire and Derbyshire at this period developments were taking place that were to affect considerably the utility of cast iron in building. The material had been used for making pipes since the middle of the seventeenth century, and the demand now for better water supply—and later for gas supply—gave the necessary stimulus to that branch of the foundry industry engaged in making cast iron service pipes, a branch which has now grown to such dimensions that it produces annually in this country some 500,000 tons of pipes. The oldest cast iron pipes of which definite records exist were laid between 1664 and 1686 in the Palace gardens of Versailles where they still supply the fountains. In *Cast Iron Pipe, Its Life and Service*, published by the Stanton Ironworks Company, descriptions are given of the condition of these pipes after excavation and examination some years ago, and also of similar pipes at St. Etienne and Marseilles. Those at Marseilles, though 100 years old, were so well preserved that it was possible to relay them. There are in Great Britain many miles of cast iron water pipes still in use that were laid some 165 years ago, towards the end of the eighteenth century.

Documents show that in 1702 furnaces were in operation at Staveley, and though these were small in scale and the output probably used mainly for

Fig. 77. A cast iron water main laid over 120 years ago in Philadelphia, Pennsylvania, U.S.A. This is still giving satisfactory service.
Reproduced by courtesy of the Cast Iron Pipe Research Association, of America.

domestic products, cast iron pipes were being produced well over a hundred years ago. To-day the foundries of the Staveley Coal & Iron Company Limited can produce more than a million-and-a-half cast iron pipes yearly, varying from 1½ inches to 72 inches in diameter. They supply these products for architectural and civil engineering works all over the world.

In 1788 iron was first made at the little village of Stanton-by-Dale in a small furnace, probably not more than thirty feet high and with a hearth about three feet in diameter. From this an undertaking developed which is now capable of producing well over 13,000 tons of pig iron a week, and employs over 14,000 workers. From the first Stanton was associated with the manufacture of cast iron pipes, and as the water supply, gas supply and sewage plants of the country were improved and extended, the Stanton foundries concentrated on this use, until they became the largest producers of cast iron pipes in the world to-day.

Until about a hundred years ago all iron pipes were cast horizontally, but in 1846, D. Y. Stewart of the Links Foundry, Montrose, obtained a patent for a method of producing vertically cast iron pipes. Most pipes and ornamental columns and lamp standards, are still cast horizontally, but the vertical method is used extensively to-day for what are termed "pressure" pipes, those used for water, gas and sewage. By this vertical method, the metal, during casting, is compressed and the pipe is thus able to withstand high pressures when in use. A compromise is often adopted in the casting of drain pipes. These are cast on the "Bank", which means on an incline, the head of the pipe being raised from 6 inches to 12 inches above the bottom. As it is poured at the higher end, a slightly closer grained metal is produced than in pipes cast horizontally. (The more recent centrifugal method of casting pressure pipes is described in Section 5, page 358.)

One other important development affecting the use of cast iron in building and equipment took place about this time. The art of enamelling on iron was discovered in Europe about the middle of the eighteenth century. In 1764 a factory in Wurttemberg was producing acid-proof enamelled cast

FIG. 78 (*Right*). Section of water main laid in London in 1810, which is still in service. This replaced the original wooden and stone pipes used by Sir Hugh Myddleton for distributing the water supplied by the New River Company.

Reproduced by courtesy of the Cast Iron Pipe Research Association, of America.

FIG. 79 (*Left*). A cast iron water main, laid in the middle of the nineteenth century, and still in service at Baltimore, Maryland, U.S.A.

Reproduced by courtesy of the Cast Iron Pipe Research Association, of America.

iron stoves, cooking utensils and chemical ware.[14] In 1799 a patent was granted to Samuel Hickling of Birmingham (Brit. Pat. No. 2296, 1799), in which is described how he intended "to improve and beautify certain vessels and utensils used for chemical, culinary and various other purposes, made of hammered iron or cast iron . . . by lining or covering them with the following vitreous compounds . . ." and complete details are then given of the materials to be used and the methods of application.[15] The variety of brilliant finishes available on cast iron equipment to-day were thus originated.

We have given the great names of the iron industry; but the list of lesser known firms is huge. In addition to actual iron producing centres, foundries sprang up all over Britain—in the Midlands, Yorkshire, Derbyshire, Lancashire, South Wales and Scotland, where production was still chiefly concerned with ordnance but where, when Government orders ceased, the owners changed over to making stoves, grates, water and gas pipes, tram rails, structural beams and stanchions, bridge components, railings, gates and general household equipment.

BRIDGES.

With the pioneer work in iron casting that was proceeding in Shropshire during the middle of the eighteenth century, it was natural that this district should be the first to encourage the use of cast iron for bridge work. When questions of headroom and foundations were considered, the lightness and strength of this material compared with stone must have greatly influenced engineers. Samuel Smiles says "The metal can be moulded in such precise forms and so accurately fitted together as to give to the arching the greatest possible rigidity; whilst it defies the destructive influences of time and atmospheric corrosion with nearly as much certainty as stone itself. The Italians and French, who took the lead in engineering down almost to the end of last century (18th century), early detected the value of this material, and made several attempts to introduce it in bridge-building; but their efforts proved unsuccessful, chiefly because of the inability of the early founders to cast large masses of iron, and also because the metal was then more expensive than either stone or timber. The first actual attempt to build a cast iron bridge was made at Lyons in 1755, and it proceeded so far that one of the arches was put together in the builder's yard; but the project was abandoned as too costly, and timber was eventually used. It was reserved for English manufacturers to triumph over the difficulties which had baffled the foreigners".[16]

We have mentioned earlier the decision to erect a bridge over the Severn at a point near Broseley where previously a ferry had been used. The advice of the great ironmaster, John Wilkinson, finally influenced the Bridge Company, and Thomas Farnolls Pritchard, a Shrewsbury architect, agreeing with Wilkinson's suggestion, was invited to make designs for a bridge in the new material, and these were eventually adopted. The whole of the work was cast at the local foundries of the Darbys, where Richard Reynolds was then

Fig. 80. The first cast iron bridge in the world, erected at Coalbrookdale, Shropshire, in 1779, across the River Severn. It was designed by Thomas Farnolls Pritchard, a Shrewsbury architect, whose portrait is shown on page 55, Fig. 45. See Figs. 81 and 82 on the pages that follow. *This illustration is reproduced from a steel engraving in the possession of the Coalbrookdale Company.*

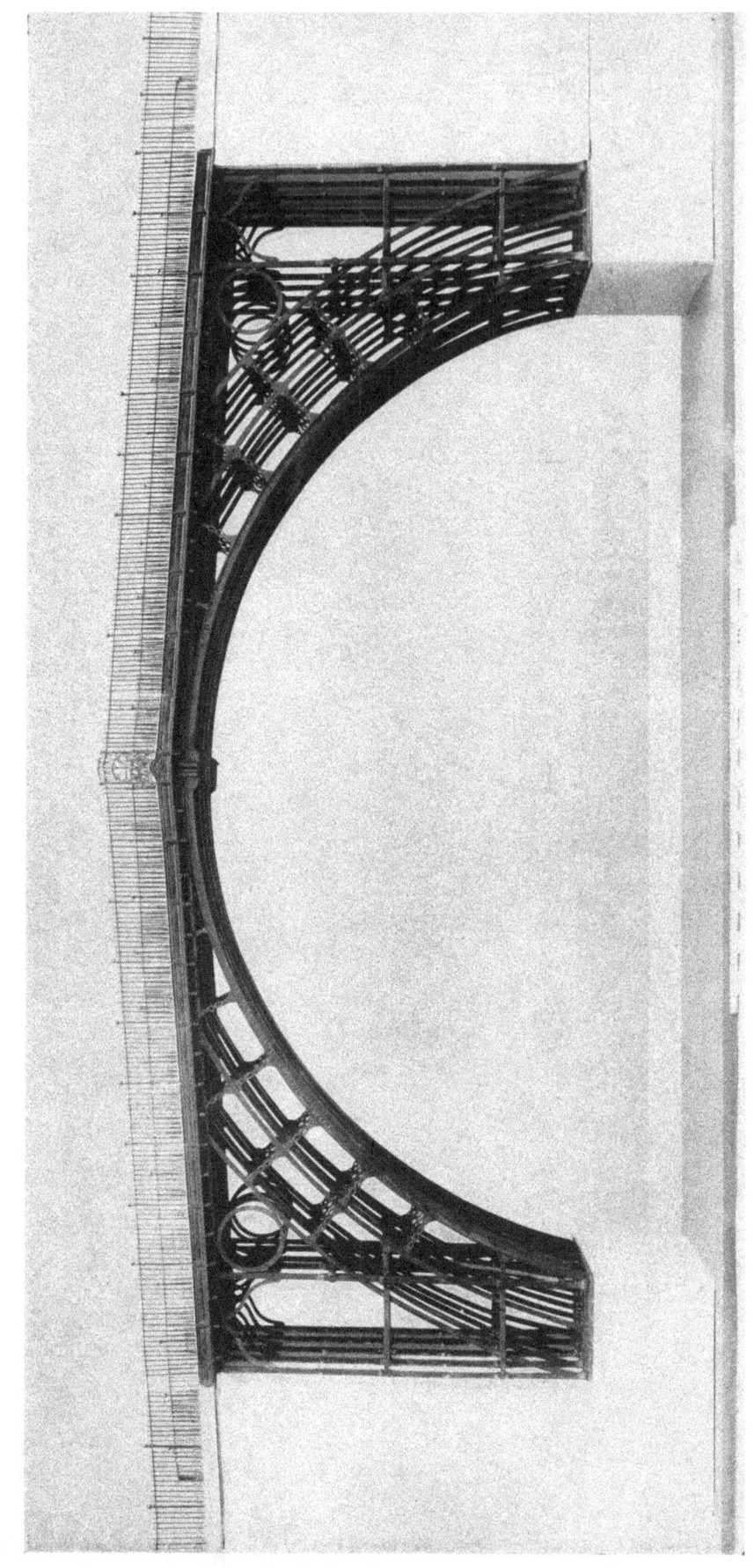

Fig. 81. The cast iron bridge at Coalbrookdale has a total length of 196 ft., a span of 100 ft. 6 ins., and a rise of 50 ft. The five main cast iron ribs, each cast in two pieces and weighing 5 tons 15 cwt. each, with masonry abutments, support a 24 ft. carriageway, carried on open sand-cast iron plates, $2\frac{1}{2}$ ins. thick and in 3 to 8 in. widths. The total weight of cast iron used in the bridge is 378 tons, 10 cwt.

Crown copyright, by courtesy of the Science Museum, London.

FIG. 82. The Coalbrookdale cast iron bridge as it is to-day. It has now been classified as an ancient monument by the Ministry of Works. See Fig. 80 on page 83, and Fig. 81 opposite.
Crown copyright, by courtesy of the Science Museum, London.

in charge, and the bridge was erected in the year 1779. It was a considerable technical achievement, the great semi-circular ribs of the arches spanning 100 feet and each being cast in two pieces only. That cultivated observer, the Hon. John Byng, viewing it in 1874, wrote "But of the iron bridge over the Severn, which we cross'd, and where we stop'd for half an hour, what shall I say? That it must be the admiration, as it is one of the wonders, of the world. It was cast in the year 1778; the arch is 100 feet wide, and 55 feet from the top of the water, and the whole length is 100 yards: the country agreed with the founder to finish it for 6000£; and have, meanly, made him suffer for his noble undertaking".[17]

This cast iron bridge—the first of its kind in the world—still stands to-day. The iron work is still comparatively sound though the bridge is closed to heavy traffic. For many years, little attention was paid to its maintenance; but now that it has been classified by the Ministry of Works, it should remain as a monument to the pioneering enterprise of the Coalbrookdale ironfounders for very many years to come.

The next designer of a cast iron bridge was Tom Paine, better known as a revolutionary politician and pamphleteer than as a pioneer in engineering. The author of the *Rights of Man*, during his stay in America, had produced a design for a cast iron bridge to span the 400 feet of the Schuylkill, and he came to England to patent his invention and to order from the Rotherham Ironworks the necessary castings for a bridge. These were actually made and fitted together on a bowling green in Paddington where this extraordinary new type of structure was exhibited to the public, at a charge of a shilling a head. On the outbreak of the French Revolution, Paine hurried to Paris and left the bridge work in the hands of his creditors. The parts were purchased and erected over the River Wear at Sunderland in 1796, to a modified design by Mr. T. Wilson. This bridge spanned 236 feet, the springing from the stone abutments being 95 feet above the river bed and having a further rise of 34 feet, and it was for many years regarded as the greatest triumph in the new art.

In the same year another cast iron bridge was being erected in Shropshire. Thomas Telford, born of humble parents at Wasterkirk in Eskdale, Dumfriesshire, in 1757, began his working life as a stone-mason at the age of fifteen, eventually becoming one of the greatest builders of roads, bridges, canals and harbours that the British Empire has known. In 1795, as county surveyor to Shropshire, he was asked to prepare designs for a new bridge over the Severn at Buildwas, a high flood having swept away the old bridge. He had studied the first iron bridge, appreciating its merits but observing some of its defects. Telford had some difficulty in getting his design accepted, for it departed from the prototype that ironfounders knew; but he eventually succeeded and in 1796 the bridge was erected. It consisted of a single arch of 130 feet span, the ribs being the segment of a very large circle. Despite the increased span, the bridge contained only 173 tons of iron as compared with the 378 tons in the earlier one at Ironbridge. Telford himself wrote:

FIG. 83. The bridge over the River Wear at Sunderland. This was a modification of a design originally made by Tom Paine, author of *The Rights of Man*, during his stay in the United States. It had been intended to span the Schuylkill, and the castings for the bridge were made and ultimately erected to a modified design, by T. Wilson, at Sunderland, in 1796. This drawing, and those on the two pages that follow, (Figs. 84 and 85) were made by Robert Clark and published in 1798. *Crown copyright, by courtesy of the Science Museum, London.*

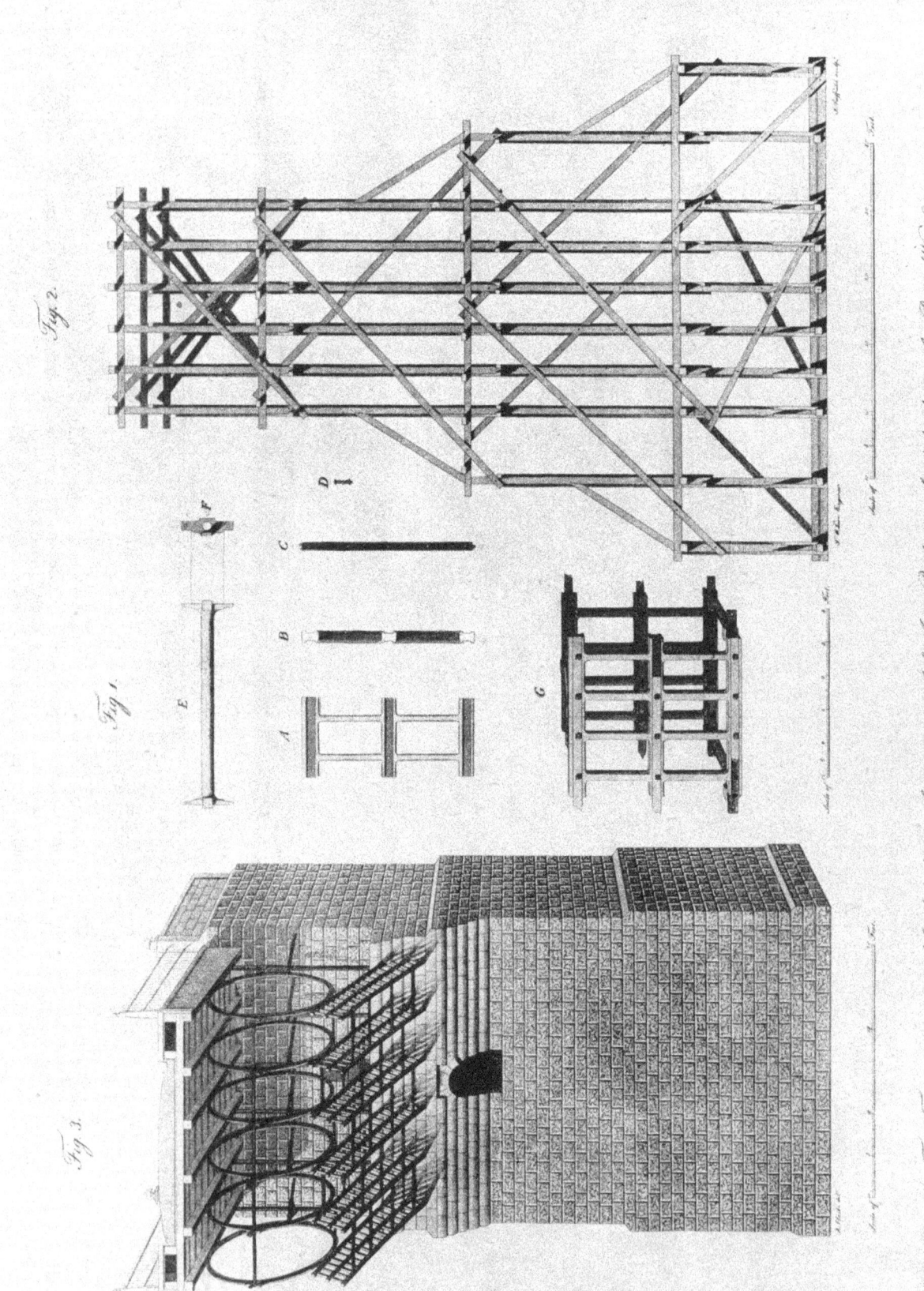

Fig. 1. The component parts of the Arch, shewing the construction of the Iron Bridge at Sunderland over the River Wear.
A. side view of a Block. B. end view. C. one of the wrought Iron Bars, by which the Blocks are united. D. one of the screw Bolts. E. length view of one of the Tubes, which unite the Ribs horizontally. F. end of the Tube. G. one of the Blocks when united, forming part of two Ribs.

Fig. 2. An elevation of one of the Frames used for supporting the Centre.

Fig. 3. A section of the Bridge, twenty feet from the northern Abutment in Perspective.

Span of the Arch 236 feet. ——— Hight from low Water 100 feet.

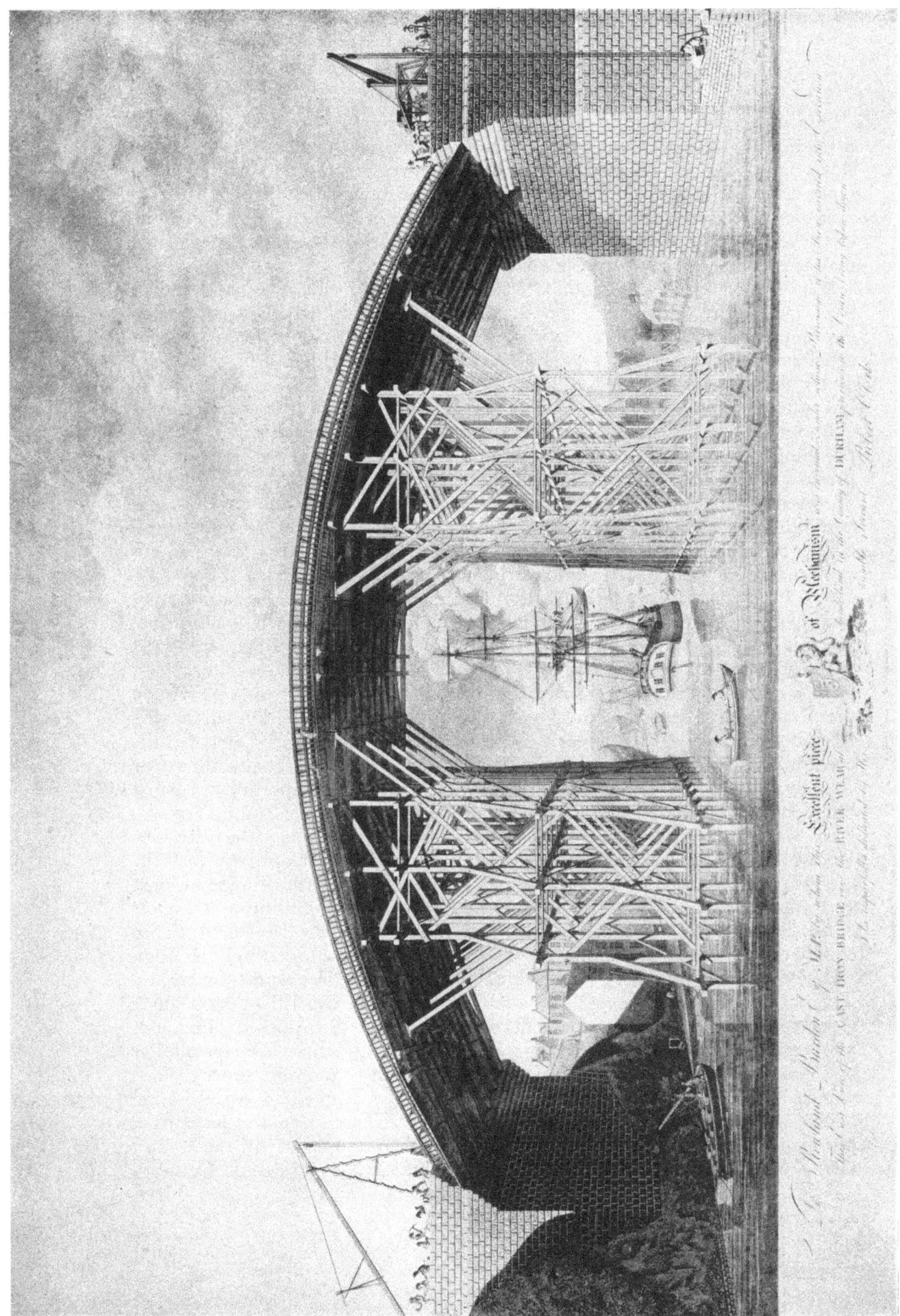

FIG. 84 (*Opposite*) and FIG. 85 (*Immediately above*). On the opposite page are drawings showing the construction of the iron bridge at Sunderland, and immediately above is a view of the bridge in course of erection. The complete bridge is shown in Fig. 83, on page 87.

Crown copyright, by courtesy of the Science Museum, London.

"The bridge was cast in an admirable manner by the Coalbrookdale ironmasters in 1796, under contract with the County magistrates. The total cost was £6,034 . 13 . 3." He also pointed out that previous builders had confined themselves to imitating in cast iron the shape of a stone bridge—just as many architects and engineers to-day imitate in steel and concrete the shape of earlier stone buildings—instead of using to the full the possibilities of the new material. (See Fig. 87, page 92.)

Telford's principal application of cast iron to building was in the construction of road bridges. In Shropshire alone, as County surveyor, he erected five and his success encouraged him, when in 1801 old London Bridge was considered unsafe, to propose a single arch cast iron bridge of 600 feet span across the Thames. The opposition to this revolutionary idea was strenuous; but after examination by a Select Committee, the proposal had to be abandoned, not because of any structural difficulty but because of the town planning problem created by the approaches, and the almost certain depreciation of adjacent property and sites.

Among many other iron bridges erected by Telford, one of the most beautiful was undoubtedly that at Craigellachie, where a light cast iron bridge of four ribs, spanning 150 feet across the river, still exists. Two famous examples of his work are Waterloo Bridge over the Conway river at Bettws-y-Coed, built in 1815, and the Galton Bridge over the Birmingham Canal, cast by the Horseley Bridge Company in 1829, and still in excellent condition. (During the 1939-45 World War it was passed by the military transport authorities as capable of carrying the heaviest tanks.)

The Horseley Bridge Company has exerted an important influence upon the production of architectural castings. A blast furnace, known as the Horseley Furnace, existed as early as 1770, on a site opposite the works of the present Horseley Bridge and Engineering Company Limited. The works at one time were under the management of Isaac Dodds, a contemporary of George Stephenson, and who married a sister of Robert Stephenson. In the *Diary of a Tour through the Midlands in* 1821, Joshua Field, describing the Horseley Ironworks, says: "The three Furnaces tap into a large Foundry where the largest casting can be made. Here were large columns for the London Docks and many castings for the Calcutta Mint Rolling Mill for Rennie . . .".[18]

In 1801 Telford was asked for a design for bridging the Menai Straits to enable the main road from Shrewsbury and Chester to be taken right through to Holyhead. He submitted two designs, one using cast iron and stone arches in combination, and the other, which he favoured, consisting of a single cast iron arch of 500 feet span. This scheme, owing to supposed difficulties in forming the necessary centering, was finally abandoned. The possibility of spanning the Straits was again discussed in 1815, and this time Telford worked on the lines of a scheme he had previously evolved for a suspension bridge at Runcorn. His suggestions were finally adopted and the suspension bridge over the Menai Straits was completed and opened in 1826. Cast iron was not used extensively in this or the later similar bridge at Conway,

Fig. 86. A perspective view of Thomas Telford's proposed design for a new London Bridge, 1801. The bridge was to be made in cast iron, with a span of 600 ft., and was abandoned only because of the difficulty created by the approaches. This illustration is reproduced from a contemporary aquatint.

Crown copyright, by courtesy of the Science Museum, London.

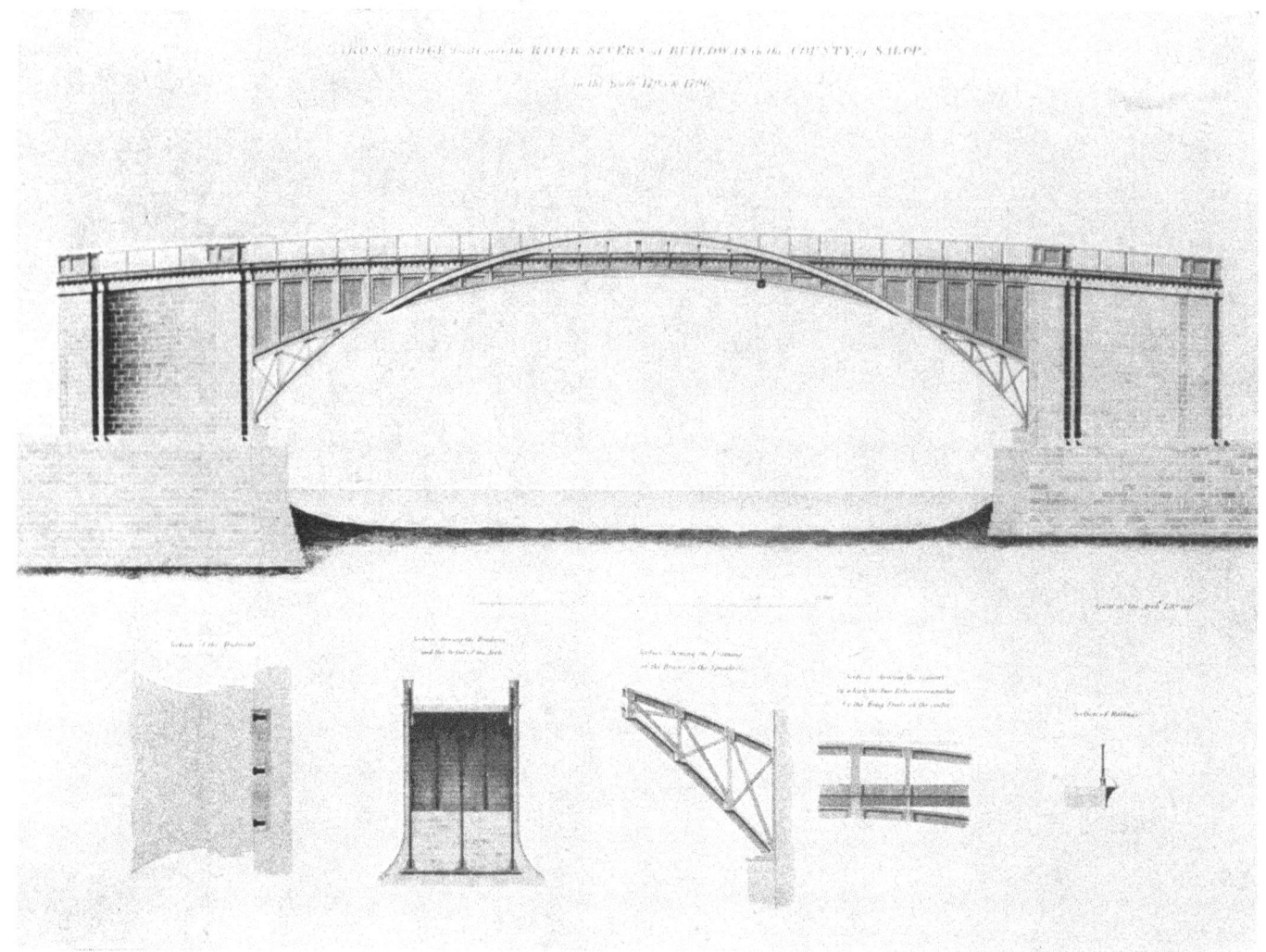

Fig. 87. Buildwas Bridge, over the Severn in Shropshire. This was the first iron bridge designed by Thomas Telford. It had a span of 130 ft., and although he was clearly influenced by the design of the first iron bridge at Coalbrookdale, Telford was much more economical of material than Pritchard; for he used only 173 tons of cast iron compared with the 378 tons Pritchard employed for a smaller span bridge.

Reproduced by courtesy of the Birmingham Central Reference Library.

but Telford realised its strength and used it for members in compression, particularly for the saddles on the tops of the piers to bear the suspension chains.

Contemporary with the work of Telford was that of John Rennie, another great civil engineer and architect. Rennie was the youngest son of a farmer at Phantassie, Haddingtonshire, where he was born in 1761. Like Telford, he rose from humble circumstances, and at an early age showed a great mechanical aptitude. He was taught by Andrew Meikle (1719-1800) who invented, in 1787, the first successful thrashing machine. (It appears that Meikle's ancestors were well esteemed, as the Scots Parliament passed a special Act in 1686 to help John Meikle, founder, who was probably the first person to introduce the art of ironfounding to Scotland, and was certainly one of the earliest founders.) Rennie started in business as a millwright, and when only nineteen, proved both his courage and his knowledge of the new

Fig. 88 (*Above*) and Fig. 89 (*Below*). Craigellachie Bridge, near Banff, 1814. Designed by Thomas Telford. This carries the road over the River Spey, and has a total length of 195 ft., a span of 152 ft., a carriageway of 315 ft. 7 ins. with no footways, and is a cast iron arch structure on masonry abutments.

Fig. 88 is reproduced by courtesy of the Public Works, Roads and Transport Congress, and Fig. 89 is reproduced by courtesy of the Birmingham Central Reference Library.

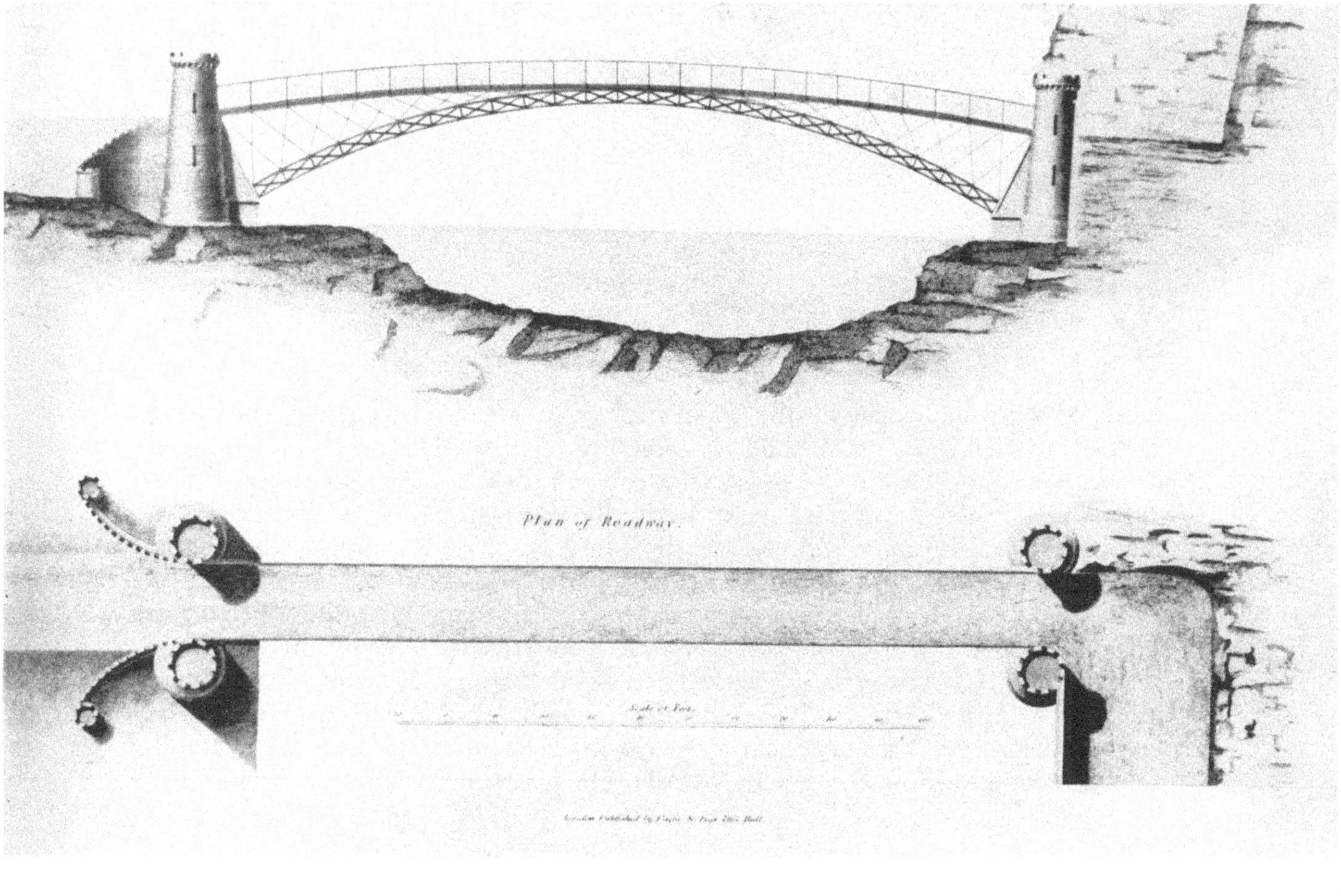

FIG. 90. The cast iron Galton Bridge at Smethwick, Birmingham, 1829. Designed by Thomas Telford. It has a span of 150 ft. and carries the road over the canal.

Reproduced by courtesy of Muir White, Esq.

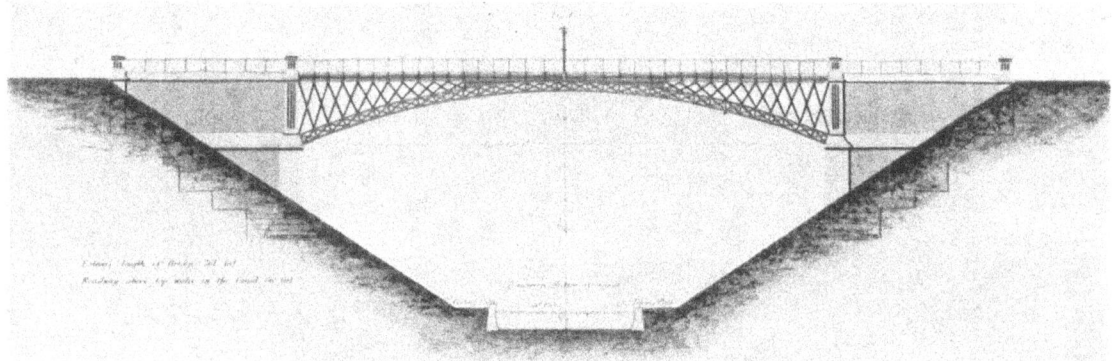

FIG. 91. Contemporary drawing showing the Galton Bridge.

Reproduced by courtesy of the Birmingham Central Reference Library.

material by using considerable quantities of cast iron in place of the wood previously used in machinery he was erecting. Later, when he was introduced to Watt in Birmingham, he undertook much mill work for the Soho firm.

Rennie's reputation grew: he was equally successful whether dealing with machinery or designing and erecting buildings, and he was finally to become famous for his bridges, factories, canals, drainage schemes, harbours and great dockyards. His first cast iron bridge was that spanning the Witham at Boston, Lincolnshire, erected in 1803. It consisted of a single arch of segmental iron ribs, and was especially successful through Rennie's foresight in allowing considerable width for the road which was kept almost level as it crossed the water. Two years earlier—in 1801—Rennie had been asked to design a bridge to cross the Menai Straits. He proposed a single great arch of

FIG. 92 (*Above*). The Waterloo Bridge at Bettws-y-Coed, Carnarvonshire, 1815, designed by Thomas Telford. This crosses the Conway River, having a total length of 125 ft., a span of 105 ft., a 20 ft. carriageway, and 5 ft. footways. It is a cast iron structure supported on masonry abutments, and in 1923 the centre cast iron ribs were encased in concrete to strengthen them.
Reproduced by courtesy of the Public Works, Roads and Transport Congress.

FIG. 93 (*Below*). Contemporary drawing of the Waterloo Bridge over the Conway River.
Reproduced by courtesy of the Birmingham Central Reference Library.

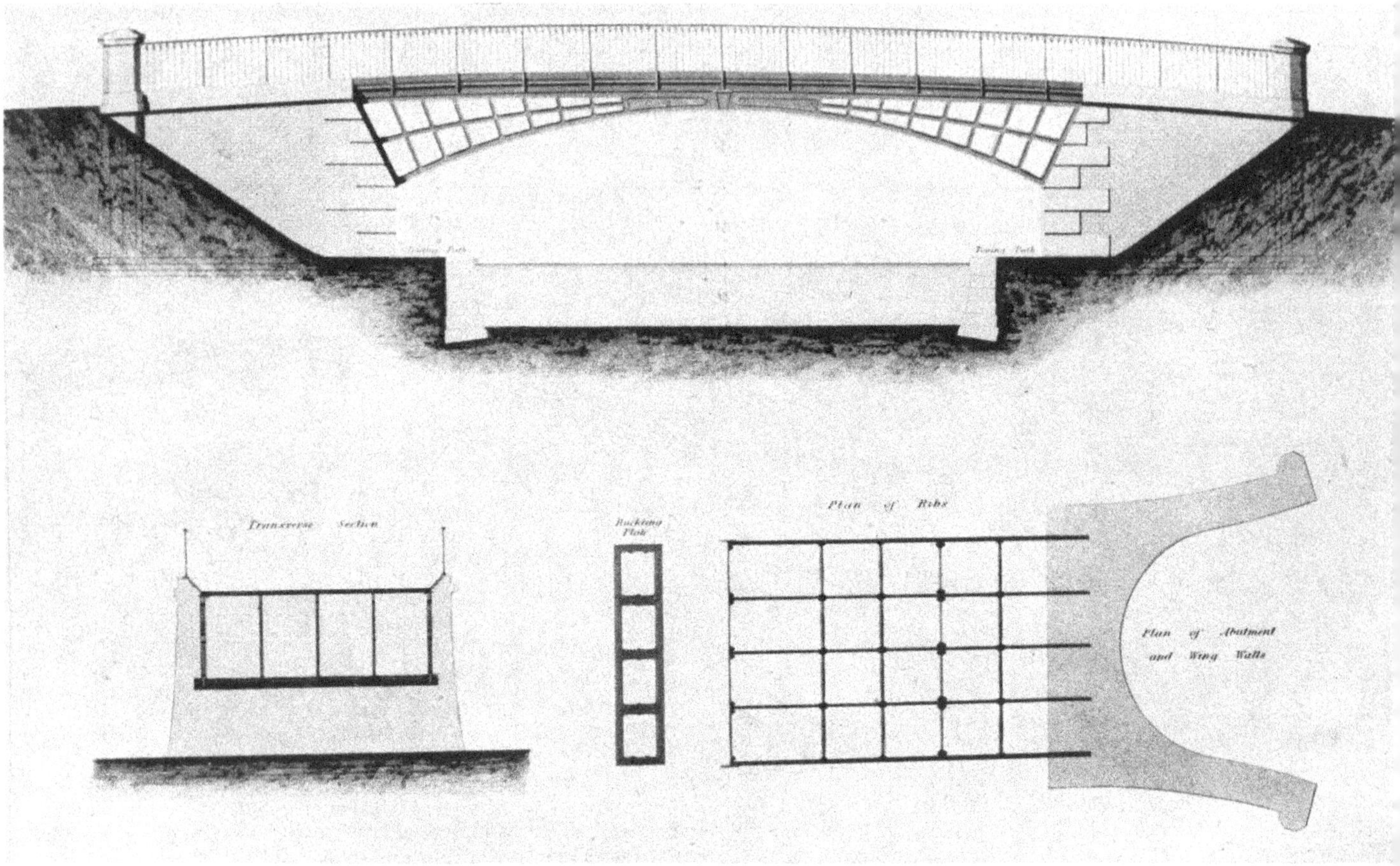

Fig. 94 (*Above*). The Ickneild Street Bridge, at Birmingham, 1828. A cast iron bridge designed by Thomas Telford, to carry the road across the canal.

Fig. 95 (*Below*). Another of Telford's cast iron bridges, the Mythe Bridge, Tewkesbury, which crosses the River Severn, 1823 - 1826. Total length 276 ft., span 170 ft., carriageway 18 ft. 2 ins., and footways 3 ft. The cast iron structure has six main ribs, standing on masonry abutments, with cast iron plates resting on cast iron bearers to carry the road. In 1923 the bridge was strengthened by the addition of a reinforced concrete slab over the decking.

Both these illustrations are reproduced by courtesy of the Birmingham Central Reference Library.

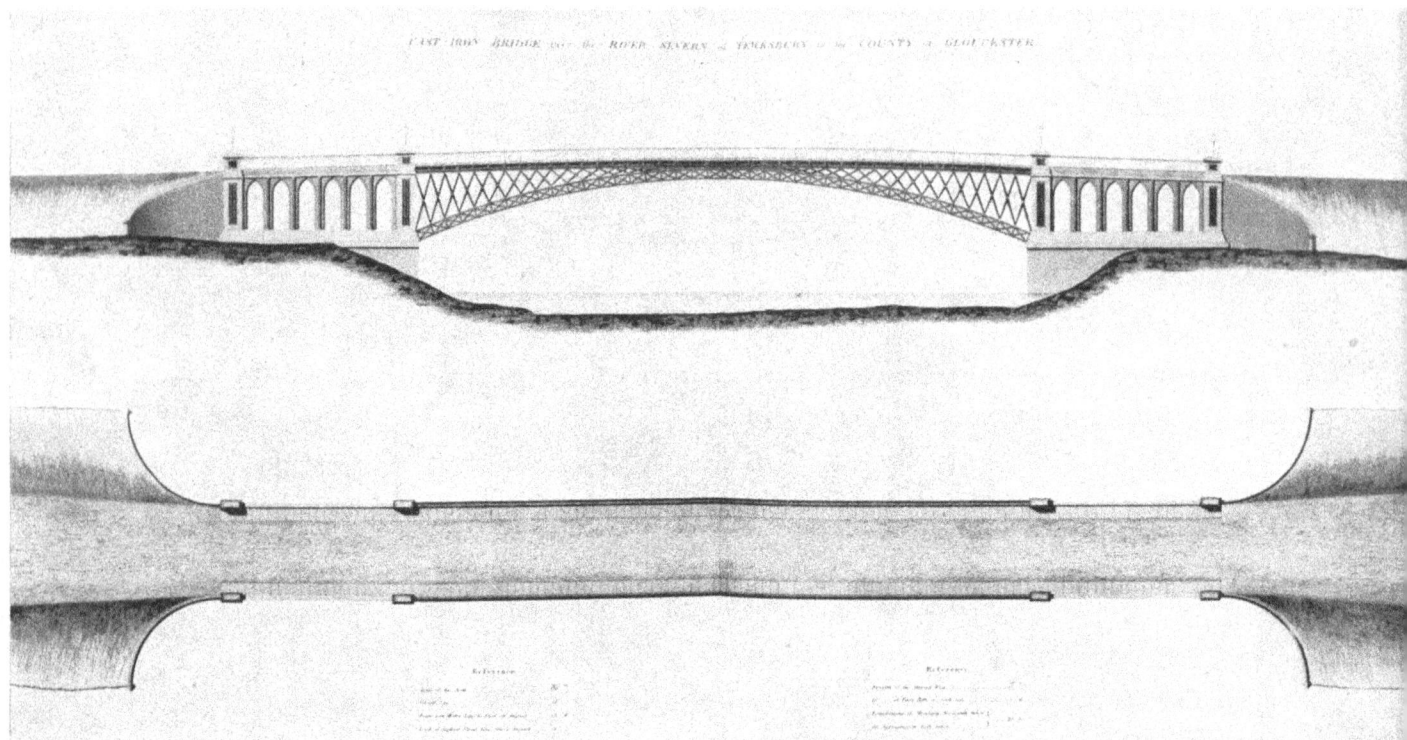

Fig. 96 (*Above*). Stokesay Bridge, Shropshire, over the River Onny, 1823. Designed by Thomas Telford. Total length 101 ft. 6 ins., span 54 ft. 9 ins., carriageway 20 ft. A cast iron rib structure on masonry abutments.

Fig. 97 (*Below*). Chepstow Bridge, Monmouthshire, 1816. Designed by John Rastrick. Crosses the River Wye. Total length 490 ft. There are 5 arches, the centre span 112 ft., two spans of 70 ft., and two spans of 30 ft. Carriageway 14 ft., footways 3 ft. The road superstructure is carried on cast iron ribs, braced together on stone piers. In 1889 the central span was strengthened by three steel box girder ribs, 10½ ins. deep, erected below the three cast iron ribs.

Both these illustrations are reproduced by courtesy of the Public Works, Roads and Transport Congress.

FIG. 89 (*Above*). Boston Town Bridge, Lincolnshire, across the River Witham. The original bridge, 1802, was the first in England designed by John Rennie; also one of the first in which wrought iron was used, the cast iron weighing 208 tons and the wrought iron just over 3 tons. This lasted until 1912 - 1913, when it was replaced by a steel structure erected on the original foundations. The present bridge has a span of 86 ft. 6 ins., a 25 ft. carriageway, 9 ft. footways, and cast iron parapets pierced in a trefoil pattern.

FIG. 99 (*Below*). Cleveland Bridge, Bath, Somerset, which crosses the Wiltshire Avon, 1833. Designed by John Hazzledine. Total length 164 ft., span 110 ft., carriageway 23 ft. 5 ins., footway 6 ft. 8 ins. This was originally a cast iron arch rib bridge with a cast iron deck and parapets. In 1930, a reinforced concrete structure was erected within the old bridge, which now remains in a state of rest, retaining its former appearance. (See Fig. 100, on opposite page.)

Both these illustrations are reproduced by courtesy of the Public Works, Roads and Transport Congress.

Fig. 100. A contemporary illustration of the New Iron Bridge over the Avon at Bathwick, Bath. This illustration was published in 1829 in a series of views of Bath and Bristol engraved by Thomas H. Shepherd, with historical and descriptive notes by John Britton, F.S.A. Of this bridge, Mr. Britton said: "Iron bridges possess the advantages of lightness, strength, and durability, combined with a superior elegance of form; and as the termination of a great public road, as it enters the precincts of a first-rate city, this erection, in every respect of a superior description, is peculiarly appropriate".

cast iron 450 feet in span, and suggested a similar one of 350 feet span to cross the Conway Ferry. Rennie held that his designs were invented as early as 1791 and was completely satisfied that they were superior to all others. But, as with Telford's designs of the same period, they were thought too revolutionary, too daring and too expensive, and were dropped.

In 1819 Rennie completed Southwark Bridge across the Thames. This consisted of three cast iron arches with stone piers and abutments, the centre arch being of 240 feet span, slightly more than the span of the Sunderland bridge, which until then was the largest cast iron arch erected. The great Doric cornice and the plinth to take the side rails were cast hollow, and immense care was taken with every detail of the foundry work, totalling some 5,000 tons, which was cast by Walkers, of Rotherham. This firm sub-let some of the work to Thorncliffe and to William Yates, the owner of the Gospel End Ironworks in Staffordshire.

Fig. 101. The North Parade Bridge, Bath, Somerset, crossing the Wiltshire Avon, 1836. This is sometimes attributed to John Hazzledine. Total length 135 ft., span 111 ft., carriageway 20 ft., footways 5 ft. 6 ins. The bridge is of cast iron rib construction with masonry abutments.
Reproduced by courtesy of the Public Works, Roads and Transport Congress.

John Rastrick, born in 1780, was engineer under Richard Trevithick, in 1808, and later joined John Hazzledine's foundry business at Bridgnorth, Shropshire, where many of the castings for Trevithick's engines were produced. Previously he had been employed at the Ketley ironworks, in the same county, under the management of William Reynolds, and had learnt much about the moulding of cast iron and its properties, and had seen how it was used in the bridges at Ironbridge and Buildwas. When he was employed to design a bridge over the Wye at Chepstow he chose the new material, and produced a bridge with a centre arch of 112 feet span, flanked by gradually diminishing arches. This bridge still stands to-day.

In the closing years of the eighteenth century, John Nash was becoming interested in cast iron: it was new, it was modish, his clients had heard about it, and one of them, Sir Edward Winnington, actually wanted to build a cast iron bridge across the Teme. In his admirable biography of Nash, John

Fig. 102. Abermule Bridge over the Severn in Montgomeryshire. This is one of the late cast iron bridges, being erected in 1852, which fact is recorded in rather corpulent lettering on the arch ring of the bridge. Span 109 ft., total length 150 ft., carriageway 20 ft. 10 ins. The structure is cast iron with masonry abutments.

Reproduced by courtesy of the Public Works, Roads and Transport Congress.

Fig. 103 (*Above*). A design for a bridge over the Thames between London and Blackfriars Bridges, by R. Dodd, Engineer. The centre arch was designed for cast iron with a 300 ft. span. This illustration is from the Report of the Select Committee upon the Improvement of the Port of London, 1800.

Fig. 104 (*Below*). Two designs for cast iron bridges over the Thames, by Telford and Douglass, included in the Report of the Select Committee upon the Improvement of the Port of London, 1800.

These illustrations are reproduced by courtesy of the Royal Institute of British Architects.

FIG. 105 (*Above*). Vauxhall Bridge from the Thames. Designed by James Walker, 1816. Total length 806 ft. There are nine arches, each with a span of 78 ft. Width between the parapets 36 ft.

FIG. 106 (*Below*). Design for a cast iron bridge over the Thames at Kingston, Surrey. This was the subject of a public competition in 1824, and the prize of 100 guineas went to an architect, J. B. Watson. His design, which was not carried out, was for a bridge of three arches, each with a span of 108 ft., the width between the parapets being 36 ft.

These illustrations are reproduced by courtesy of the Royal Institute of British Architects.

103

Summerson records the great architect's early interest in cast iron as a material. He suggests that, as evidence exists of Nash's connection with Broseley, he probably had relatives living there and conceivably studied cast iron technique at the foundries of Coalbrookdale and Bersham. "Indeed, it might be possible to claim for him a place of some distinction among the early engineers," writes Summerson, "if only there were more facts to go upon".[19] His use of the new material for a single span bridge over the Teme at Stanford was unfortunate. The bridge was completed, but it was hardly erected when it slowly collapsed. There are no records showing its design or construction; and Summerson contents himself with the moderate statement that Nash's client, Sir Edward Winnington, "was disappointed".[20] But he trusted his architect, and Nash, undaunted by the failure, built another bridge in cast iron which was opened in September, 1797, and stood until 1905. Nash claimed to have had some hand in the designing of the cast iron bridge across the Wear at Sunderland, but though he was called in for advice, it seems that the design was almost wholly the work of Tom Paine.

Nash actually took out a patent on the type of construction to be used for cast iron bridges. John Summerson has expressed the following views on this aspect of Nash's work: "The patent for the 'Construction of Plate Iron Bridges' is interesting as part of the history of the use of iron as a structural material. Nash's idea represented an ingenious attempt to avoid the worst treacheries of cast iron construction. It must be remembered that cast iron in compression is admirable; in tension, as unreliable as stone. The invaluable tensile strength of steel, which enables us to span vast distances with superb efficiency, was unknown till long after Nash's time. Thus, except for very light short-span bridges, iron was not really an economic substitute for brick or stone. Nash's 'plate iron' bridge merely substituted cast iron boxes, made of iron plates bolted together, for voussoirs (*i.e.* blocks of stone forming an arch), and as the boxes were to be filled with 'earth, sand, stone, gravel, or other materials to make solid bodies', the patent bridge presents nothing much more original than an arch, each of whose voussoirs has a casing of cast iron plates. The bolting together of the voussoirs is a somewhat unscientific safety device making a very flat arch practicable".[21]

CANALS AND AQUEDUCTS.

In the last half of the eighteenth century and the first quarter of the nineteenth, canals accelerated the growth and development of the first industrial revolution. The vast increase in trade found the road system quite unable to deal with the new volume of traffic, and inland water transport ultimately cheapened the movement of goods and raw materials. Canals connected various river highways, and greatly increased the value of the adjoining land and sites. Brindley had made such a success of the Duke of Bridgewater's canal scheme that others quickly followed.

Even before his design for the Buildwas bridge, Telford was thinking of structural uses for cast iron. In 1793 he was, with the strong backing of John

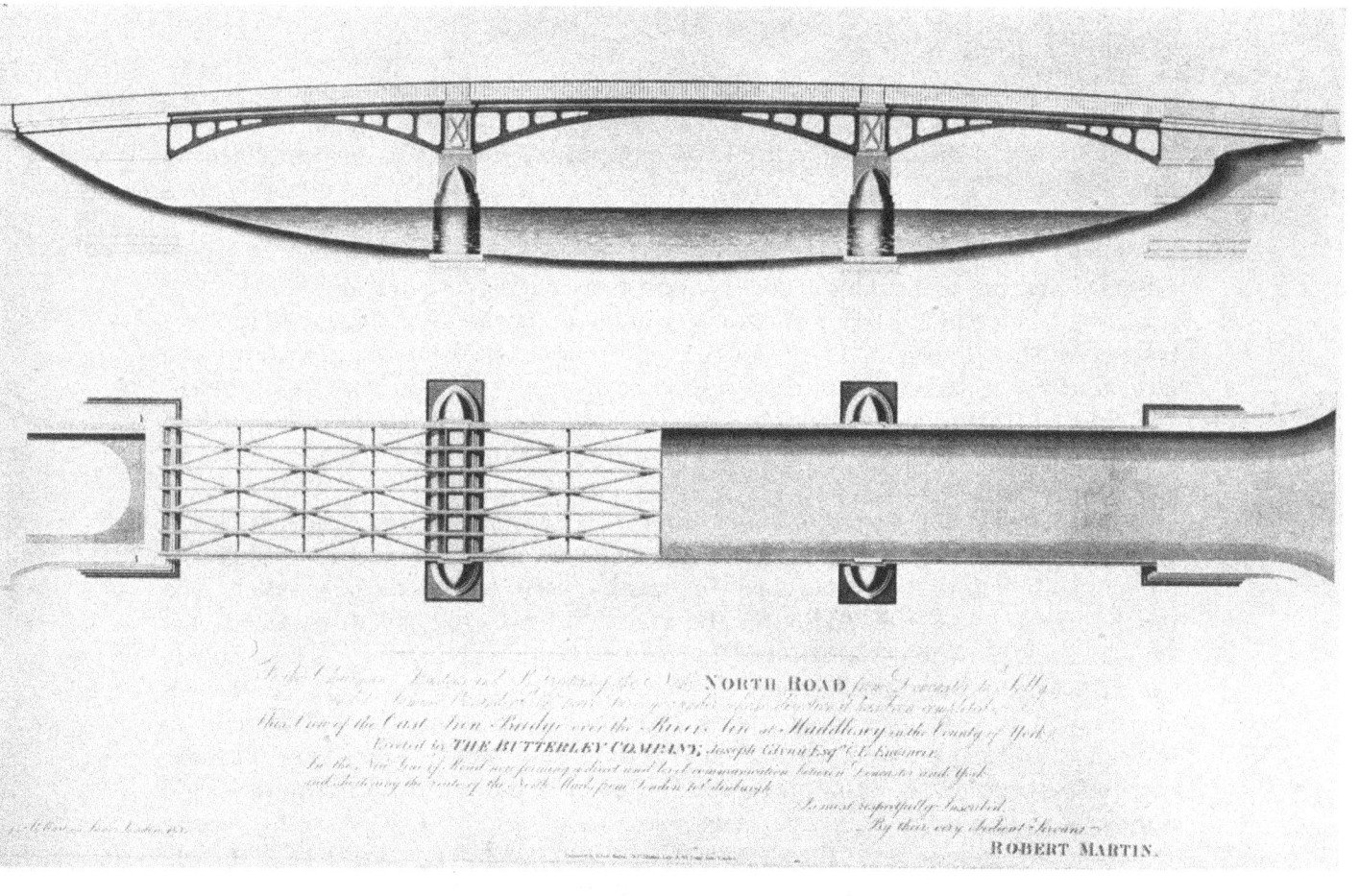

FIG. 107. Cast iron bridge crossing the River Aire at Haddlesey, Yorkshire. Erected by the Butterley Company, Joseph Glynn, Engineer. From an engraving by Robert Martin, 1834.
Reproduced by courtesy of the Royal Institute of British Architects.

Wilkinson, appointed to prepare a scheme for connecting the Mersey, Dee and Severn, which became known as the Ellesmere Canal System. Telford realised that although the initial cost would be increased, it was necessary to keep the number of locks to a minimum in order to speed and cheapen traffic. Apart from the engineering feat of cutting these canals, this decision meant that some means had to be devised for carrying the canal across the great valleys of the Dee and the Ceriog. Telford did this with two great aqueducts at Chirk and Pont-Cysylltau, described in wonder and admiration by Phillips as "among the boldest efforts of human invention in modern times".[22]

Previously canals had been carried on masonry piers and arches of sufficient breadth and strength to allow room for a puddled water tray. Before Telford had started work on the Chirk and Pont-Cysylltau aqueducts, he had been appointed to work on a canal connecting Shrewsbury with the collieries near Ketley, where in 1795, he successfully used a cast iron trough, carried on cast iron columns, to bear the canal some sixty yards across the river Tern. He adapted this idea for the new and impressively large aqueducts. The Chirk aqueduct consists of ten masonry arches, each of 40 foot span, carrying the

105

water some 70 feet above the river below, the canal bottom consists of cast iron plates bolted together, while the sides are formed in masonry. (The plan and elevation of this aqueduct are shown opposite, with a section which shows the placing of the cast iron plates in relation to the slightly sloping masonry sides.)

The Pont-Cysylltau aqueduct is greater in scale, having nineteen masonry arches, covering a distance of over 1,000 feet, with the canal carried 97 feet above the river level. Here Telford used metal even more extensively than before, and set on top of the masonry a complete cast iron trough with towing path and side rails, all accurately bolted together, the water way being about 12 feet; 4 feet 8 inches of which were occupied by a towing path supported on cast iron columns, resting on the bed of the canal. (See Figs. 110, 111 and 112, on pages 108 and 109.)

Telford was concerned with more than thirty canal schemes, and with each year's experience he learnt more of the uses of cast iron and introduced it where timber and stone had been generally used before. Smiles says: "On the Ellesmere, and afterwards on the Caledonian Canal, he introduced cast iron lock-gates, which were found to answer admirably, being more durable than timber, and not liable like it to shrink and expand with alternate dryness and wet. The turnbridges which he introduced upon his canals, instead of the old drawbridges, were also of cast iron; and in some cases even the locks themselves were of the same material. Thus, on a part of the Ellesmere Canal opposite Beeston Castle, in Cheshire, where a couple of locks, together rising 17 feet, having been built on a stratum of quicksand, were repeatedly undermined, the idea of constructing the entire locks of cast iron was suggested; and this extraordinary application of the new material was successfully accomplished, with entirely satisfactory results".[23]

MACHINERY.

One of the earliest successful uses of cast iron for machinery has already been mentioned (Section 1, page 48), namely the rebuilding of the London Bridge Works for pumping water from the Thames in 1704. In 1754 John Smeaton had introduced a cast iron shaft for a windmill, and in 1760 John Murdoch, father of the celebrated William Murdoch, is said to have had a cast iron cog wheel made at the Carron Works. But these and other examples were just early experiments and of somewhat inferior workmanship, and it was not until 1784 that Rennie successfully used cast iron for nearly all the parts of the machinery in the Albion Flour Mills on the banks of the Thames near Blackfriars Bridge. James Watt designed the special steam engines to drive this new machinery, and Rennie was employed by Boulton & Watt to design and supervise the lay-out of the mill which proved a landmark in the history of mechanical improvements, and subsequently influenced the planning and designing of all factories.

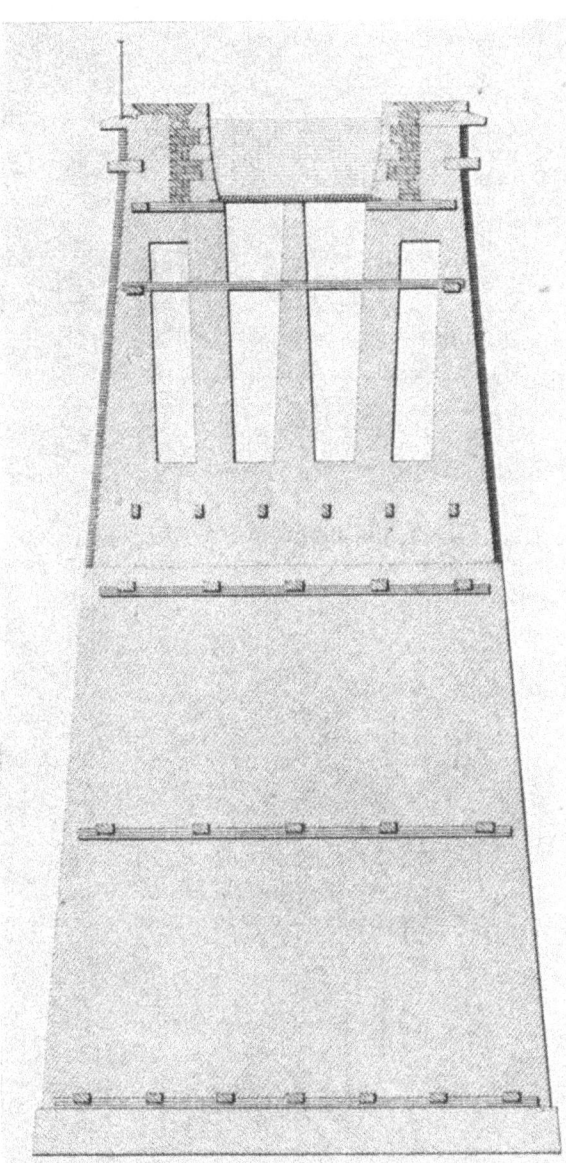

Fig. 108 (*Above*). Plan and elevation of the Chirk Aqueduct, designed by Thomas Telford, carrying the canal across the valley of the Ceriog. Ten masonry arches, each of 40 ft. span, carry cast iron plates, which form the bottom of the canal. 1796 - 1801.

Fig. 109 (*Right*). Section, showing the structure and the relative position of masonry and cast iron plates.

Reproduced by courtesy of the Birmingham Central Reference Library.

Fig. 110. The **Pont-Cysylltau Aqueduct**, designed by Thomas Telford, carrying the canal across the River Dee at the bottom of the Vale of Llangollen. See Figs. 111 and 112, on page 109 for detail drawings.
Reproduced by courtesy of the Institution of Civil Engineers

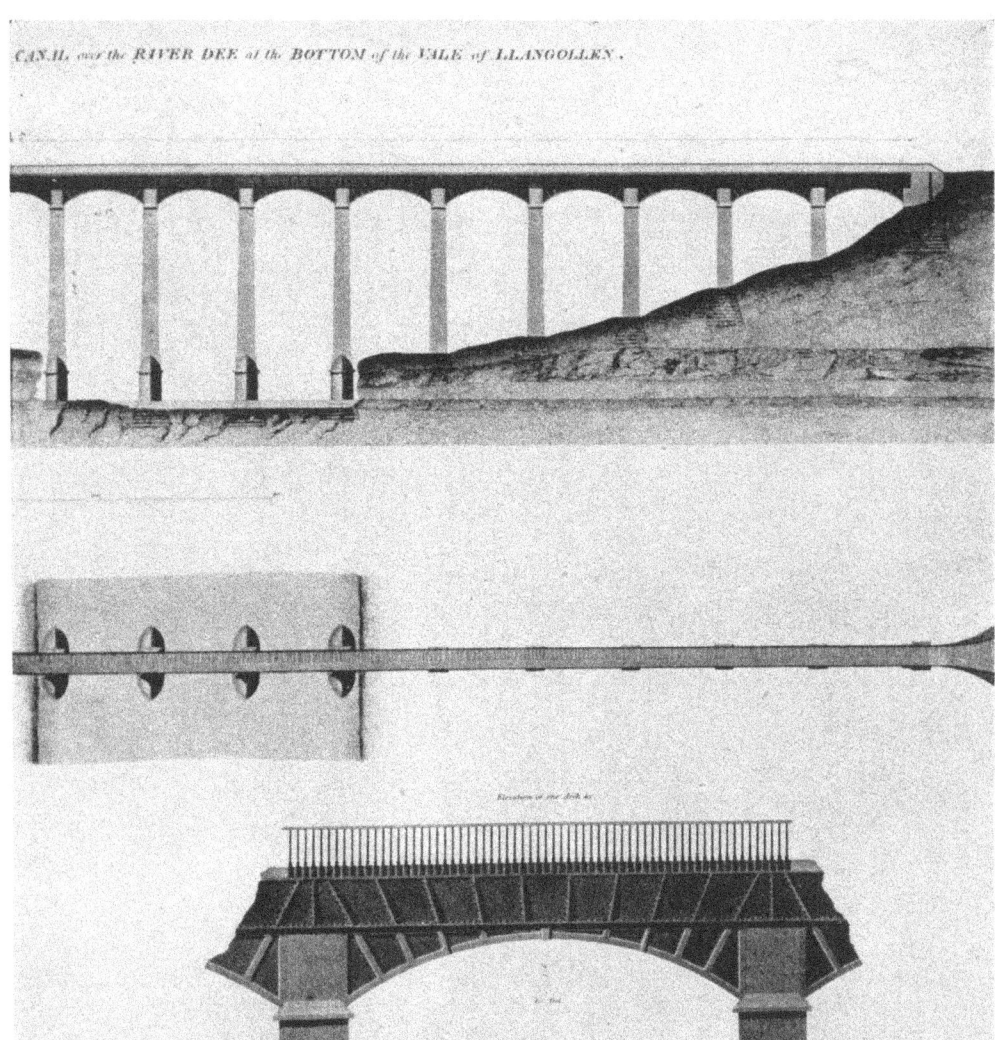

Fig. 111 (*Above*). Elevations and plan of the Pont-Cysylltau Aqueduct. The nineteen masonry arches which carried the canal across the Dee valley supported a complete cast iron trough, with a towing path and side rails of the same material. Designed by Thomas Telford, 1795 - 1803. A complete view of the Aqueduct is shown in Fig. 110 opposite.

Fig. 112 (*Left*). Section through the aqueduct, showing the cast iron structure of the trough.

Reproduced by courtesy of the Birmingham Central Reference Library.

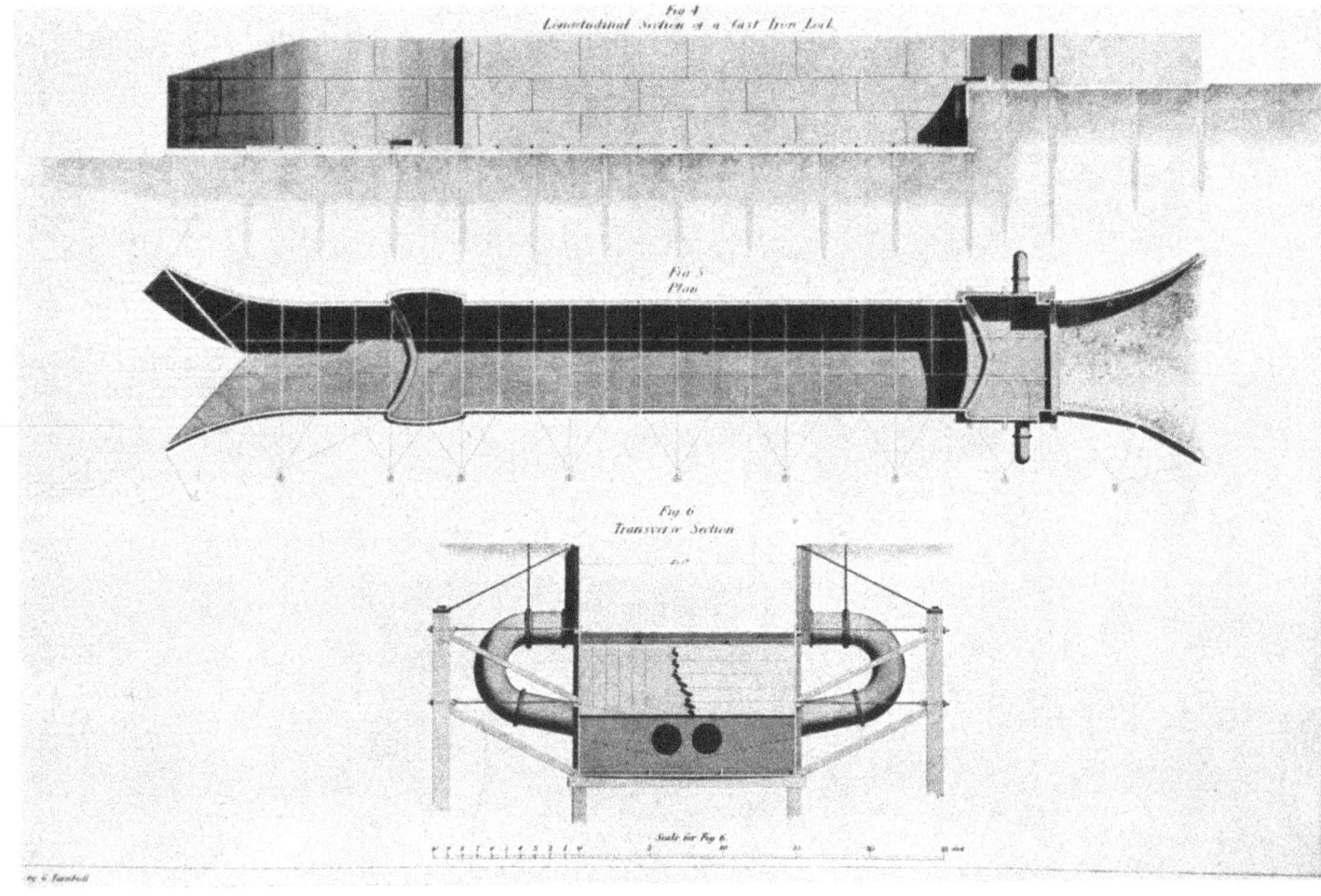

FIG. 113. Plan and longitudinal and transverse sections of a cast iron lock designed by Thomas Telford.
Reproduced by courtesy of the Birmingham Central Reference Library.

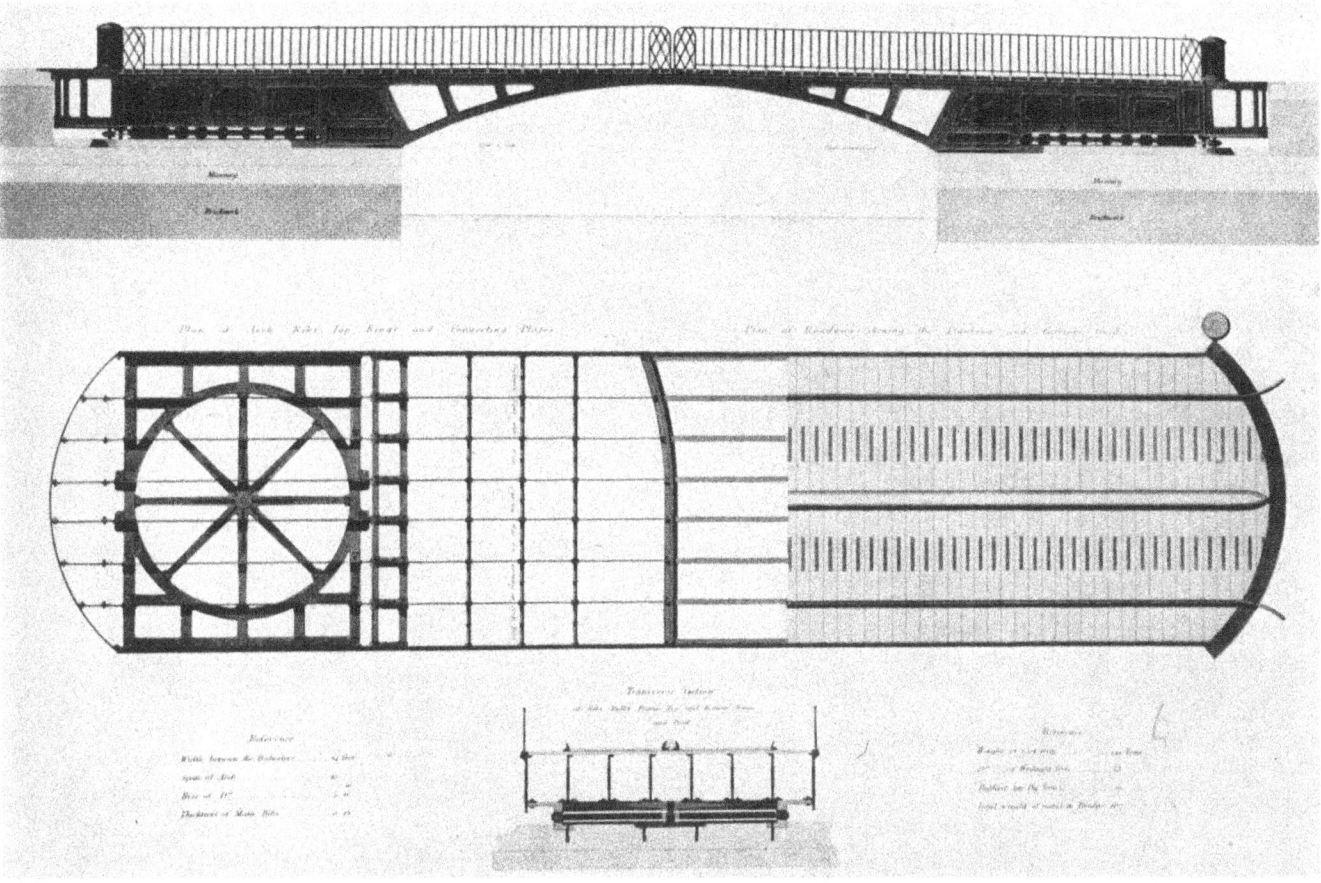

Fig. 114. Elevation, plan and transverse section of cast iron swivel bridge over the entrance lock at St. Katharine's Docks, London. Designed by Thomas Telford, 1824 - 1828.

Reproduced by courtesy of the Birmingham Central Reference Library.

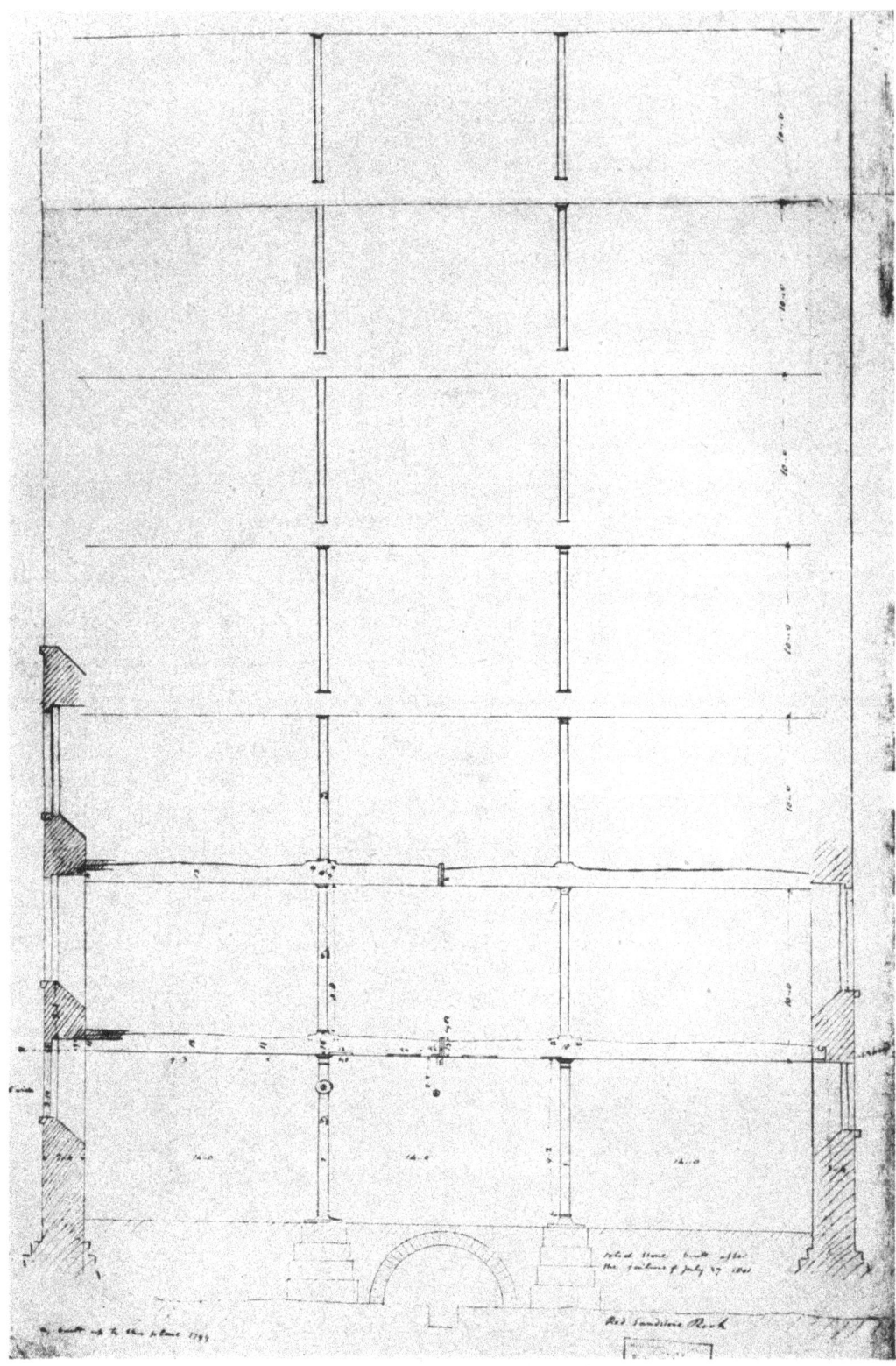

Fig. 115. Cross section of the cotton mill erected in 1801 by Messrs. Phillips, Wood & Lee in Manchester. It was designed by Boulton & Watt, and was probably the first successful use of cast iron beams as structural units in a building of this scale. The mill was 140 ft. long, 42 ft. wide, and seven storeys high. The illustration above, and those in Figs. 116 and 117 (opposite and on page 114) are reproduced from the original drawings of Boulton & Watt.

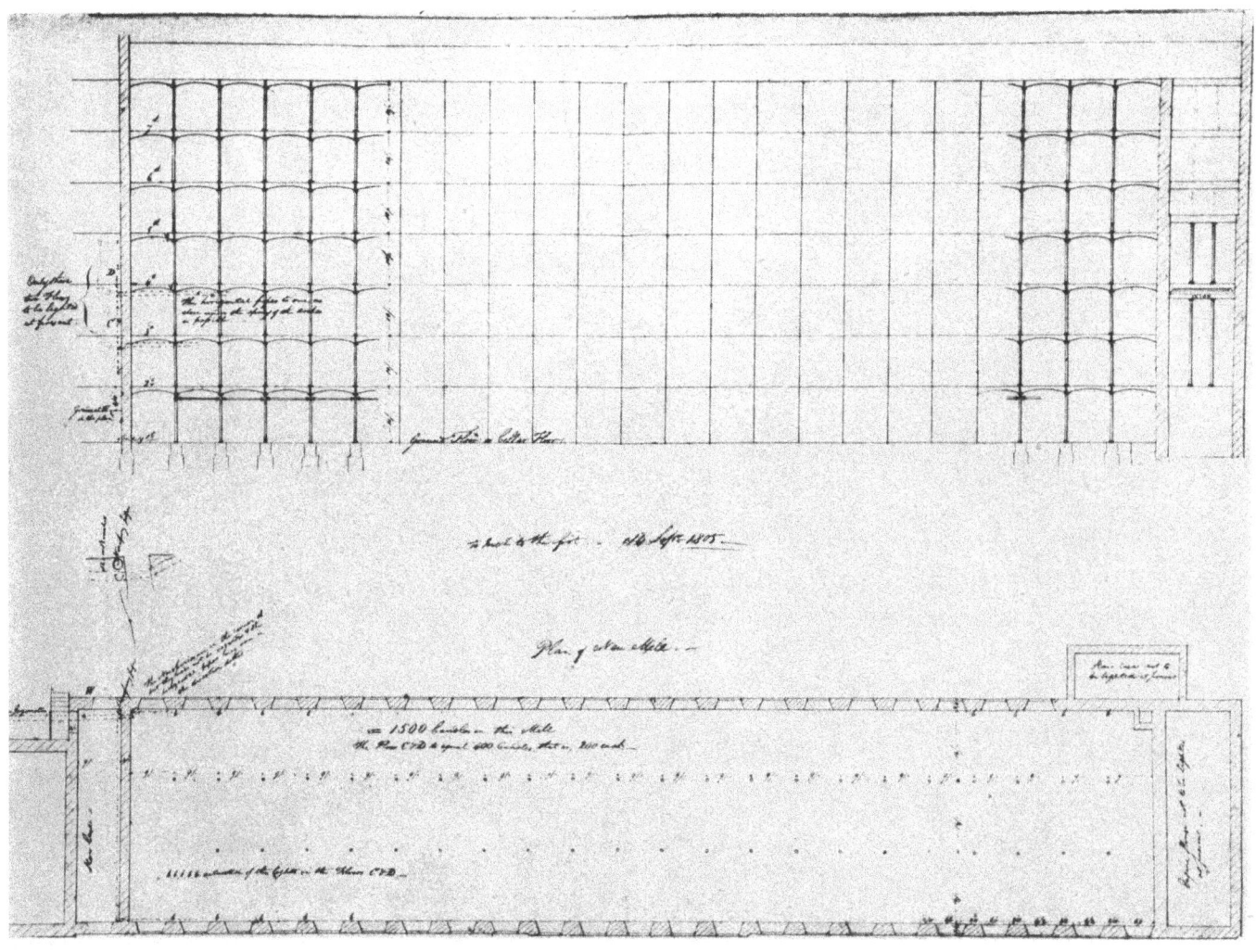

Fig. 116. Plan and section of the Manchester cotton mill of Messrs. Phillips, Wood & Lee. See Fig. 115 on the opposite page for the cross section, and Fig. 117 on the page that follows for structural details of columns and bases.

BUILDINGS.

The first successful use of cast iron beams as structural units in a building was in a cotton mill erected in 1801 by Messrs. Phillips & Lee in Manchester. These beams were designed by Boulton & Watt, and, as William Fairbairn points out, considering that they were the first of their kind and the limited state of knowledge on the subject at the period, they reflect great credit upon the skill of the designer.[24] The mill was 140 feet long, 42 feet wide and seven storeys high. The width of the building was divided into three bays, each of 14 feet span, by two rows of cast iron stanchions at 9 feet intervals, the floor being carried on $13\frac{1}{2}$ ins. by $3\frac{1}{4}$ ins. inverted T type cast iron beams 14 feet long. For the next twenty years this mill was the model for many

113

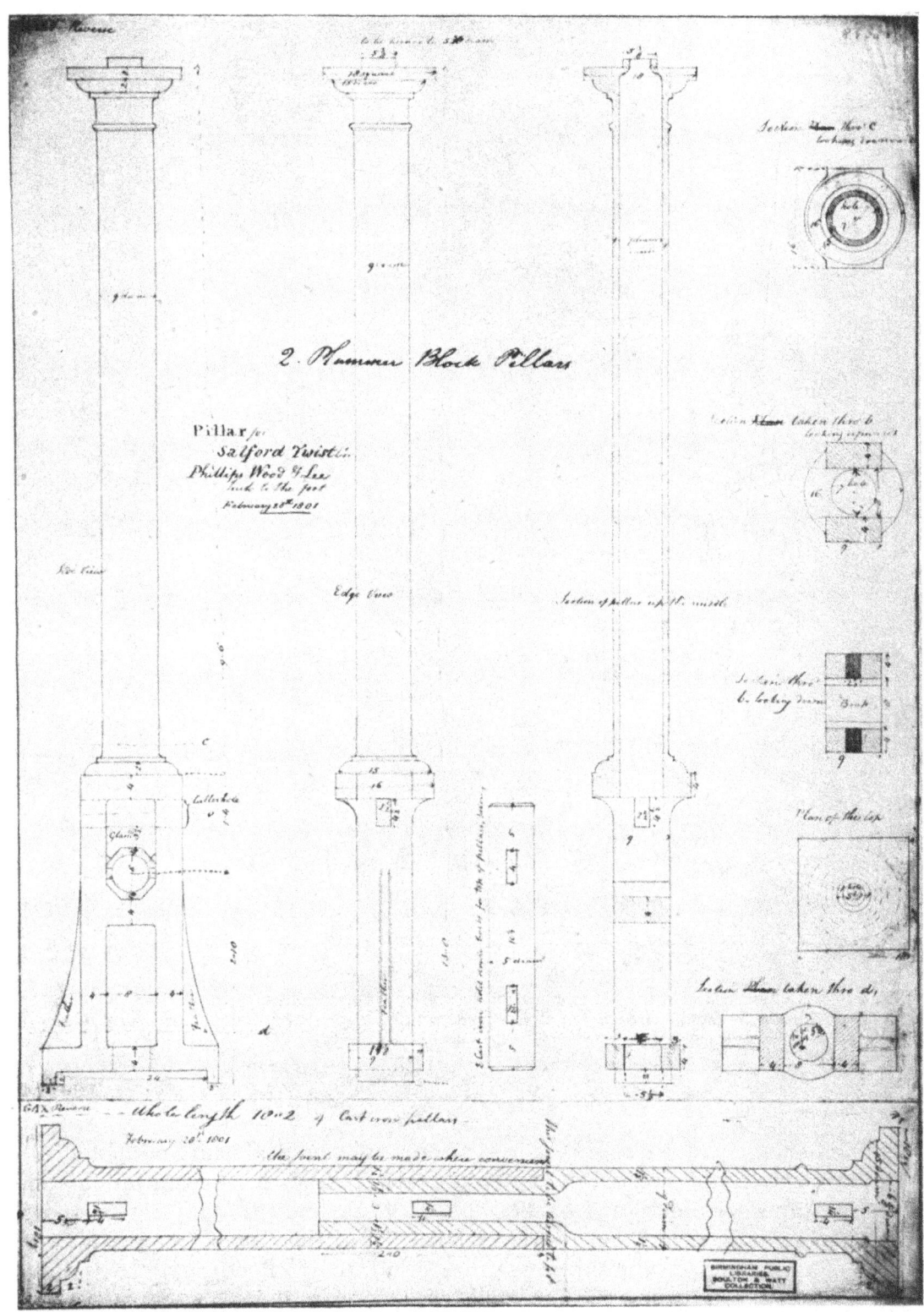

Fig. 117. Details of columns and bases in the Manchester cotton mill of Phillips, Wood & Lee, designed by Boulton & Watt. A cross section of the building is given in Fig. 115 (page 112), and plan and section in Fig. 116 (page 113).

This illustration, and those shown in Figs. 115 and 116, are reproduced from drawings in the Boulton & Watt collection of the Birmingham Central Reference Library.

others, and the form of building and form of beam, having proved satisfactory, varied very little until Tredgold, in 1824, published the second edition of his work on the strength of cast iron. Professor Hodgkinson carried out his well-known investigations into the strength and design of iron beams in 1827, and William Fairbairn, after a series of practical experiments, increased the area of the bottom flange of beams he employed in mill buildings in Leeds and Bradford.

Decoratively, cast iron had a considerable influence on late eighteenth century architecture. In the lay-out of such large London estates as the Bedford, Portland, Grosvenor, Berkeley and Portman, and in similar building schemes in towns all over the country, the terrace house created a demand for miles of railings of a repetitive design, which foundries could produce easily and economically. These basemented houses needed protective rails round the areas which lighted the lower floors, and apart from the decorative value of a railed balcony at first floor level, there was its safety value in case of fire. A fashion had arisen in this period for balconies to all town dwellings. There were cast iron railings and balconies to houses in London, Liverpool, Birmingham, Edinburgh, and in fact all over Great Britain, particularly in the holiday and health resorts, such as Brighton, Bath and Hastings. Many of the earlier balconies and railings were of wrought iron, but as the custom and demand increased, cast iron was more frequently used. The designers were generally craftsmen of the foundry, who seemed to realise instinctively the repetitive and inherent qualities of the material in which they designed, often with admirable results. W. R. Lethaby wrote of such railings: "They show indeed that, before the age of style mongering tightened its grim hold, the English manufacturers really could do pleasant and reasonable things. Some of the railings were formed of simple uprights, some of open pilaster-like strips, and others, which were the most characteristic, were all-over lattice panels. For natural and unashamed treatment of the method of iron-casting, these things could not be bettered. We never think of comparing them with wrought work to their disadvantage; both are excellent in different categories. Ambitious modern work in cast iron most usually fails by raising the comparison with wrought iron, imitating in infinitely careful pattern-making just those effects in which to succeed is most of all to fail".[25]

The first cast iron railings recorded were those used round St. Paul's Cathedral, which were fixed in 1714 (see page 37). These railings were cast at Lamberhurst on the Sussex border of Kent. The rails are of baluster form with a stronger member of similar profile at seven or eight foot intervals. There were over 2,500 of the main type and 150 of the heavier type, the whole, with all the accessories, weighing about 200 tons and costing £11,000. Sir Christopher Wren disapproved of them. Lethaby, in writing of their design, said: "I do not see how the railings could have been better, they are heavy and rather blunt as befits the situation and the material of which they are made".[26] Some are still in place, and a few are preserved in pleasant surroundings at Lewes Castle by the Sussex Archæological Society.

In 1874 the iron railings at the west end of the Cathedral, including the western gates, were sold by public auction for £349 . 5 . 0. They were sent to Toronto and suffered shipwreck, but some were bought to enclose the tomb of Mr. John G. Howard, architect of that city, and bear the following inscription:

> "S. Pauls Cathedral for 160 years I did enclose
> Oh! stranger look with reverence.
> Man! Man! unstable man!
> It was thou who caused the severance.
> November 18, 1875. J. G. H."

Similar railings were designed by James Gibbs for the Cambridge Senate House in 1722-30. These are magnificent in scale and bold in detail. They have a wrought bar between each cast baluster, like the rails round the statue of Henry VI at Eton. (See Figs. 42 and 43, pages 46 and 47.) Gibbs used cast iron railings to enclose St. Martin's-in-the-Fields Church about 1726 and also for the altar rails in the same church, and in 1740 George Dance designed cast iron railings to surround St. Leonard's Church, Shoreditch. But the introduction of cast iron was gradual, and often the new material was used only for parts that were difficult or costly to produce in wrought iron. The most characteristic feature of all iron work of the period is perhaps the cast iron finials to the standards of railings. The pine- or acorn-shaped finial, and the vase-shaped, either fluted or heavily gadrooned, are typical.

Isaac Ware in *The Complete Body of Architecture*, 1756, says "that cast iron is very serviceable to the builder and a vast expense is saved in many cases by using it; in rails and balusters it makes a rich and massy appearance when it has cost very little, and when wrought iron, much less substantial, would cost a vast sum".[27]

In Carter's *Builders' Magazine* of 1774, many designs are shown for gates and railings using cast iron extensively, and nearly fifty years later, about 1823, L. N. Cottingham's famous *Smith and Founders Director* was published, giving many examples of existing and new designs.

Sir Robert Taylor, in 1774, designed some fine iron railings for Stone Buildings, Lincoln's Inn, in the form of thin elongated Doric columns, having used the same form in the railings to "Ely House" in Dover Street, in 1772. But some of the most decorative of the iron work of the period was designed by Robert Adam. He was a prolific and facile designer, and through his association with the Carron Works acquired a complete knowledge of the possibilities of cast iron. He never hesitated to mix cast iron, wrought iron, brass and steel, and used the material that was best suited to his design. Unfortunately, some of his finest schemes were never carried out, such as his entrances to Hyde Park (1778) in which he used bars crossing diagonally, with rosettes at each intersection, in the manner of the contemporary French *treillage*. But he left masterly examples of the use, or partial use, of cast iron in this kind of work in the gates at Sion House, Isleworth; Lansdowne

Fig. 118. Pair of gates from the garden entrance of Lansdowne House, Berkeley Square, London. Probably designed by Robert Adam about 1770. The gates are a combination of cast iron and wrought iron units.

From the Victoria and Albert Museum, by whose courtesy this is reproduced. Crown copyright reserved.

House, London; Chandos House, Queen Anne Street, London; Boodles Club, St. James's; and houses in Portman Square and the Adelphi, London. The brothers Adam devised a type of design which was peculiarly suited in its severity and repetition of delicate detail for reproduction in the foundry.

Apart from all these visible and perhaps obvious decorative uses, and the structural employment of cast iron in mills, factories and warehouses, architects in the late eighteenth and early nineteenth centuries discovered that it was a convenient and adaptable material and made it perform a variety of unusual and now almost forgotten tasks. For example, John Summerson records that when, in 1942, he measured the lantern on the roof of the Middle Temple Hall "a light and graceful piece of early nineteenth century Gothic" he "firmly believed the whole thing was made of timber, until, in 1944, the roof was burnt and down came great chunks of the neatest cast iron cusping imaginable".[28] The same authority has described how the blitzed Temple disclosed the extent to which the later Georgian architects used iron. "Sir Robert Smirke, in particular, had a most knowing appreciation of the material. In his Inner Temple Library, 1824, and the rather later Harcourt Buildings there were iron girders to all the floors and iron dwarf stanchions concealed behind the stone mullions of the windows. The girders, which bore the name of 'Dewar, London,' were of ⊥ section with a cambered flange. In the library the floors had all been 'fireproofed' with iron sheeting, a method which might have stood up well to a homely Georgian conflagration, but which crumpled up pathetically in the blitz. Some of the garden and area railings in the Temple are also cast iron, admirably appropriate in design. In the interests of safety rather than amenity they have been allowed to survive.

FIG. 119. Detail of cast iron railings, Stone Buildings, Lincoln's Inn, London, designed by Sir Robert Taylor, 1774. (See page 116.)

FIG. 120. Cast iron railings designed by Sir Robert Taylor for Ely House, Dover Street, London. These, like the railings shown on page 117 in Fig. 119, are really elongated Roman Doric columns. The cast iron lions are later additions, and are designed by Alfred Stevens. (See Figs. 268 and 271, on pages 216 and 217.)

Photograph: Quentin Lloyd.

"When we recorded Southwark Cathedral we found that the roof of the choir (above the stone vault) was supported by really magnificent cast iron trusses, introduced, I imagine, by Gwilt during his restoration of 1822-5; each truss was cast in pieces and bolted together, recalling, in its scale and boldness the cast ironwork of Rennie's Southwark Bridge, built a very few years earlier".[29]

It was left to John Nash to use cast iron not only in railings, as he did so well in Regents Park, but structurally and decoratively, as he did in Carlton House Terrace, Buckingham Palace and the Pavilion at Brighton. Nash's experiments with cast iron as a material for bridges were elementary from an engineering point of view, compared with the work of Rennie and Telford, and it was not until he had the opportunity to design on a much larger scale that he realised how cast iron could help him in his great building schemes in London and the South of England. Then he was, to quote Summerson's critical phrase, "quite merciless in his use of iron as a general factotum".[30]

In 1815 Nash began his designs for the extension of the Pavilion at Brighton, and by the end of the following year the main gallery was completed with its two famous staircases of cast iron in the Chinese manner, with bamboo so successfully simulated that many people to this day do not detect the imitation. The *Sussex Weekly Advertiser* for September 4th, 1815, says: "The smiths who attended to fit the elegant cast iron staircases at the Pavilion finished their work to-day and returned to town". And in the *Journal of Mary Frampton*, 1779-1846, in a quoted letter of 2nd February, 1816, we find: "All the rooms open into this beautiful gallery, which is terminated at each end by the lightest and prettiest Chinese staircases you can imagine, made of cast iron and bamboo with glass doors beneath, which reflect the gay lanterns, etc., at each end".

Nash also used the new material for elegant, thin, decorated pillars, carrying the roof of several rooms, with particular success in the great Kitchen where they took up much less floor space than would have been necessary for the normal type of construction.

Nash began the rebuilding of the Regent Street Quadrant in 1818. In its original form it must have possessed singular dignity, with its 145 great Roman Doric columns carried through the ground floor and mezzanine floor. These columns were all of cast iron carrying a stone entablature; so were the later columns to the wings at Buckingham Palace (to the surprise of most people who saw them after the "blitz"); and still later, in 1827, cast iron was used for the Greek Doric columns carrying the terrace to Carlton House.

Sir Robert Smirke introduced iron in domestic building during this period, and some facts about his use of cast iron were given by his brother, Sidney Smirke, in the course of a discussion which followed the reading of a memoir at an ordinary general meeting of the Royal Institute of British Architects held on June 17th, 1867. This memoir of Sir Robert Smirke was read by Edward Smirke, another of Sir Robert's brothers. In the discussion it was said "no architect employed cast iron largely as a building material" until it had

Fig. 121. Detail of cast iron and lead railings, Lincolns Inn, London, late eighteenth century.

been introduced by Sir Robert Smirke. Sidney Smirke stated that Sir Robert had used "huge cast iron bearers" in Lord Bathurst's house, "where he had some practical difficulties to overcome, and heavy walls to retain and undermine without injury to the fabric: and cast iron beams were employed of as large scantling as they were now using".[31]

Cast iron had "arrived" as a building material: it had been accepted and used in architecture as wood and stone and brick had been used, and was brought into the framework of contemporary architectural design. Only in large scale engineering had its use stimulated a new technique of design.

On this and the opposite page, cast iron verandahs and balconies of the late eighteenth and early nineteenth centuries show how the classical tradition was maintained and comparative simplicity of form was consistently favoured.

FIG. 122 (*Above*). Cast iron balcony at Grand Parade, Brighton.

FIG. 123 (*Right*). Nos. 17, 18 and 19 Grand Parade, Brighton.

Both these illustrations are reproduced by courtesy of the National Buildings Record.

Fig. 124 (*Above*). Montpelier Road, Brighton.
Reproduced by courtesy of the National Buildings Record.

Fig. 125 (*Left*). House in Edwardes Square, Kensington, London.
Photograph: E. R. Jarrett.

Fig. 126 (*Below*). Litfield Place, Clifton, Bristol.
Photograph: E. R. Jarrett.

Fig. 127. Balconies in Charlotte Street, Bristol. Design here is growing ornate; loss of elegance is already foreshadowed by these balcony railings. Compare these designs with the work that was being produced half a century later, and which is illustrated in Figs. 341, 342 and 343 on pages 280, 281 and 282.

Photograph: *A. F. Kersting.*

Fig. 128. Savile Place, Clifton, Bristol. The balcony railings here resemble the character of those illustrated in Fig. 121, on page 121. In this period of the nineteenth century the most appropriate use was made by designers of this most obliging material. Only as the century progressed was it required to perform complicated and almost acrobatic feats.

Photograph: E. R. Jarrett.

Examples of early nineteenth century cast iron balcony railings in Bloomsbury, London, are shown on this and the opposite page.

Fig. 129 (*Above*). Has a faintly Gothic flavour. Some lingering traces of the taste that emanated from Horace Walpole's antiquarian exercises, which were given a new lease of life by the eccentric but talented William Beckford at Fonthill Abbey. This interlaced, pointed arch motif occurred in the glazing bars of windows and bookcases, and here it is translated in cast iron.

Fig. 130 (*Right*). Here is a forerunner of a much later form of decorative cast iron work; in the early years of the nineteenth century it still retained classical affinities. Later it was to lose its lightness and, in common with most Victorian forms, to become gross.

Fig. 131 (*Above left*). The shell motif below the top rail has become rather bloated: the ornamental treatment at the base of the railings is beginning to thicken up, to lose the characteristic grace of Georgian design.

Fig. 132 (*Above right*). Here, the influence of the Greek Revival is apparent, and a geometric severity characterises both this design and that shown below in Fig. 133, below.

Fig. 133 is reproduced by courtesy of F. R. Yerbury.

Fig. 133.

Fig. 134. This early nineteenth century design again rings the changes on geometric motifs, and is a less rigid variation of the patterns shown in Figs. 132 and 133 on the previous page.

Reproduced by courtesy of F. R. Yerbury.

Fig. 135 (*Right*). Early nineteenth century cast iron balcony railings at Downshire Hill, Hampstead, London.

Photograph: E. R. Jarrett.

Fig. 136 (*Above*). A simple geometric pattern, although broken to secure privacy for individual balconies, gives horizontal continuity to this terrace of houses in Kentish Town Road, London.

Fig. 137 (*Below*). A close geometric pattern with emphasis on the vertical elements, is used for these balcony railings at Earls Terrace, Kensington Road, London.

Photographs: *E. R. Jarrett*.

Fig. 138. Balcony railings in cast iron which suggest the influence of Robert Adam, although they lack the rigid certitude associated with the work of that architect. There are more concessions here to the nature of the material than Robert Adam generally allowed in his designs for cast iron. This is No. 43 Claremont Square, Pentonville Road, London. The ubiquity of the particular design with the double anthemion motif was astonishing: it is found in houses throughout England, and was one of the favourite Cottingham patterns. See Fig. 284, bottom left, on page 227.

Photograph: E. R. Jarrett.

Fig. 139 (*Above*). Cast iron balcony railings in Victoria Terrace, Weymouth.

Fig. 140 (*Left*). Cast iron balcony railings in Claremont Square, Pentonville Road, London. The designer has gained an easy mastery of the material, and has produced an intricate pattern which creates the impression of elegant simplicity.

Photographs: E. R. Jarrett.

Fig. 141. Cast iron balconies in Regent Terrace, Edinburgh. A straightforward geometric pattern that provides horizontal continuity and becomes a semi-transparent architectural feature echoing the lines of the cornices that terminate the upper storeys.

Photograph: E. R. Jarrett.

FIG. 142 (*Right*). Balcony railings at Brunswick Place, Southampton.
Photograph: E. R. Jarrett.

FIG. 143 (*Left*). Cast iron balcony railings at Clarence Terrace, Regents Park, London, designed by Decimus Burton, under the direction of John Nash, *circa* 1822. The design on the single balcony is shown in Cottingham's *Director*. See Fig. 285, top right, on page 228.

Reproduced by courtesy of the National Buildings Record.

Fig. 144 (*Above*). Balcony railings at Chester Terrace, Regents Park, London, designed by Decimus Burton and John Nash, 1825.
Photograph: *A. F. Kersting*.

Fig. 145 (*Below*). Railings outside No. 3 Chester Terrace, Regents Park, London. This is a simple treatment in which a rounded arch motif is repeated.
Photograph: *E. R. Jarrett*.

Fig. 146 (*Above*). Railings at York Terrace, Regents Park, London, designed by John Nash, 1822. The lamp post in the foreground is a variation of a design in Cottingham's *Director*. See Fig. 281, right, on page 224.

Fig. 147 (*Left*). Another view of the cast iron railings at York Terrace, Regents Park, London. The posts are similar to those illustrated in Cottingham's *Director*. See Fig. 281, centre, on page 224.

Photographs: *E. R. Jarrett*.

Fig. 148. Balconies and railings at Dorset Square, London. Many of the early nineteenth century balconies represented associations of wrought and cast iron, but as the century progressed the tendency to use cast iron patterns exclusively increased, particularly when cheapness and ease of repetition became important in large scale building schemes.
Photograph: E. R. Jarrett.

FIG. 149 (*Right*). Balcony railing of house in the Adelphi, London, designed by Robert Adam, *circa* 1770. The principal motif is based on the Greek anthemion ornament, adroitly adapted by a master designer for casting in metal: a far more inventive treatment of cast iron by Adam, than his normal habit of using decorative forms that were common alike to wood, plaster, stone or iron, and were interchangeable, deriving neither originality of form nor distinctive character from the material employed. Here, Adam is obviously conscious of the properties and possibilities of cast iron as a new material.

Reproduced by courtesy of F. R. Yerbury.

FIG. 150 (*Left*). Railings and lamp standards at Chandos House, Chandos Street, Cavendish Square, London, designed by the brothers Adam, 1771. Again the anthemion motif appears in the lamp standards, though subordinated in the general pattern of the design, not dominating it, as it does in the balcony railing shown above in Fig. 149.

Photograph: Quentin Lloyd.

Fig. 151 (*Left*). Railings outside a house in Portman Square, London.
Reproduced by courtesy of F. R. Yerbury.

Fig. 152 (*Below*). Railings at No. 5 Upper Harley Street, London. The cast iron lamp post in the foreground bears the monogram of George IV.
Photograph: E. R. Jarrett.

FIG. 153. Railings and cast iron balcony at No. 4 Cleveland Square, London.
Reproduced by courtesy of J. D. M. Harvey.

Fig. 154. Detail of cast iron balcony at No. 5 Columbia Place, Winchcombe Street, Cheltenham, west side, which is shown opposite. This is a free treatment of Greek ornamental forms, making a striking use of the properties of cast iron. Compare this with the more rigid handling of a similar subject, in the balcony designed by Robert Adam, shown in Fig. 149, on page 137.

Reproduced by courtesy of the National Buildings Record.

Fig. 155. Cast iron balcony at No. 5 Columbia Place, Winchcombe Street, Cheltenham, west side. See Fig. 154 on page 140 for enlarged detail.

Reproduced by courtesy of the National Buildings Record.

Fig. 156 (*Above*). Details of cast iron railings and urn finial, at Bryanston Square, Marylebone, London.

Fig. 157 (*Below*). Another variation of the urn device used as a finial, in railings at Manchester Square, Marylebone, London.

These illustrations are reproduced by courtesy of the National Buildings Record.

Fig. 158. Early nineteenth century gate at Campden Hill Square, London. There is a family likeness between all these early nineteenth century railings: compare the example above with the detailed photographs on page 142 (Figs. 156 and 157), and the railings shown on the two pages that follow. All these variations, based on designs that were suggested by the details of classic architecture, were ultimately recorded in L. N. Cottingham's *Smith and Founders Director* (see page 116, and Figs. 276 to 292 inclusive, on pages 221 to 235).

Fig. 159 (*Above*). Detail of railings designed by John Nash at Park Crescent, Regents Park, London. The posts and railings are illustrated in Cottingham's *Director*. See Fig. 281, page 224.

Photograph: E. R. Jarrett.

Fig. 160 (*Above right*). Detail of railings at Norfolk Square, Paddington, London. The influence of the Greek Revival is apparent here, as in the examples given in Figs. 154, 155 and 159.

Fig. 161 (*Below right*). Detail of cast iron railings at Gloucester Square, Paddington, London.

These illustrations are reproduced by courtesy of the National Buildings Record.

Fig. 162. Early nineteenth century cast iron gates in Portman Square, London. Compare the treatment of the gate posts with those at Surrey Street, Norwich, shown in Fig. 163, which represented a slightly simpler treatment. Again, comparison should be made with the elaborations set forth in Cottingham's designs for piers and lamps, reproduced in Fig. 279, on page 223.

Reproduced by courtesy of the National Buildings Record.

Fig. 163. Cast iron railings and gate posts in Surrey Street, Norwich.

Reproduced by courtesy of the National Buildings Record.

Fig. 164 (*Above*). Cast iron railings to Marlborough House, Brighton.

Reproduced by courtesy of the National Buildings Record.

Fig. 165 (*Left*). Cast iron nameplate to Hall's warehouse, Church Street, Worcester, *circa* 1820. The lettering used still retains the letter forms based on the incised lettering of the Trajan column. Compare this lettering with that on the steam cylinder shown in Fig. 52, on page 67, and with the blunt, clumsy lettering on the arch ring of the bridge in Fig. 102, on page 101, which was cast over thirty years later. The lettering on the bridge shown in Figs. 92 and 93 had not become debased and was as clear and well proportioned, with perfectly formed serifs, as the plates on this warehouse door. (See page 95.)

Reproduced by courtesy of the National Buildings Record.

Fig. 166. Early nineteenth century cast iron bollards, probably made from old patterns for cannon. Whether they were thus adapted, or were designed specifically as bollards, the simplicity and effectiveness of the design are obvious.

Reproduced by courtesy of the Architectural Press.

Fig. 167 (*Above*). Cast iron balusters used in the hall gallery of Bedford Hotel, Brighton.

Reproduced by courtesy of the National Buildings Record.

Fig. 168 (*Right*). Cast iron balusters to the gallery in the Octagon, Victoria Rooms, Clifton, Bristol. Designed by Charles Dyer, 1794-1848.

Photograph: E. R. Jarrett.

FIG. 169. Cast iron stair balusters at the De Grey Rooms, York. Here a naturalistic motif is allowed considerable freedom, but without loss of control. John Summerson has described this particular design as "One of the most exquisite examples I know of cast iron used in such a way as to elicit its capacity for small-scale elegance . . ." ("Records of an Iron Age", by John Summerson, F.S.A., A.R.I.B.A. *Official Architect*, Vol. VIII, No. 5, page 235.)

Reproduced by courtesy of the National Buildings Record.

FIG. 170 (*Above*). Ornamental fluted cast iron columns designed by John Nash to support the ceiling of one of the reception rooms at the Royal Pavilion, Brighton.

FIG. 171 (*Below*). The cast iron columns supporting the ceiling of the kitchen in the Royal Pavilion, Brighton. The leaves forming the rather extravagant capitals are made of copper. Designed by John Nash.

These illustrations are reproduced by courtesy of the Brighton Art Gallery.

FIG. 172. The cast iron staircase at the Royal Pavilion, Brighton. The balustrade, which is only partly visible in this view, was made to imitate bamboo. Designed by John Nash.
Reproduced by courtesy of the Brighton Art Gallery.

Fig. 173. The head office of the Essex and Suffolk Equitable Insurance Society Limited, at Colchester, Essex. This is described in *Picturesque Beauties of Great Britain*, by Vertue, published in 1831, as follows: "On the ground floor of this elegant building there was formerly a public edifice called Red Row, and afterwards named Exchange which whilst the bay trade flourished here, was daily frequented by great numbers of substantial merchants, and over it was Dutch Bay Hall. The new building was erected about the year 1820 by subscription for the use of the corn merchants and farmers. The builder was Mr. Hayward, of Colchester, and the Architect, Mr. David Laing. The Corn Market occupies the basement ground storey which is an open colonnade of well formed fluted cast iron pillars. The upper rooms are occupied as the Essex Fire and Life Insurance Office".

The second floor was added in 1914. Enlarged views of the cast iron Doric columns are shown in Fig. 174 on page 153 and in Fig. 175 on the page that follows.

Reproduced by courtesy of the Essex and Suffolk Equitable Insurance Society, Limited.

Fig. 174. This shows the delicate moulding of the Greek Doric detail and the cast iron base of the columns. See Fig. 173 on previous page and Fig. 175 on the following page.
Reproduced by courtesy of the Essex and Suffolk Equitable Insurance Society Limited.

Fig. 175 (*Above*). Cast iron Doric columns in the head office of the Essex and Suffolk Equitable Insurance Society Limited, at Colchester. See Figs. 173 and 174 on the previous pages. Compare these with Nash's designs shown opposite.
Reproduced by courtesy of the Essex and Suffolk Equitable Insurance Society Limited.

Fig. 176 (*Above next page*). Cast iron Doric columns at Carlton House Terrace, London, designed by John Nash, 1827. These were damaged in an air raid during the second world war, 1939 · 1945.

Fig. 177 (*Below next page*). Cast iron Greek Doric columns at the North Lodge, Buckingham Palace, London. Designed by John Nash, 1825. Damaged in an air raid during the second world war.
Photographs: Central Press.

FIG. 176 (*Right*).

FIG. 177.

155

SOURCES OF REFERENCES IN SECTION TWO

[1] *The Industrial Development of Birmingham and the Black Country*, by G. C. Allen, M. Com., Ph.D. George Allen & Unwin Ltd., 1929. p. 31.

[2] *Papers on Iron and Steel*, by David Mushet. John Weale, London, 1840. pp. 42, 43, 44.

[3] *Iron and Steel in the Industrial Revolution*, by T. S. Ashton, M.A., University Press, Manchester and Longmans, Green & Co. 1924. Appendix B, p. 235.

[4] "A sketch of the Industrial History of the Coalbrookdale District", by Rhys Jenkins, from the *Transactions of the Newcomen Society*, Vol. IV, 1923-34. Printed for the Society by the Courier Press, Leamington Spa, 1925. pp. 103, 104.

[5] *Ibid*, pp. 103, 104.

[6] *The Torrington Diaries*. Eyre and Spottiswoode, Ltd. 1934. Vol. I, p. 185.

[7] *Die Geschichte des Eisens*, by Dr. Ludwig Beck. Vol. III. Braunschweig; Druck und Verlag von Friedrich Vieweg und Sohn. 1897. p. 1,078.

[8] *Iron and Steel in the Industrial Revolution*, by T. S. Ashton, M.A., University Press, Manchester and Longmans, Green & Co. 1924. p. 86.

[9] *Lives of the Engineers*, by Samuel Smiles. Vol. II. John Murray, London. 1863. p. 356.

[10] *Recreations in Agriculture, Natural-history, Arts*. Vol. VI, p. 106, by James Anderson, LL.D. Printed by T. Bensley, Fleet Street, London, 1802, from chapter on "Travels in Scotland".

[11] *Annals of Commerce*, by D. Macpherson. Vol. III, 1805. p. 609.

[12] "Yesterday and Today—Designing for Cast Iron", by Grey Wornum, F.R.I.B.A. *The Official Architect*, Vol. VIII, No. 5. Special Cast Iron Number. May, 1945. p. 243.

[13] *The Story of the Forth*, by H. M. Cadell, published by James Maclehose & Sons, 1913.

[14] *Historical development of enamelling technique*, by R. Aldinger. Glashutte 68. 1939.

[15] "Research and the Vitreous enamelling industry", by G. H. Abbott, B.Sc., paper given to the Institute of Vitreous Enamellers, November 24th and 25th, 1944.

[16] *Lives of the Engineers*, by Samuel Smiles. Vol. II. John Murray, London, 1863. p. 355.

[17] *The Torrington Diaries*. Eyre & Spottiswoode Ltd., 1934. Vol. 1, p. 184.

[18] *Diary of a Tour through the Midlands in 1821*, by Joshua Field. Original in Science Museum, South Kensington. Extracts in Newcomen Society Transactions, 1925-26. Printed by The Courier Press, Leamington Spa, 1927. Vol. VI, pp. 20, 21.

[19] *John Nash*, by John Summerson. George Allen & Unwin Limited, 1935. Chap. II, p. 44.

[20] *Ibid*, p. 46.

[21] *Ibid*, p. 54.

[22] *A General History of Inland Navigation, Foreign and Domestic*, by J. Phillips, 4th Edition, London, 1803.

[23] *Lives of the Engineers*, by Samuel Smiles. Vol. II. John Murray, London, 1863. p. 361.

[24] *On the Application of Cast and Wrought Iron to Building Purposes*, by William Fairbairn, C.E., F.R.S., F.G.S., 3rd edition. Longman, Green, Longman, Roberts & Green, 1864. p. 2.

[25] "English Cast Iron, II", by Professor W. R. Lethaby. *The Builder*, Nov. 5th, 1926. p. 742.

[26] *Ibid*, p. 741.

[27] *A Complete Body of Architecture*, by Isaac Ware of His Majesty's Board of Works. Printed by T. Osborne and J. Shipton in Gray's Inn, J. Hodges near London Bridge, L. Davis in Fleet Street, J. Ward in Cornhill, and R. Baldwin in Pater-Noster-Row, 1756. p. 89.

[28] "Records of an Iron Age", by John Summerson, A.R.I.B.A. *The Official Architect*, Vol. VIII, No. 5, May 1945, p. 235.

[29] *Ibid*.

[30] *Ibid*.

[31] "A memoir of the late Sir Robert Smirke", by Edward Smirke. Read at the ordinary general meeting of the Royal Institute of British Architects, June 17th, 1867. London: Published at the rooms of the Institute, 9 Conduit Street, Hanover Square, W., 1867.

SECTION THREE

SECTION THREE

THE INDUSTRIAL EXPANSION OF THE CAST IRON INDUSTRY: 1820-1860

IN the opening sentences of the introduction, it was pointed out that iron was the basic material for industrial development; and the first industrial revolution was largely the result of improved methods of using iron and coal, while steam power gave it momentum. The period of forty years between 1820 and 1860 may be regarded as the "heroic age" of British engineering; an adventurous spirit stimulated enterprise, not unlike the spirit that inspired the Elizabethan adventurers and merchants, who risked their lives and liberty and capital to open up markets in the new world, and incidentally to found the United States of America. Telford and Rennie had in the previous thirty years improved the road system and helped to create the network of canals; but the following period of industrial and mechanical progress made a new pace for the affairs of men, and saw the beginning of those future developments of transport and communication which, in the words of Professor J. B. S. Haldane "are limited only by the velocity of light." We had begun our approach to the condition "when any two persons on earth will be able to be completely present to one another in not more than one twenty-fourth of a second." Haldane says: "We shall never reach it, but that is the limit which we shall approach indefinitely".[1]

To the men and women of the eighteen twenties and thirties, the new methods of locomotion achieved by the great engineers seemed both astonishing and alarming. The first public railway was opened in 1803, and ran from Croydon to Wandsworth. It was known as the Surrey Iron Railway; but it was not until 1830 that mechanised railway traffic was established, when the Liverpool and Manchester line was opened. The railway solved the problem of steam traction: iron rails allowed hundreds of tons of traffic to travel at high speed.

"Steam coaches for road traffic had been tried out on paper, and some experimental models had been made during the last two decades of the eighteenth century. William Murdock, who was employed by Boulton & Watt at Birmingham, had built experimental models, and did much work on a steam carriage. William Symington produced a working model of a steam carriage in 1786. The machine ran on four wheels and consisted of a carriage with a locomotive at the rear end. But Symington's interests were diverted from road transport to marine engineering. It was not until the eighteen-twenties that steam carriages appeared on the roads. They were then regarded with the greatest alarm and hostility by many people, their uncouth

Fig. 178. George Stephenson, 1781 - 1848, from an engraving by W. Holl, after the portrait by John Lucas. From Samuel Smiles' *Lives of the Engineers*.

appearance repelling even the most progressive and adventurous souls. The road improvements made in Britain between 1750 and 1830, and the increased speed of stage coaches and private vehicles, had aroused the greatest popular interest in traffic. The stage coach became a national institution; it held the affection and admiration of townsmen and countrymen alike, of passengers and people who never travelled. As the railway age began, so did the age of steam carriages. Some of them operated successfully for a short time, but they never captured public imagination and their designers seemed to be incapable of discarding the forms of horse-drawn vehicles. The ghost of the horse galloped before them, as eighty years later it was to gallop before the motor car, and only towards the middle of the present century has that ghost been effectually laid.

"Although steam coaches and railway locomotives possessed mechanical originality, they began with many handicaps. The form of the early locomotive still bore a strong family likeness to its parent, the steam pumping engine; it seemed to be constructed for vertical rather than horizontal movement. Only when The Rocket was designed by George and Robert Stephenson in 1829, were later developments foreshadowed, for in this engine the up-and-down motion which had hitherto communicated the thrust of the piston rods to the wheels was abandoned. The Rocket is the great ancestor of the modern railway engine: earlier models, such as the Killingworth locomotive which

PLATE II. Exterior view of the Crystal Palace, designed by Sir Joseph Paxton. From *Dickinsons Comprehensive Pictures of the Great Exhibition of 1851*, painted by Messrs. Nash, Haghe and Roberts. Published by Dickinson Bros., 114 New Bond Street, 1854.

George Stephenson designed in 1816, represented a *cul de sac* in development, as Neanderthal man represented a dead end in human evolution. But there is as much difference between The Rocket and one of the latest London, Midland and Scottish or Great Western locomotives, or the Pennsylvania Railroad type, in whose design Raymond Loewy has collaborated, as there is between the Java ape man and modern man. Unfortunately the scale of railway development was determined before steam locomotives were used. The gauge of the track was that of the horse-drawn cart. Certainly the first industrial revolution was hampered by many quite needless limitations. The men who saw so clearly what applied science could do for civilisation, even men with original creative minds like George Stephenson, could not make a clean break with precedents".[2]

The railways, or the "iron roads" as they were called, evolved from the tram roads, which had been in use for many generations in the collieries. As early as 1630 wooden planks were embedded in the track on which horse-drawn coal waggons ran. In 1791 Faujas de Saint-Fond, the French geologist and traveller, described the wooden rails at a Newcastle mine. They were formed with a rounded upper surface like a projecting moulding, and, so that they might fit the rounded surface, the wheels were "made of cast iron and hollowed in the manner of a metal pulley". This custom spread throughout Britain and improvements were gradually made. Metal strips were nailed to the upper surface of the wooden rails, and cast iron rails were tried, and

Fig. 179. Robert Stephenson, 1803-1859, from an engraving by W. Holl, after a photograph by Claudet.

From Samuel Smiles' *Lives of the Engineers*.

it is probable that the first to be used were those at Whitehaven in 1738. It has been mentioned earlier that cast iron rails were laid at Coalbrookdale by the Darbys in 1767; and in 1776, at a Sheffield colliery, a cast iron tramway was nailed to wooden sleepers and was cast as an L section to afford a guide for wheels. In 1789 a line at Loughborough, Leicestershire, had a cast iron edge rail with flanges upon the cast iron tyre of the wheels to keep them on the track, and these wheels may well be the forerunners of present-day railway wheels.

By 1820 it was common practice to move waggons with heavy loads on rail tracks. Although the use of steam power had enormously increased the manufacture of commodities, without the introduction of steam traction, the easy and rapid transit of raw materials and economic distribution of goods would have been impossible. The population of Britain was doubled between 1801 and 1851; and this increase was accompanied by a growing output of domestic iron products, and the erection of new factories, warehouses, offices and churches, in which iron was used structurally. This expansion of consumer demand and industrial productivity accounted for a great increase in the number of blast furnaces, forges, mills and foundries.

Iron production and foundry practice had been profoundly affected in the early eighteenth century by the invention of coke smelting by Abraham Darby I; and in the early nineteenth century an invention of comparable significance increased the production of iron. Before 1826 it was generally accepted that the colder the air current injected into the blast furnace, the more efficient would that furnace be; a belief so strongly implanted, that ironmasters even incurred the expense of passing the air over cold water or through pipes packed with ice. It seemed a revolutionary, and even a ludicrous notion, that a bigger production of iron could be obtained by a hot blast; but this method was patented by J. B. Nielsen; in 1828 his patent was tried at the Clyde Ironworks; and to the surprise of everybody, he was able to reduce fuel consumption from 8 tons to 5 tons per ton of iron. An even greater economy of fuel was later secured by using the waste gases from the furnace to heat the blast current. Nielsen's first experiment was with a blast temperature of 150°C. To-day temperatures of 600°C. to 850°C. are common, and with suitable ore a ton of iron can be smelted with a ton of coke or less.

The most striking results of the use of the hot blast were obtained in Scotland where there were large deposits of iron ore known as black band ironstone. This iron ore, with cold blast, had been almost unworkable; but David Mushet reduced it with the hot blast and thus saved the considerable imports of iron ore from England that had hitherto been essential. In 1830 the production of pig iron in the Scottish industry was 37,500 tons; but by 1840, as a result of extensive use of the hot blast, the figure had risen to 200,000 tons. Similarly, in South Wales it was found that anthracite coal previously unusable, could be successfully employed with the hot blast. In America, where anthracite is abundant in certain districts, the same results were achieved, leading ultimately to the development and expansion of the world's great iron-producing centre of Pennsylvania.

CAST IRON IN RAILWAY ARCHITECTURE.

Apart from the extensive use of cast iron for the early permanent ways and bridges that carried the new rail traffic, the material was used extensively for a variety of new architectural requirements: stations, protective railings, signal cabins, foot-bridges, and such things as signs, lamp-posts and drinking fountains. George Stephenson had devised a new form of cast iron rail for permanent ways. These rails had overlapping joints and improved cast iron chairs to carry the junctions. During the period of early railway development, George Stephenson and his son Robert planned and carried through an immense programme of work. The Liverpool to Manchester railway was an amazing feat of engineering, involving a tunnel 2,200 yards long, through rock at Liverpool, a cutting about two miles in length and in places 100 feet deep, driven through solid sandstone, the crossing of Chat Moss, a huge peat bog, the construction of sixty-three bridges under or over the railway, and the building of one viaduct. Stephenson found it impossible to use the normal arch construction for many of the bridges, because the railway had to be kept level. "In such cases he employed simple cast iron beams," writes Samuel Smiles, "by which he safely bridged gaps of moderate width, economising headway, and introducing the use of a new material of the greatest possible value to the railway engineer".[3]

FIG. 180. An example of cast iron railway architecture, where the material has been used to form a completely independent structure. Full use has been made of its properties, and the vertical columns and the treads and risers of the stairs are of cast iron.

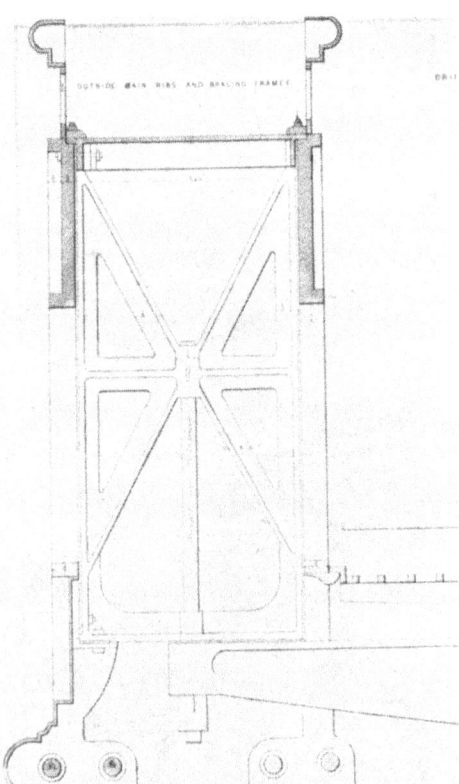

FIG. 181 (*Above*). Cast iron bridge designed by Robert Stephenson, which crosses the Regent's Canal near Chalk Farm, carrying the old London and Birmingham Railway, which was later to become the London and North Western, and is now the London, Midland and Scottish. From a lithograph by John C. Bourne, dated 1838.

FIG. 182 (*Below*) and FIG. 183 (*Left*). Detailed drawings of the bridge shown above. From *The First Series of Railway Practice*, by S. C. Brees, C.E., published by John Williams & Co., 1847.

Reproduced by courtesy of the London, Midland & Scottish Railway.

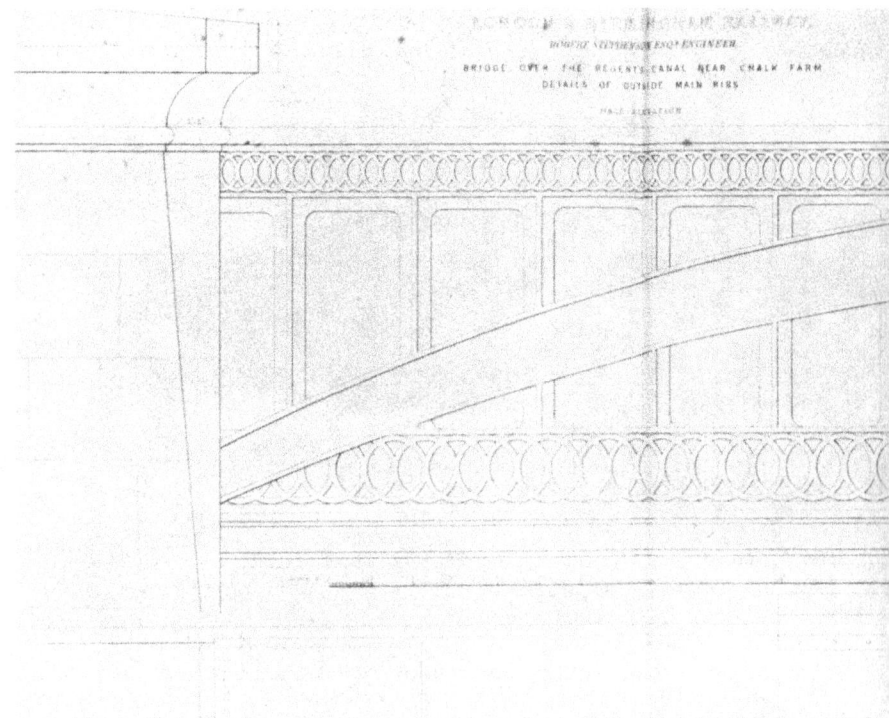

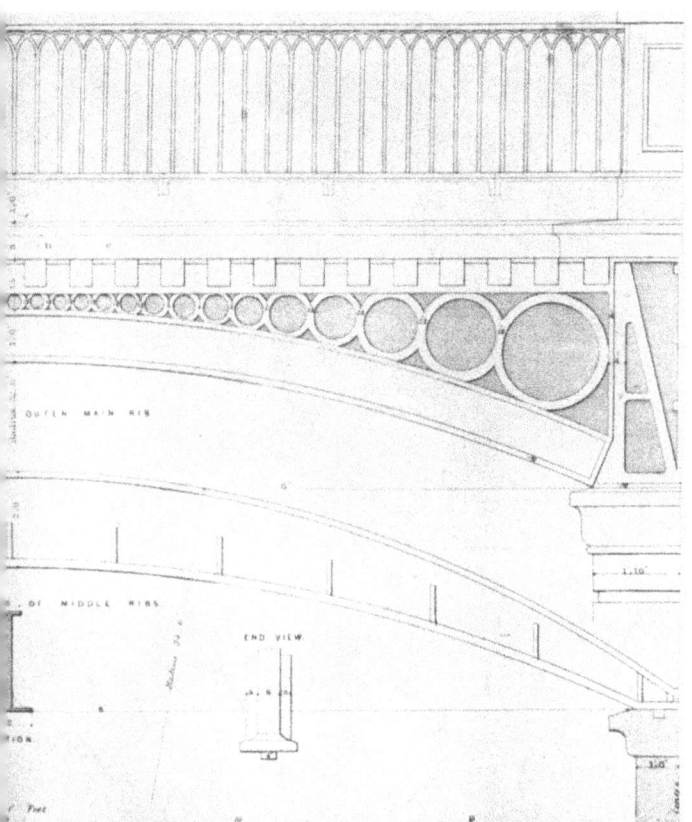

Fig. 184. Cast iron arched girder bridge designed by Robert Stephenson. Known as the Hampstead Bridge, and carrying the London and Birmingham Railway. From a lithograph by John C. Bourne, dated 1837.

Fig. 185 (*Left*). Detail showing part of the elevation of the outer main rib of the Hampstead Bridge, shown above. The cast iron parapet railings are similar to those which are still in use on the London, Midland & Scottish line immediately outside Euston Station. From *The First Series of Railway Practice*, by S. C. Brees, C.E., published by John Williams & Co., 1847.

Reproduced by courtesy of the London, Midland & Scottish Railway.

165

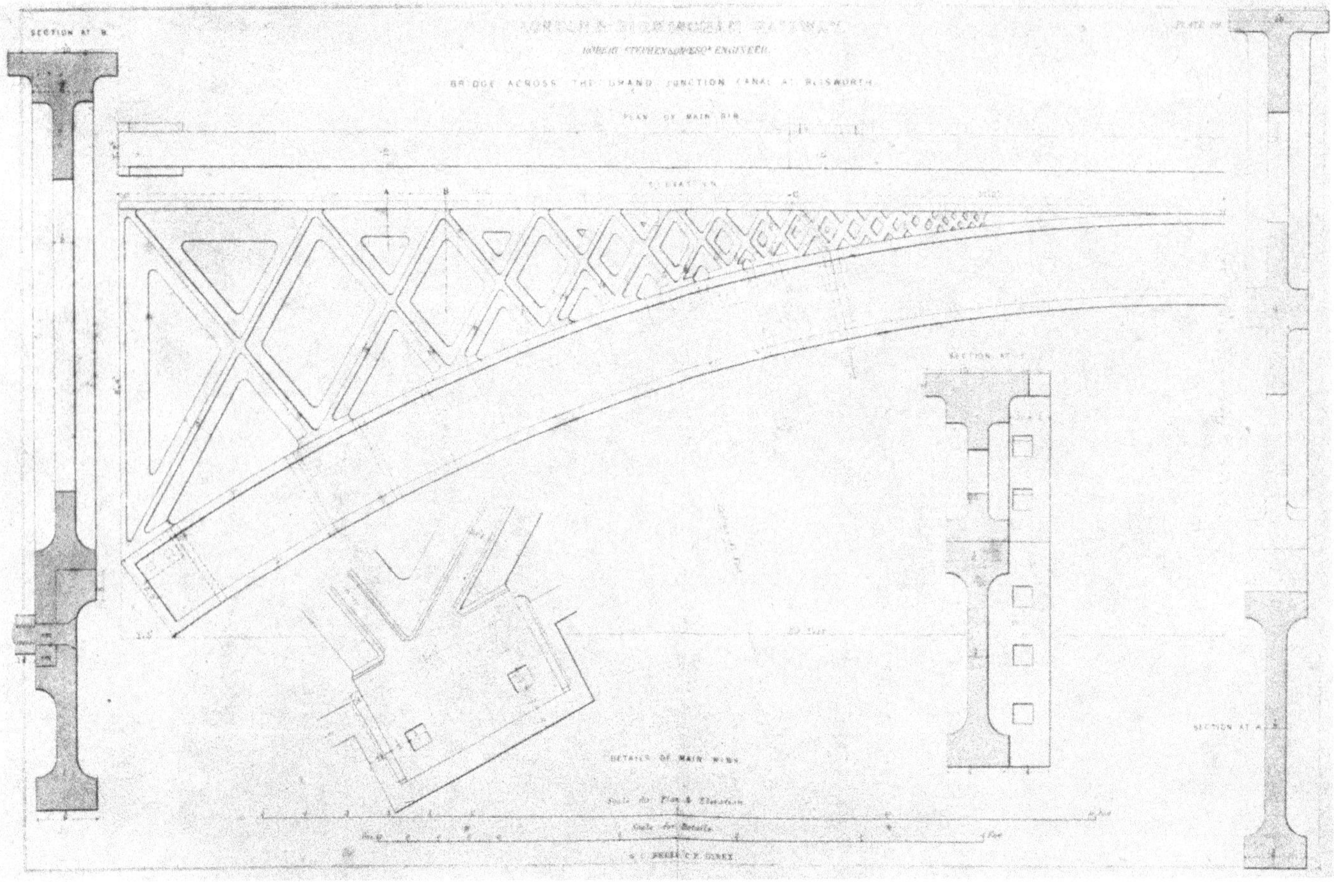

Fig. 186 (*Above*). Detail of the main outer rib of the cast iron bridge over the Grand Junction Canal at Blisworth, shown in Fig. 188 on the opposite page. From *The First Series of Railway Practice*, by S. C. Brees, C.E. published by John Williams & Co., 1847.

Fig. 187 (*Right*). Part of the cross section, and detail of the cast iron railings of the bridge over the Grand Junction Canal at Blisworth. See above and opposite From *The First Series of Railway Practice*.

Reproduced by courtesy of the London, Midland & Scottish Railway.

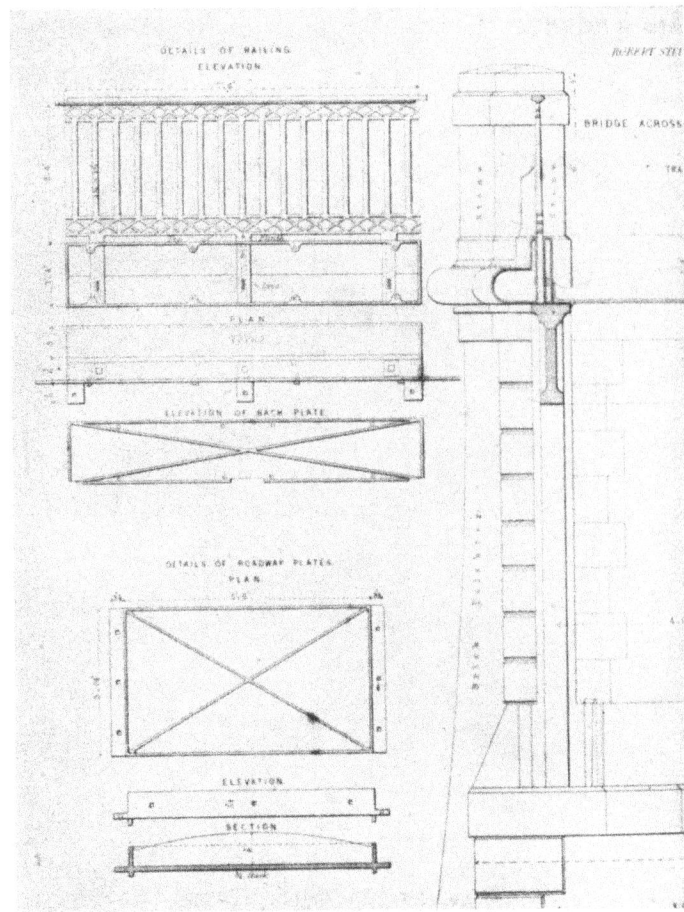

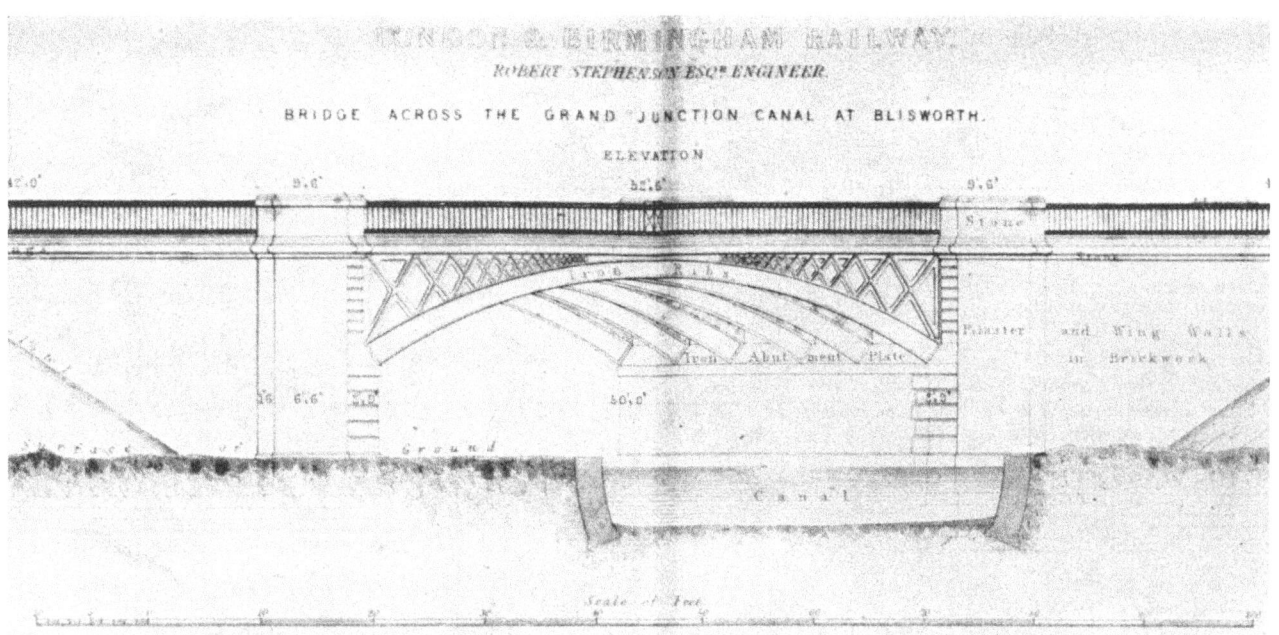

Fig. 188. Elevation of the cast iron bridge designed by Robert Stephenson, which carried the London & Birmingham Railway over the Grand Junction Canal at Blisworth. 1833 - 1838. See details on the opposite page. From *The First Series of Railway Practice.*
Reproduced by courtesy of the London, Midland & Scottish Railway.

In 1830 a railway was projected between London and Birmingham, and George Stephenson and his son were appointed as engineers. The line, on which work began in 1833, was just over 112 miles in length, with great cuttings at Tring and Blisworth, and eight tunnels, the greatest being at Primrose Hill, Watford and Kilsby. There were numerous bridges, and again the Stephensons made use of the cast iron girder type which had been successfully employed on the Manchester-Liverpool line. The line was opened in 1838 and was immediately followed by the opening of many more by the same engineers: the Birmingham and Derby line in 1839; the Sheffield and Rotherham in 1839; the Midland; the York and North Midland; the Chester and Crewe; the Chester and Birkenhead; the Manchester and Birmingham; and the Manchester and Leeds in 1840. The Stephensons, in carrying out this enormous programme relied on a highly competent staff of engineers, many of them trained originally as students under George Stephenson—men like Vignolles, Locke, Dixon, Gooch, Swanwick, Birkenshaw and Cabrey. Meanwhile, in 1833, another great engineer, I. K. Brunel, had been appointed at the age of twenty-seven to construct the Great Western Railway. He carried out this work in the grand manner: its scale was magnificent and impressive, and he introduced the famous seven-foot "broad gauge". Many fine masonry bridges bore aloft the permanent way of the Great Western, and at Chepstow and Saltash, Brunel designed iron bridges.

"The early architecture of the railways, like that of the early industrial buildings, preserved its connection with tradition. The first railway stations were designed unpretentiously; they were perfectly fitted for the function they had to perform; but soon they became influenced by the 'battle of the styles', which disturbed their functional character and tricked them out in the most unsuitable costume. This aspect of industrial architecture did not escape criticism; but the critics generally wrote or spoke from the point of

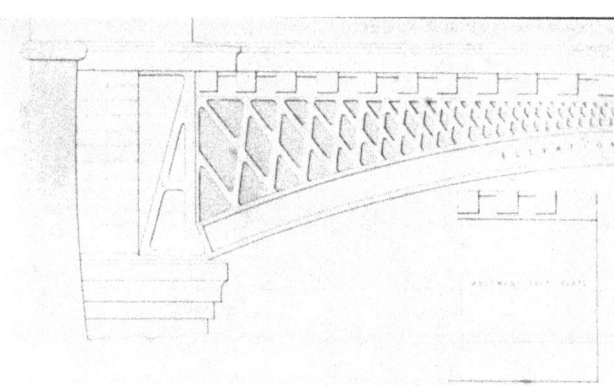

Fig. 189 (*Above*) and Fig. 190 (*Below*). Details of Park Street Bridge, London, designed by Robert Stephenson for the London & Birmingham Railway. From *The First Series of Railway Practice*, by S. C. Brees, C.E., published by John Williams & Co., 1847.

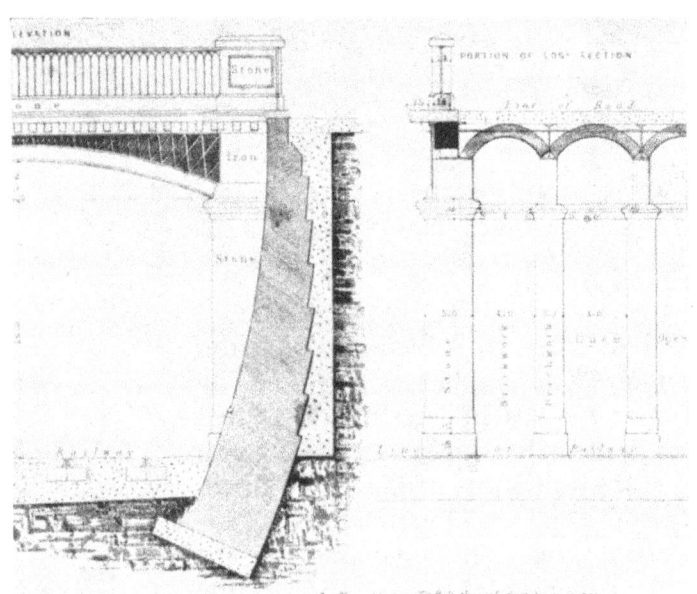

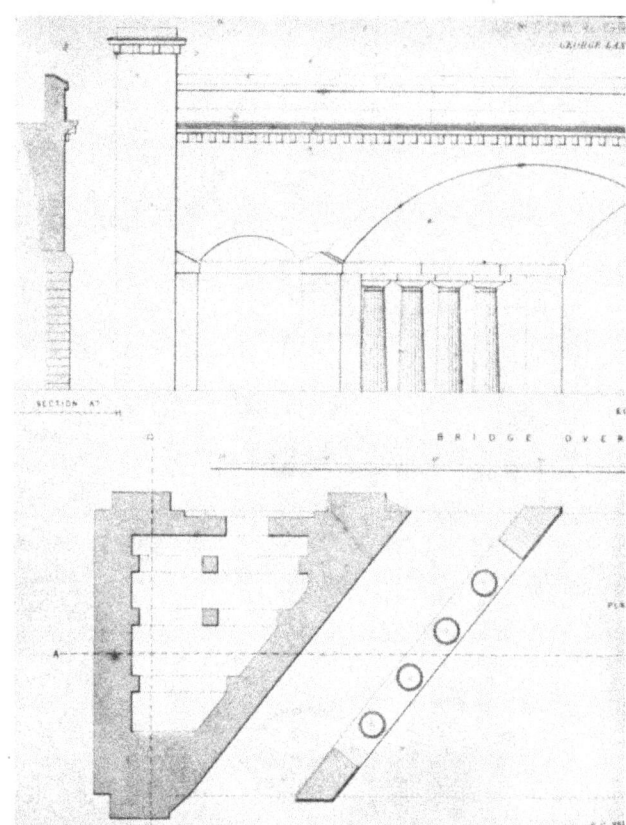

Fig. 191 (*Above*). Bridge over Spa Road, London & Greenwich Railway, designed by George Landmann, Engineer. This shows part of the elevation and plan of the cast iron Greek Doric columns supporting the masonry arches over. From *The First Series of Railway Practice*, by S. C. Brees, C.E., published by John Williams & Co., 1847.

Fig. 192 (*Below*). Cast iron girder bridge designed by Robert Stephenson, intended to cross the River Nene, near Wisbech, circa 1848.

Reproduced by courtesy of the London, Midland & Scottish Railway.

Fig. 193. Nash Mill cast iron bridge, designed by Robert Stephenson. This carried the London & Birmingham Railway, near Kings Langley, Herts. From a lithograph by John C. Bourne, 1838. *Reproduced by courtesy of the London, Midland & Scottish Railway.*

view of one or other of the protagonists in the battle of the styles. A railway station, for example, that retained a classical elegance of design, would be ruthlessly condemned, almost branded as immoral, by a Gothic Revivalist, who had read his Ruskin and had taken to heart the winged words which dismissed the classic orders and all the architectural works of Greek and Roman antiquity as visible evidence of an improper and subversive paganism".[4]

In most of the stations, cast iron was used extensively in the structure; but no new forms were evolved for structural members: piers were merely elongated Doric or Tuscan or Corinthian columns of cast iron, carrying entablatures of the same material which in turn carried cast iron trusses, brackets or cantilevers. Obscured by layers of brown paint, which has blinded their original mouldings, they still stand to-day, evidence of the normal inability of designers to discard prototypes. When the great engineers used a material such as cast iron, they were uninfluenced by past styles and mannerisms: they were the real, though largely unrecognised, architects and industrial designers of the nineteenth century, and they produced appropriate and fine solutions to the problems of railway architecture. Much of Robert Stephenson's

SEC. AT A
(half real size)
SEC. AT C
SEC. AT B. B
DETAILS OF SKYLIGHT
Plans of Bolts
half full size
DETAILS OF ROOF
Scale for Plan
Feet

PLAN OF COLUMN

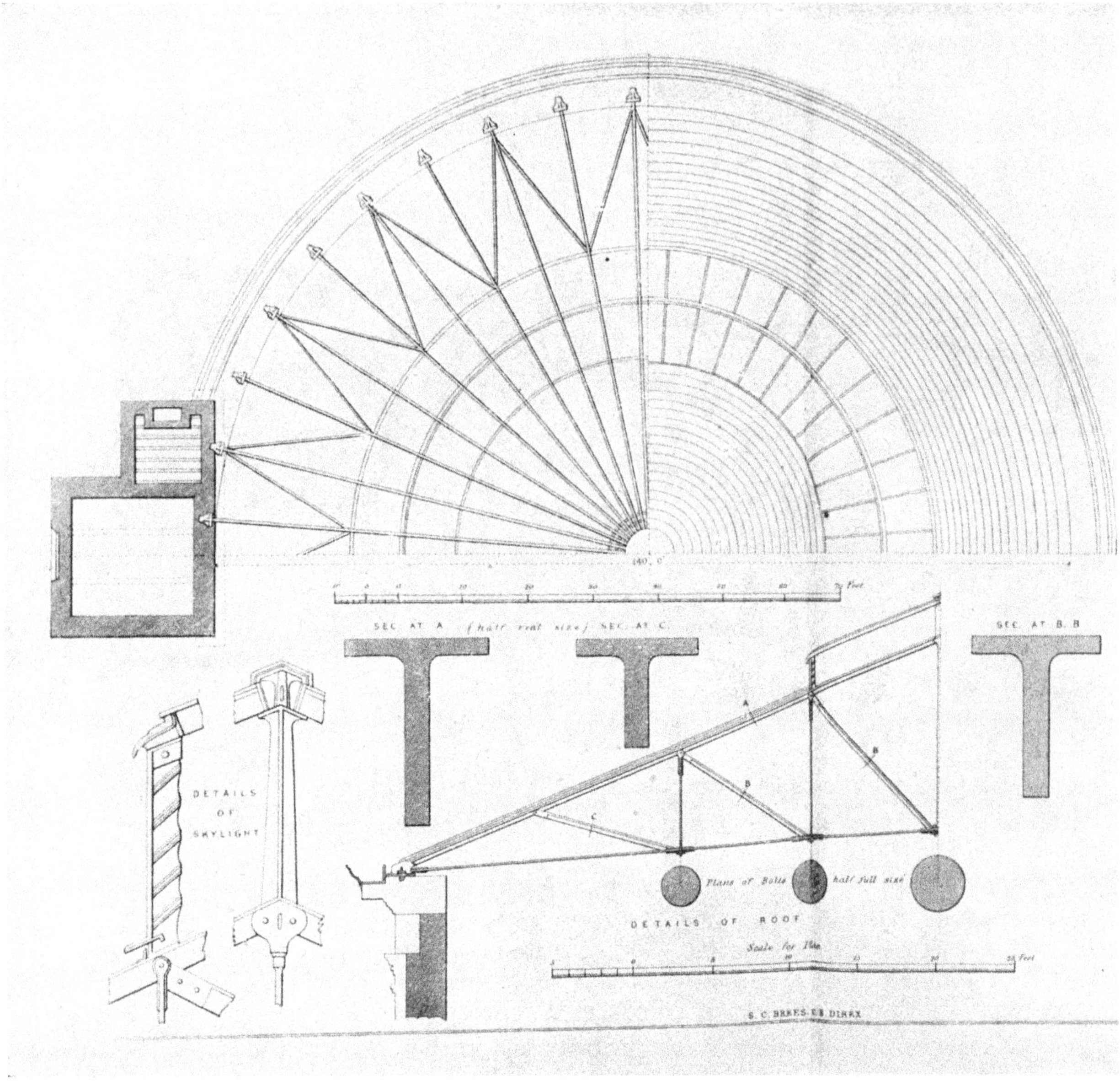

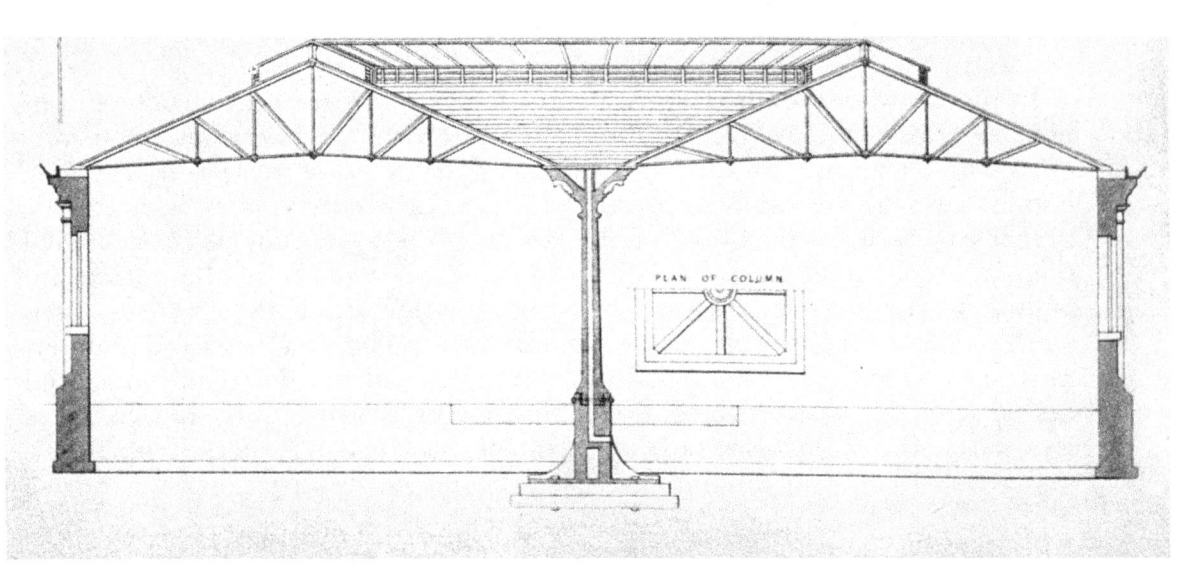

Fig. 194 (*Opposite page*). Plan, section and details of circular engine house at Gorton Depôt on the Sheffield & Manchester Railway. Weightman & Hadfield, Architects: A. S. Jee, Engineer. From *The First Series of Railway Practice*, by S. C. Brees, c.e., published by John Williams & Co., 1847.
Reproduced by courtesy of the London, Midland & Scottish Railway.

Fig. 195 (*Above*). Circular engine house, North Western Railway, Camden Town Depôt, 1847. Diameter 160 ft. Roof supported on 24 cast iron columns in a circle of 40 ft. diameter. From a print in the Public Library of the Metropolitan Borough of St. Pancras.
Reproduced by courtesy of the St. Pancras Public Library.

work in cast iron has great distinction: and it would be difficult to better the repeating bays of railings he designed to run for miles, bordering the track outside Euston station, or the hand rails on many of his bridges.

Often the early stations were merely open sheds, with a road driven straight in, the roof upheld on slender cast iron columns, supporting decorated brackets and the roof members. But the railway companies soon became conscious of the importance of stations, and even if the platform space was not improved and enlarged, impressive buildings were erected at the approach. Some of the smaller stations were designed without any conscious attempt to accommodate some style; and they were unpretentious, neat and fitted into rural surroundings appropriately enough. Cast iron was used successfully for the thin, not inelegant supports for the wooden roofs of such structures. Even when this cast iron and woodwork became ornate, such early station architecture was not defaced and disrupted by the indiscriminate placing of advertisements upon nearly every available surface, which occurred later in the nineteenth century. In the design of stations, the architect and the engineer often worked together, sometimes with agreeable results; but often the architect merely added the trimmings, and they were generally borrowed from some traditional style. Writing in the middle of the century, Frederick S. Williams in his book, *Our Iron Roads*, reveals the typical approach of the contemporary architect to the problem of station design. "The styles of architecture which have been adopted in the construction of railway stations are very various," he says. "Sometimes they are heavy and massive, or large and handsome; in other places they are neat or picturesque; and occasionally they have no one good quality to apologize for their existence. The characteristics of the neighbourhood in which they are erected have in some instances appropriately determined the style. On the Tunbridge Wells and Hastings Railway, the Battle station is Gothic, and is built of native stone, with Caen stone dressings, the roof being covered with alternate bands of plain and ornamental tiles".[5]

The same writer describes in some detail, the accommodation and character of Chester station, which represented an uneasy compromise between the architect and the engineer. "It consists of a noble pile of buildings in the Italian style," he tells us, "the facade fronting the city being 1,050 feet long. The centre of the station, which is two storeys in height, contains on the ground-floor the usual offices for passengers; and in the upper part are those required for the management of the business of the Companies whose lines here unite. These apartments are more than fifty in number. The wings are formed of projecting arcades, with iron roofs, and are appropriated to private and public vehicles waiting the arrival of trains. On the inner-side of the office-buildings is a large platform 750 feet long, by 20 wide, and chiefly used for departure trains. It is covered, as are three lines of rails, by an iron roof of sixty feet span, which is one of the most elegant yet constructed; and, as the height of the walls on which it rests is twenty-four feet from the platform, the entire structure has an imposing appearance. Behind this shed, and only divided from it by a series of pillars and arches, is a space for spare carriages, 450 feet long by 52 in width, which is also covered by a beautifully constructed iron roof. There are likewise two sheds for arrival trains".[6]

When engineers used cast iron to solve an engineering problem, and were unconscious of or ignored any obligation to respect the canons of some traditional style of architecture, they produced such remarkable examples of original design, as Brunel's station roof at Paddington. The great terminal station which Brunel and W. D. Wyatt designed in 1854, replaced a mere shed, with cast iron columns bearing a wooden roof. The new station had great cast iron columns, and they still exist together with the decorated, curved trusses and the detail round the offices. The structure and material sustained comparatively little damage when Paddington had a direct hit from a bomb during the second world war. A length of curved truss and some of the adjoining work disappeared; but the main structure suffered little, and traffic was almost uninterrupted.

At Euston, Hardwicke adorned his vast Greek Doric entrance gateway, with superb cast iron gates and railings; but behind this impressive portal

Fig. 196 (*Above*). Interior of the Nine Elms Goods Depôt, which was the original terminus of the old London & Southampton Railway, which became the London & South Western Railway, and is now the Southern Railway. Designed by Sir W. Tite in 1837. This view shows the timber roof supported on cast iron columns carrying girders.

Fig. 197 (*Left*). Detail of column and girders in the Nine Elms Goods Depôt.

These illustrations are reproduced by courtesy of 'The Builder'.

FIG. 198 (*Above*). A contemporary drawing of the interior of Paddington Station, Great Western Railway, designed by I. K. Brunel and W. D. Wyatt, 1854. Cast iron columns were used, but wrought iron was principally employed in the roof, together with some cast iron work. From *The Builder*, Vol. XII, No. 591, page 291.

FIG. 199 (*Below*). Cast iron and plaster decoration at the entrance to the offices at Paddington Station. From *The Builder*, Vol. XII, No. 593, page 323.

These illustrations are reproduced by courtesy of 'The Builder'.

the station itself was just a series of sheds. Cast iron columns and trusses supported the roof, and many of them are in use to-day. At the other end of the line that ran from Euston, Birmingham also had a dignified arched entrance with great cast iron entrance gates, and mere sheds behind. Only at Kings Cross, which was built in 1851, and where the two great arches flanking the central clock tower suggested the form of a station, was there any indication of what went on behind the facade. (The clock was taken from the Great Exhibition buildings when they were moved from Hyde Park to

FIG. 200. The entrance to Euston Station, London, designed by Philip Hardwick, and opened in 1838. The gates and railings are of cast iron. From a lithograph by T. Allom presented to the Royal Institute of British Architects by the architect in 1837.

Reproduced by courtesy of the Royal Institute of British Architects.

Sydenham.) Contemporary drawings show how the front elevation originally appeared, before it was hidden by a jumble of mean buildings and shacks.

In nearly all the large terminal stations and junctions, there was a general tendency to bedeck the bold and dignified results of courageous innovation in structure, with inappropriate architectural trimmings. Rainwater gutters became Doric entablatures and down pipes Corinthian columns. The architect was becoming an antiquary rather than an original designer; and too often the engineer felt obliged to be "architectural". The development of an

Fig. 201 (*Above*). Statue of Robert Stephenson outside Euston Station, London. Cast iron gates, railings, gate posts and bollards appear in this photograph.
Photograph: A. *Newton & Sons.*

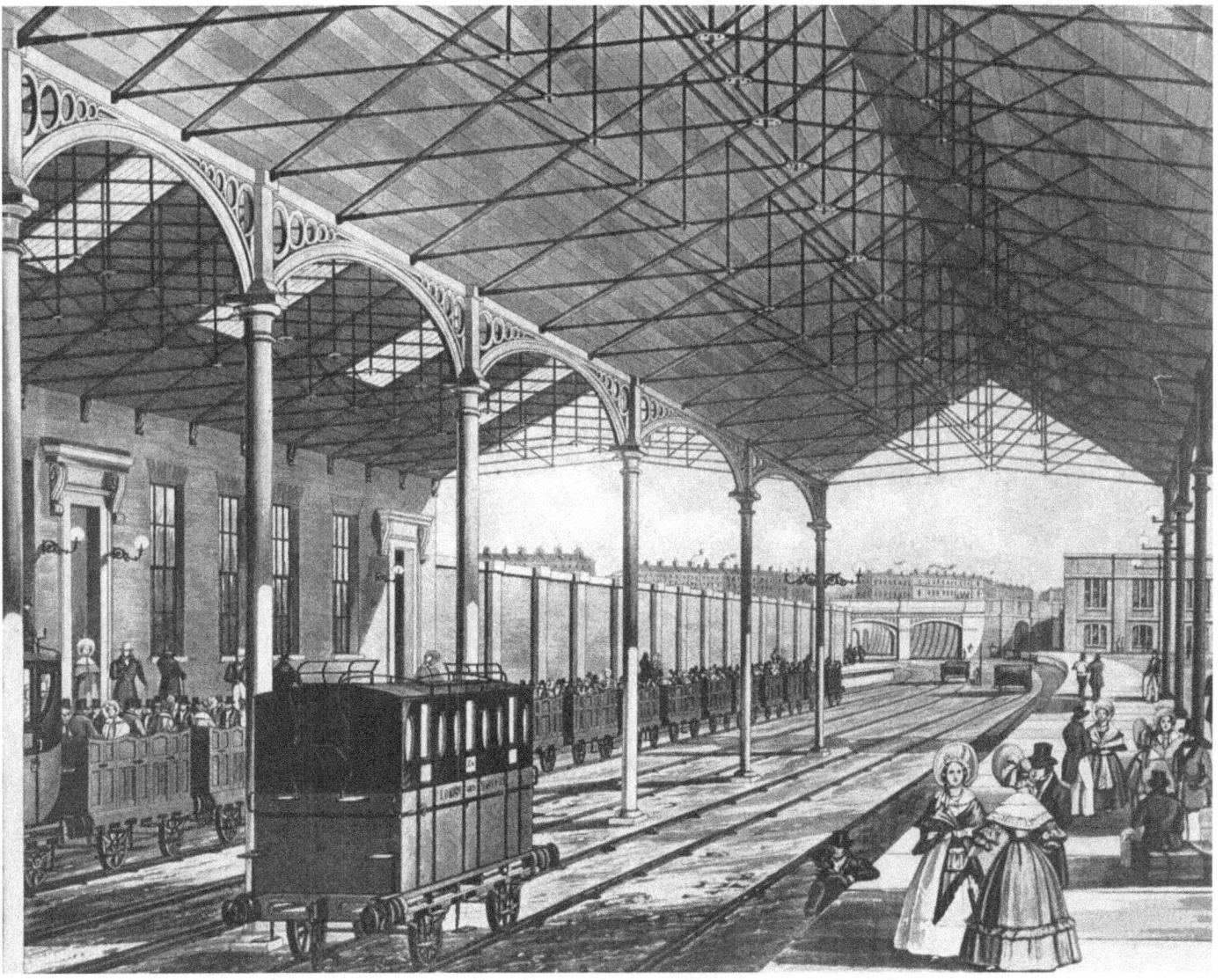

Fig. 202 (*Above*). The interior of Euston Station, from a drawing by T. T. Berry, published by Ackerman in 1837. This shows the cast iron columns and arched connecting girders, many of which still remain in the station today.

Reproduced by courtesy of the British Museum.

Fig. 203 (*Opposite left*). Euston Station, entrance and cast iron gates, designed by Philip Hardwick.

Fig. 204 (*Opposite right*). Euston Station. Detail of cast iron gates.

These illustrations are reproduced by courtesy of John Summerson.

Fig. 205. An early print showing the roof of St. Pancras Station, London. Architect: W. H. Barlow, 1868. Cast iron was used with wrought iron and glass to enclose a vast space.

Reproduced by courtesy of the Butterley Company Limited.

original technique for the use of cast iron in architecture was arrested. But occasionally the promise of such a development was revived, as in a scheme proposed in 1841 for a walk with an overhead railway, supported by cast iron columns of delicate and beautiful design, to run along the bank of the Thames from Hungerford market to London Bridge. This early attempt to plan the area round St. Paul's and to open a vista from the river to the south portico of the Cathedral, exemplified a singularly appropriate use of cast iron in architecture.

CAST IRON GIRDER BRIDGES.

The extent to which cast iron girder bridges were used in building railways during the first half of the nineteenth century has been revealed by a research conducted in 1944 by a group of engineers. As a result of the Road and Rail Traffic Act, 1933, the Railway Companies calculated and tabulated the restrictions which they considered should be applied to certain bridges, and after conferring with the Ministry of Transport and other interested bodies, a priority list was drawn up for the reconstruction of all those bridges which were deemed unsuitable for loads exceeding twelve tons. It was realised that many bridges, which by the ordinary theoretical standards were considered capable of carrying only light loads, were known to carry much heavier routine loads without signs of distress. Tests, taken sometimes to breaking point, were carried out on many bridges of this type, and the results were published in a paper by C. S. Chettoe, B.Sc.(Eng.), M.Inst.C.E., N. Davey, B.Sc.(Eng.), M.Inst.C.E., and G. R. Mitchell in 1944.[7]

The cast iron girder and jack arch system of construction was first used, as mentioned on page 113, in 1801 in the building of the cotton mill of Phillips & Lee in Manchester; and it was quickly realised that here was a method of bridging roads and railways without the use of arched construction, which gave a level bridge surface and good headroom beneath. There are to-day, still in existence, more than 4,000 of such bridges in Great Britain; none of them less than fifty years old and the majority built before 1860.

The research work of Professor Hodgkinson, William Fairbairn and Tredgold influenced the character and use of beams for mill and warehouse construction and had a comparable influence upon girder bridges. Various types of beams were used extensively as the main bearers spanning between abutments of bridges. The space between the beams was bridged by brick jack arches which carried the filling and road surface. The spans were under fifty feet, and the brick arches usually of nine-inch thickness, spanning about seven-feet between girders and resting on the bottom flanges. Sometimes cast iron plates or stone slabs were used instead of the brick arches. Wrought iron tie rods were used to couple the girders and take the thrust of the brick arches. On the outside girders, strengthened for the purpose, the brick, stone or cast iron parapet was usually built.

Chettoe, Davey and Mitchell in their report describe fully the types of girders, the various bricks, mortars, fillings and road surfaces met with during their investigations, and show that the strength of such heterogeneous structures cannot be assessed by ordinary analysis and rules. They also reported that in the tested bridges the quality of the cast iron varied considerably, the castings not always being "clean"; a common defect being slag inclusions and blow-holes at junction points between flanges and webs, these defects being the primarily responsible factor in most cases of failure.

Apart from the straight cast iron girder bridge, nineteenth century engineers experimented with arched beams or bowstring girders, the arched beams being held together by horizontal ties to resist the thrust, instead of abutments. Robert Stephenson erected several of this type on the original London and Birmingham railway, and another fine specimen was the great High Level Bridge at Newcastle, built in 1849. The problem was to throw a bridge carrying a road and a railway across the valley that separated Newcastle and Gateshead. Stephenson solved it by building enormous masonry piers carrying a double bridge, the railway above and the road beneath. The upper level was borne on cast iron arched members, the lower road taken on the horizontal ties to the arches.

In 1844 an Act was passed authorising the extension of the railway from Chester to Holyhead, and the Stephensons had the problem of carrying a bridge across the Menai Straits between the mainland and Anglesey.

FIG. 206 (*Left*). Aldridge Road Bridge, Birmingham, which carried the road over the Birmingham Canal. The span is 55 ft. 1 in. Observe the 10 in. camber on the beams.

Examples of cast iron girder bridges that were tested in the Survey mentioned on pages 178 and 179.

Figs 206 to 225 inclusive are reproduced by permission of the Director of Building Research. Crown copyright reserved.

FIG. 207 (*Right*). This shows the bottom flange of the beams, which are of I section. Cast iron plates rest on the bottom flanges and a second layer rests on the top flanges and carries the roadway. Girders 27 ins. deep overall, 12 ins. wide bottom flange, 3¼ ins. thick, and 5½ ins. wide top flange.

Fig. 208 (*Above*). The Town Bridge at Thetford, Norfolk. The span is 32 ft. 9 ins.

Fig. 209 (*Left*). The under side of the Town Bridge at Thetford, which carries the main road over the Little Ouse. This shows the under side of the cast iron ribs, and the curved cast iron plates forming the soffit. Each rib has a bolted joint at mid-span, and is 15½ ins. deep, with a 5 in. by 1½ in. bottom flange, and no top flange.

Fig. 210 (*Above*). The Spey Bridge at Fochabers, near Elgin, Morayshire, built in 1853 to replace a wooden bridge which in turn was built in 1832 to replace two of the stone arches of the original bridge built in 1803. The steelwork for the carriageway was added in 1911. The span is 183 ft., and the rise about 21 ft.

Fig. 211 (*Right*). The cast iron shoes for the ribs of the Spey Bridge, Fochabers.

Fig. 212 (*Below*). The ribs of the Spey Bridge, Fochabers; also the typical joint between segments, the ties and tapered bracing struts.

Fig. 213 (*Left*). Here, the steelwork deck and sub-deck of the Spey Bridge, which were built in 1911, are shown. This view illustrates the cast iron spandrel columns, and ties, the tapered cast iron diagonal bracing struts, and the four main cast iron ribs. The ribs are 36½ ins. deep, the top and bottom flanges are 8 ins. and 11 ins. wide respectively, and the metal is 2 ins. thick.

Fig. 214 (*Above left*). Gingerbread Hall Bridge, Gt. Baddow, Essex, which was built in 1870 and carried the Chelmsford - Southend road over a small stream. Here, the bridge is seen after the removal of the parapets, and the camber on the under side of the outside girders can be observed, while the inside girders are not cambered. The span is 13 ft.

Fig. 215 (*Above right*). This view of the Gingerbread Hall Bridge illustrates how the outer girder has fractured after loading to breaking point. The architectural treatment, and the gap in the web at the end for holding down bolts are of interest.

Fig. 216 (*Right*). Here, the fractured ends of an inside girder on the Gingerbread Hall Bridge, can be seen, after loading to breaking point; also the flaw at the junction of web and flange. (The hole at this point was a drilled hole made for test purposes.)

Fig. 217 (*Left*). Stafford Bridge, Oakley, Beds. This shows the fracture of an inside girder after loading to breaking point. Observe the form of the fracture; the inverted T section of the girder with no top flange; the small flaw at the junction of web and flange, and the holes for the bars. The web is 25 ins. deep, the flange is 20 ins. wide, and the thickness about 1¾ ins.

Fig. 218 (*Left*). Babraham Bridge, Babraham, Cambs., which was built in 1847, probably under the direction of Robert Stephenson. It carried the Cambridge - Haverhill road over the Newmarket and Chesterford Railway, which was abandoned in 1858. The tie bar and brick skewback for jack arches are shown here, also the fractured girder which has resulted from loading to breaking point. The girder is 36 ft. long, 24 ins. deep and has a 12 in. flange. The camber on the top flange is 6 ins.

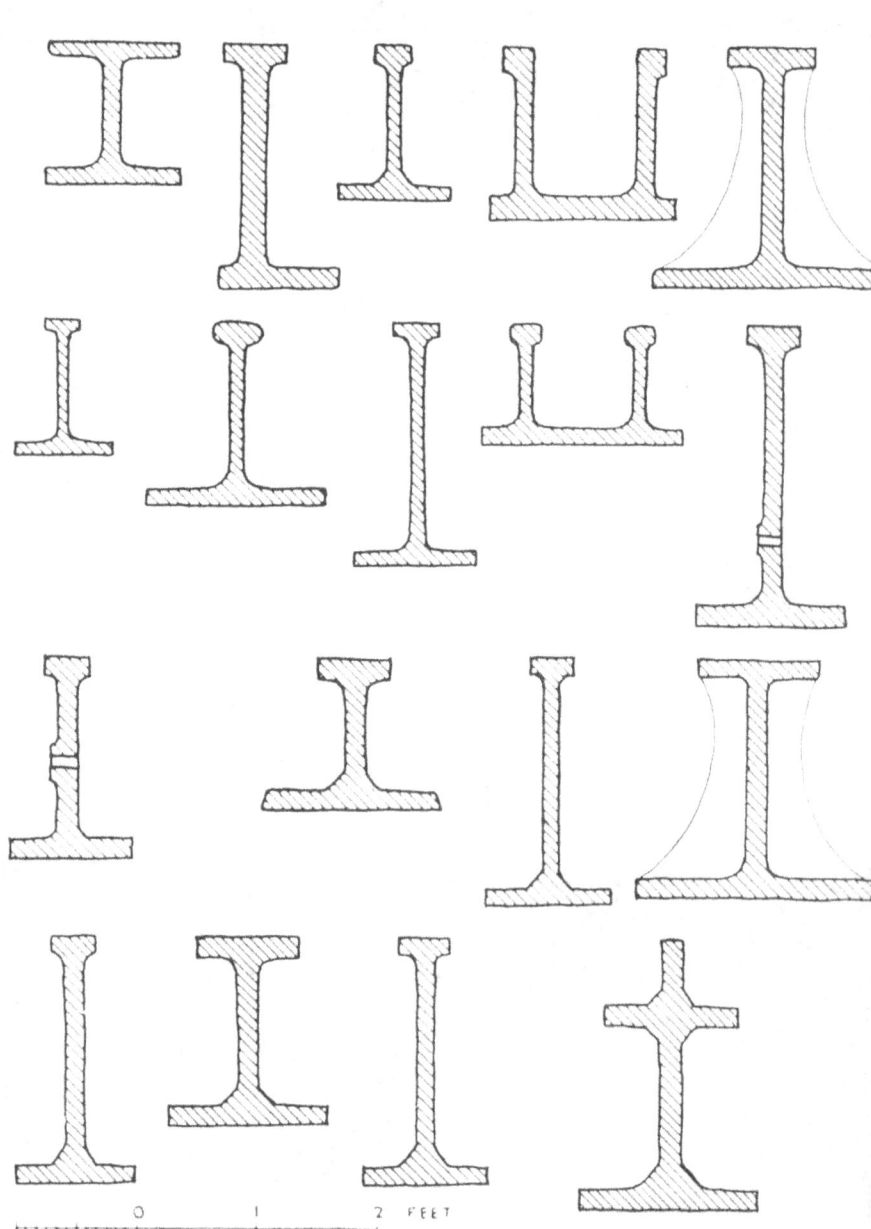

Fig. 219 (*Right*). Here are a few of the many various sections of cast iron beams found in the Survey referred to on pages 178 and 179.

Fig. 220 (*Above*). Gt. Barr Street Bridge, Birmingham, over the Grand Union Canal. Span 32 ft. 7 ins. Arched cast iron troughing, with ornamental fascia beam. Troughing is 3 ft. 6 ins. wide, the outer horns 10 ins. high, the midrib 6 ins. high, and the metal 1½ ins. thick.

Fig. 221 (*Below*). Islington Row Bridge, Birmingham, over the Worcester Canal. Span 22 ft. Typical cast iron girder and brick jack arch construction.

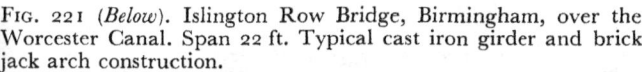

Fig. 222 (*Below*). Tindal Bridge, King Edward's Road, Birmingham. Span 42 ft. 3 ins. This view shows I section construction similar to that shown in Fig. 207, but with only one set of plates. The bottom flange is 10 ins. wide and 2½ ins. thick, while the top flange is 6 ins. wide and 1½ ins. thick. Overall depth is 19½ ins.

Fig. 223 (*Above*). Millington Hall Bridge, Retford, Notts. Span 29 ft. 10 ins. This carries the Great North Road over the old Great Central line at Retford. The cast iron railings have been covered by a timber hoarding.

Fig 224. (*Above*). Warrington Road Bridge, Culcheth, Lancs., which carries the Warrington road over the London North Eastern Glazebrook - St. Helens and Wigan line, and is typical of hundreds of railway overbridges. Span 25 ft. 8 ins.

Fig. 225 (*Above*). Eastleigh Bridge, Eastleigh, Hants., 1865 - 1868, which carried a secondary road over the Eastleigh - Romsey line. Observe the skew, camber, and architectural treatment.

FIG. 226 (*Above*). Cast iron arches, originally proposed for the Britannia Bridge, North Wales. From *The Britannia and Conway Tubular Bridges*, by Edwin Clark, resident engineer, with the sanction and under the immediate supervision of Robert Stephenson. Published by John Weale, 1850.

The first idea was for an iron bridge of two arches, each of 350 feet span, much as Rennie had suggested in 1801 and Telford in 1810, though the latter had envisaged a single arch of 500 feet. But as the necessary centering and scaffolding for erection would temporarily interfere with navigation, the scheme was abandoned. Finally the problem was solved by the construction of two continuous tubular wrought iron beams, each 1,511 feet long, carried on cast iron frames inserted as bearings on masonry towers. This, the famous Britannia Bridge, was opened in 1850, and meanwhile a similar but smaller edition had been erected over the estuary of the Conway.

Fairbairn and Hodgkinson were responsible for most of the calculations affecting these enormous undertakings and their work was again checked by Stephenson's very able assistant, Edwin Clark. As we have seen, Fairbairn was, despite his brilliant research work on the use of cast iron, inclining strongly towards wrought iron as a structural material, and it was natural, especially in view of the design of the Britannia Bridge, that this material

FIG. 227 (*Below*). Bridge over the River Trent, on the London, Midland & Scottish main line, erected in 1850 and of cast iron construction.
Published in the catalogue of the Butterley Company Limited, by whose courtesy this is reproduced.

FIG. 228 (*Above*). Lambeth Suspension Bridge, London, designed by Peter W. Barlow, F.R.S., Engineer, in 1836, and demolished in 1929. The pier cylinders are formed of cast iron plates cast in segments 12 ft. in diameter, in which is built the brickwork forming the support for the pylons. From *A Record of the Progress of Modern Engineering*, edited by William Humber, 1863 - 1864.

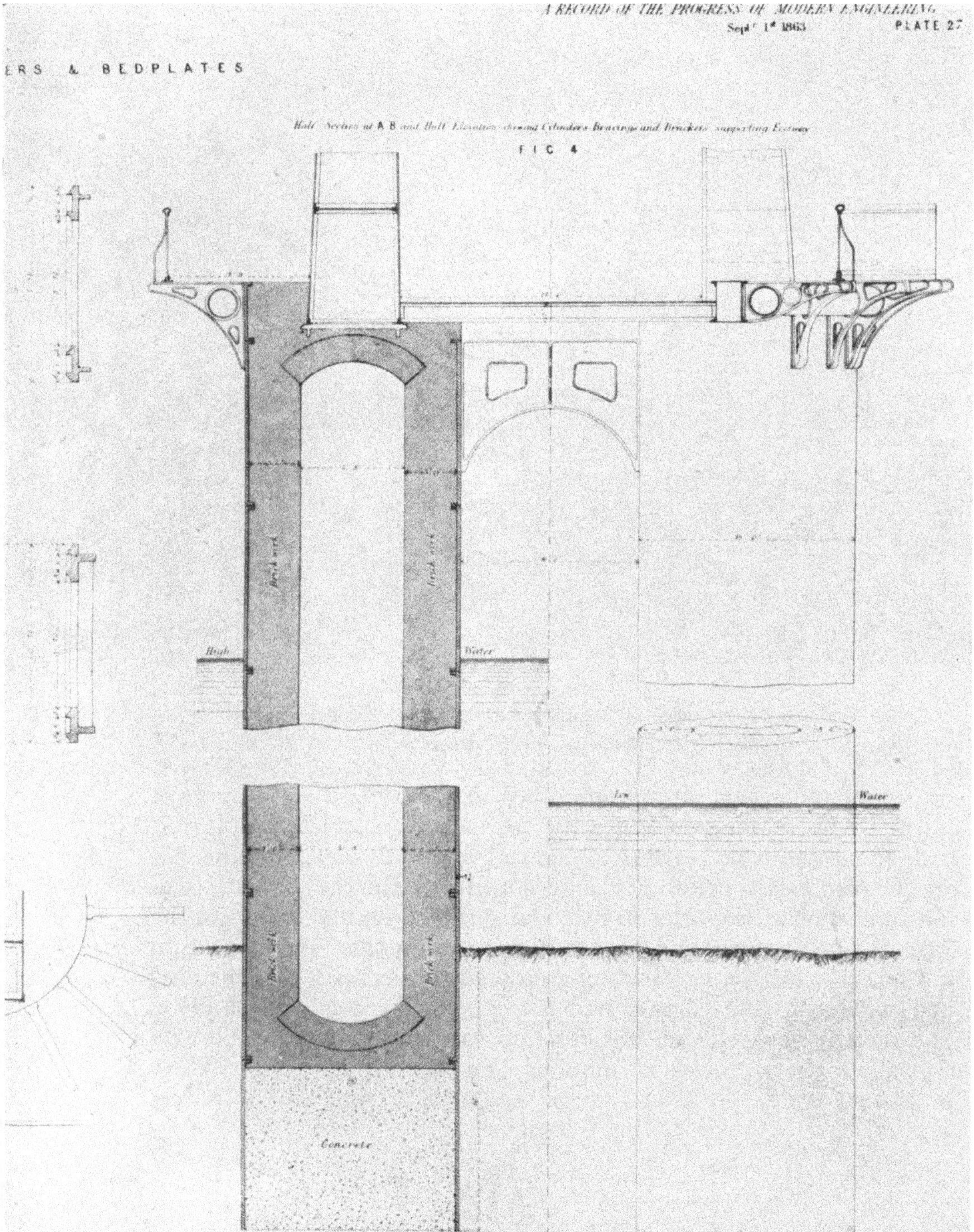

FIG. 229 (*Left*). Details of the cast iron cylinders, brackets, footway, of the Lambeth Suspension Bridge shown above. From *A Record of the Progress of Modern Engineering*.

These illustrations are reproduced by courtesy of the Royal Institute of British Architects.

should be used, though cast iron was successfully employed for those members in compression.

Of the Conway bridge, Williams gives some instructive details in *Our Iron Roads*, and describes an unusual function performed by cast iron. "The Conway tubular bridge," he writes, "which has deservedly attracted much attention is, in reality, a rectangular tunnel, or hollow square box, having top, bottom and sides. Around each end is a great deal of wrought-iron work, for the purpose of giving strength to the whole structure, the work at the top,

Fig. 230. Cast iron used for the towers of Hammersmith Suspension Bridge, London.
Reproduced by courtesy of the Public Works, Roads and Transport Congress.

bottom, and sides having each a separate office to perform. It is this part of the work in which Mr. Stephenson's scientific knowledge is specially displayed: the iron-work above the tube consists of eight square cells or tubes, and has to resist compression; that below the tube consists of six cells, and has to resist tension; and that at the sides has to secure the combined action of the top and bottom. The Conway end of the tube is immovable, being fixed on the pier, and made to rest on two beds of creosoted timber with intermediate cast-iron bed-plates; but the Chester end is free, so that it may expand by heat and contract by cold, as the tube rests on cast-iron rollers, which give play so as to allow twelve inches of motion. The whole mass weighs 1,140 tons".[8]

Brunel's iron bridge at Chepstow, he describes as follows: "A tubular bridge has been constructed over the Wye, at Chepstow, on the South Wales Railway, to which allusion must be made. This bridge consists of four spans, three of about a hundred feet each, and one of 290 feet, extending altogether from bank to bank for 610 feet. The three smaller spans rest upon iron piers, filled with concrete, supporting cast-iron girders, on which the railway is laid. The fourth or chief span is made upon the suspension principle, the great length of the girders requiring more support than that afforded by the piers alone at each extremity. Mr. Brunel accordingly contrived that this should be accomplished by means of a tube 309 feet in length, and nine in diameter, which, having been raised to the summit of piers erected on the cast bank, and in the centre of the river, is strengthened by massive chains secured to the girders. These girders are fifty feet above high-water mark at spring tides, which here rise from fifty to sixty feet—a greater height than in any other river in the kingdom".[9]

Cast iron had secured an important and trusted reputation with designers

FIG. 231 (*Above*). Chelsea Suspension Bridge, London, designed by Thomas Page, Engineer, in 1854. Cast iron was used extensively for the towers and foundation piers.

Reproduced by courtesy of 'The Builder'.

FIG. 232 (*Left*). Chelsea Suspension Bridge, as rebuilt. Cast iron is still used in the towers and foundation piers.

Photograph: Herbert Felton.

FIG. 233. Westminster Bridge, London, from a contemporary drawing. See Figs. 234 and 235 on the opposite page.
Reproduced by courtesy of 'The Builder'

FIG. 235 (*Below*). Westminster Bridge, London, designed by Thomas Page in 1857: carries the road over the River Thames. The total length is 811 ft. 6 ins., the carriageway is 58 ft. 5 ins., the footways 13 ft. There are seven arches, partly of cast iron and partly of wrought iron, upon piers and abutments of brickwork faced with granite.

Reproduced by courtesy of the Public Works, Roads and Transport Congress.

FIG. 234 (*Above*). Westminster Bridge, London. Detail of piers, from a contemporary drawing. See Fig. 233 opposite.

Reproduced by courtesy of 'The Builder'.

of bridges. The durability of the material has preserved much of this fine, adventurous work by the nineteenth century engineers.

BUILDINGS

Section II described how experiments made about 1825 affected the design of cast iron beams. Tredgold had advocated a beam with equal top and bottom flanges; Fairbairn and Lillie favoured one with no top flange but with a wider bottom flange; and finally Professor Hodgkinson, by his research of 1827, proved the superiority of a beam with a wide bottom flange and a smaller top flange. This improvement in the capabilities of the cast iron beam, together with the desire to raise the standard of fire-resistance in buildings, increased the general use of these beams and incidentally rendered practical an increase in the size of the fourteen-feet bays into which mills and factories had, till then, been divided, thus allowing greater areas of open floor space between stanchions for accommodating machinery. Fairbairn actually constructed one seven-storied building fifty-two feet in width, with one row of columns down the centre and cast iron beams spanning the twenty-six feet between the columns and the outside walls.

Beams of this revised section were used extensively to carry the floors of buildings and also as the main girders in bridges up to forty feet in span, and in one instance a British firm used such girders for a bridge on the Haarlem railway up to seventy-six feet span. Fairburn pointed out, that although such girders were satisfactory when used by experts, failures occurred, sometimes with catastrophic results, which could nearly always be traced to inefficient and ignorant design, or faults in the metal through lack of control in the foundry work. The research work of Hodgkinson and Fairbairn in the first half of the nineteenth century influenced the whole cast iron industry, and in spite of his proclaimed opinion that wrought iron was the finest structural material and overcame all the drawbacks of cast iron, Fairbairn did realise that there was the possibility, if the necessary experiments and research were carried out, of altering and controlling the characteristics of cast iron to suit particular uses. He experimented with "toughened" cast iron, and foresaw many of the results of present-day research and the improvements that were necessary in foundry technique before our own contemporary metallurgical standards could be attained. In 1854 he wrote: "On the subject of the mixtures of the different kinds of British iron, for the purpose of producing castings suitable to the purposes for which they are intended, we have no fixed or determined rule from which we can obtain anything like correct results. Every ironfounder appears to exercise his own judgment in those matters; and it is very difficult to obtain castings under conditions where the proportions are specified, unless prepared under a strict and rigid surveillance. Iron-founders, managers and furnace-men appear to work under the impression of the non-importance of attention to the mixture of metals; and hence arise all the anomalous conditions of soft and hard, strong and weak iron, and many other disqualifications which might easily be avoided

PLATE III. Interior view of the Crystal Palace, showing the supporting cast iron columns and the simple unit construction. From *Dickinsons Comprehensive Pictures of the Great Exhibition of 1851*, painted by Messrs. Nash, Haghe and Roberts. Published by Dickinson Bros., 114 New Bond Street, 1854.

Fig. 236. The Market Hall, Birmingham. Designed in 1833 by the Birmingham architect, Charles Edge, and erected in 1835. There were three aisles separated by rows of cast iron columns, having vertical reeding on the lower portion. Some of these columns can be seen in the photograph taken after the destruction of the Hall by enemy action in the second world war.

Reproduced by courtesy of the National Buildings Record.

Fig. 237. Bridgewater House, London. The picture gallery, designed by Charles Barry in 1847, shows a characteristic use of cast iron principals in the roof construction.
Copyright: The Times.

by closer attention to the quality and due proportion of each particular iron and the quantity of carbon and flux used in its liquefaction. All these are considerations of great importance in the art of founding, and we have yet much to learn in the preparation of metals, as well as in the manipulative art of moulding, and the necessary process of ventilation; a process which requires no inconsiderable degree of thought and skill. . . . We have almost every variety in the pig—soft, hard, ductile, rich and poor, as well as the white, blue, grey, etc., irons, all of which are adapted in combination to form almost any description of metal required in the useful arts. They are, moreover, calculated to effect great improvement in the quality of the castings, and from compounds which, with proper care, may be varied according to the uses for which the castings are intended".[10]

These researches encouraged an extensive use of cast iron in mills, warehouses, factories, churches and public buildings. The main uses were for vertical supports and for the beams supporting the floors of multi-storey buildings; but all over Britain we find curious examples of the employment of the material. Many of the old warehouses exist still, and in many more the cast iron work was only brought to light by air raid damage in London, Birmingham, Hull and other industrial areas. Often walls would be blown away, exposing the cast iron structure for the first time since its erection a

FIG. 238 (*Right*). Interior of the Riding School, Welbeck Abbey, showing cast iron columns and part of the trusses.

Copyright: *W. H. Smith & Son Ltd.*

FIG. 239 (*Left*). Cast iron columns in the City Temple, London. Architects: Lockwood and Mawson, 1874. This was probably the best known Congregational church. See also Fig. 240 on page 196.

Photograph: *Sport and General Press Agency.*

hundred years earlier; and when buildings were demolished by fire it was amazing to see the cast iron skeleton still standing when the steel joists of later adjoining buildings were bent and distorted. It was said when the old Alhambra was burned down years ago, that in spite of the fire and water, the cast iron columns and bressummers stood firm whereas the new lattice girders "hung in festoons like tripe".

But it was not only in Britain that cast iron played a major part in building. All over the world the cast iron column, due to its fire-resistant qualities, its cheapness, simplicity of manufacture and resistance to heavy loads, was to be found playing a significant part in architecture. British firms exported to the colonies the framework of whole buildings in pre-fabricated parts. In France cast iron was used in a lighter, more delicate and logical way. In the Bibliothèque Nationale in Paris, 1858-68, the vaulted ceiling of the Reading Room was supported on sixteen slender cast iron columns, and Henri Labrouste the architect, used simple cast iron grids for the floors in the stack room, thus admitting light to the shelves from top to bottom. Earlier,

FIG. 240. The City Temple, London. The piers of the nave, basement and galleries are of cast iron. See Fig. 239 on the previous page.

Photograph: *Sport and General Press Agency.*

FIG. 241 and FIG. 242. Two views showing cast iron columns in the Metropolitan Tabernacle, Newington Butts, London, designed by Searle and Hayes in 1898. See interior view on opposite page, Fig. 240.

Photographs: Sport and General Press Agency.

Fig. 243. Cast iron front to Nos. 219-221 Chestnut Street, St. Louis, Missouri, U.S.A.: 1877. This facade was recently removed by the National Park Service for future use in a museum.

Reproduced by courtesy of the United States Department of the Interior.

in 1824, very light cast iron columns were used to support the roof of the Market Hall of the Madeleine, Paris.

In America, as early as the eighteen forties, and later in Britain, cast iron was used to provide the frame for what was almost a pre-fabricated building, and later even the first skyscrapers had the cast iron column as the chief means of support. Sigfried Giedion describes how "commercial buildings with cast iron fronts, and often with cast iron skeletons as well, sprung up all over the United States between 1850 and 1880—the so-called 'cast iron age' ... The river front at St. Louis, was the city's oldest business area, and most of it was built up during this flourishing period".[11] Many of these water front buildings were of this frame design with the infilling of cast iron panels; and they were fine in detail, though often designed and produced by the foundry without any guidance from architect or engineer. In 1855 the old wooden dome of the Capitol at Washington was replaced by a dome with cast iron ribs, which was altered in 1870, further cast iron work being added. In the previous section we referred to John Summerson's discovery of the cast iron lantern on the roof of Middle Temple Hall. William Wilkins used cast iron columns and beams in the construction of University College,

London, in 1828. Some of the ornamental columns still exist, and the same architect used the material structurally in the National Gallery. The Gothic Revival in architecture encouraged an extensive use of the material, particularly in church building. The Ecclesiastical Commissioners actually recommended that cast iron should be used to reproduce standard Gothic details for the new churches, that were being built to serve the rapidly growing industrial and residential areas in the eighteen twenties and thirties. A fine example of this early Gothic revival work, is St. George's Church, Birmingham, designed by Thomas Rickman in 1822, with its cast iron columns, arches and gallery front.

A new and successful use of cast iron in this period was for greenhouses, conservatories, palmhouses and exhibition buildings. Such works demanded a new technique, and for the first time it could be said that the roof and walls of a structure were of glass, held in a light metal frame. As early as 1808,

Fig. 244. Cast iron gallery front and arches at St. George's Church, Birmingham, by Thomas Rickman, 1822.

Reproduced by courtesy of the National Buildings Record.

Fig. 245. The Crystal Palace, which housed the Great Exhibition of 1851, as shown in the coloured plates I, II, III and IV. Here is a contemporary drawing of the interior which clearly illustrates the type of construction.

Reproduced by courtesy of The Builder.

Humphry Repton, who only claimed to be a landscape architect, produced a design for the Prince Regent's Royal Pavilion at Brighton, which included a pheasantry, chiefly of glass supported on a thin cast iron frame. In 1814 Nash had designed a conservatory formed of cast iron trellised pilasters and glass for the Prince Regent at Royal Lodge, Windsor. In 1828, when head gardener to the Duke of Devonshire, Joseph Paxton had interested himself in the design and erection of iron and glass structures, and by 1837 he designed the famous Great Conservatory at Chatsworth. Paxton desired to make the structural part as light as possible in order to obtain all the daylight and sunshine he could, and the results he achieved affected the design of all contemporary work where a large area had to be roofed for display or exhibition purposes. The Great Conservatory at Chatsworth was 277 feet long, 132 feet broad and 67 feet high at the highest point. Paxton followed this up with a Lily house at Chatsworth, with a form of roof differing from the curvilinear roof of the Great Conservatory, and he used cast iron columns not only structurally but as rain water pipes. These experiences encouraged him to design an exhibition building in cast iron and glass—combining the two types of roof, semi-cylindrical vault and the flat ridge-and-furrow—which was adopted for the Great Exhibition of 1851. The winning design in the competition, which was submitted by the Building Committee to the Commissioners, was recognised as impracticable, involving, as it did, an enormous permanent structure; and eventually Paxton's design was sanctioned. It achieved lasting fame and the affectionate regard of the public as "the Crystal Palace".

The enormity of the undertaking was described by Charles Dickens in *Household Words*: "Two parties in London, relying on the accuracy and good faith of certain iron-masters, glass-workers in the provinces, and of one master carpenter in London, bound themselves for a certain sum of money, and in the course of some four months, to cover eighteen acres of ground with a building upwards of a third of a mile long (1,851 feet—the exact date of the year) and some four hundred and fifty feet broad. In order to do this, the

Fig. 246. Detail of the interior of the Crystal Palace, showing the construction of the galleries.

Reproduced by courtesy of F. R. Yerbury.

FIG. 247. The Crystal Palace as it was re-erected at Sydenham, London: view of the main front. Compare this with the coloured plate II, which shows it as originally erected in Hyde Park.

Reproduced by courtesy of F. R. Yerbury.

glass-maker promised to supply in the required time, nine hundred thousand square feet of glass (weighing more than four hundred tons) in separate panes, and these the largest that ever were made of sheet glass; each being forty-nine inches long. The iron-master passed his word in like manner, to cast in due time three thousand three hundred iron columns varying from fourteen feet and-a-half to twenty feet in length; thirty-four miles of guttering tube, to join every individual tube together, under the ground; two thousand two hundred and twenty-four girders; besides eleven hundred and twenty-eight bearers for supporting galleries . . ."

The entire building was erected in seventeen weeks. Here was pre-fabrication on the grand scale. The whole structure was "planned on straightforward mathematical lines, not only based on standard structural units but organised in multiples of unit dimensions".[12]   Subsequently the Crystal Palace was demolished and re-erected at Sydenham as a permanent exhibition building, until it was destroyed by fire in 1936.

Of the architects of the period, Decimus Burton was perhaps most interested in the new iron and glass construction. He designed, in collaboration with Richard Turner, a Dublin engineer, the Winter Garden in Regents Park (demolished in 1932); the Palm House at Chatsworth, and the famous Palm House at Kew. In France, Poland and Russia, similar buildings were being erected. In Paris the Jardin d'Hiver, built in 1847, contained a great ball room 100 feet by 60 feet, leading into further vast corridors, galleries and gardens. The building was composed almost entirely of cast iron and glass, the roof being supported on more than a hundred exceedingly light and elegant cast iron columns.

Everywhere the influence of Paxton's experiments spread, and the iron and glass buildings of the first half of the nineteenth century have had a much greater effect on subsequent architectural design than is generally realised. In 1891, James Ferguson, in his *History of the Modern Styles of Architecture* said, "There is, perhaps, no incident in the history of architecture so felicitous as

Fig. 248 (*Right*). Detail of the entrance to the Crystal Palace, at the end of the great barrel-vaulted centre aisle.

Fig. 249 (*Below*). Detail of the roof of the barrel vault and side galleries.

Both these illustrations are reproduced by courtesy of F. R. Yerbury.

Fig. 250. The Crystal Palace: detail of interior.

Reproduced by courtesy of F. R. Yerbury.

Sir Joseph Paxton's suggestion of a magnificent conservatory to contain that great collection. At a time when men were puzzling themselves over domes to rival the Pantheon, or halls to surpass the Baths of Caracalla, it was wonderful that a man could be found to suggest a thing which had no other merit than being the best, and indeed, the only thing then known which could answer the purpose; and a still more remarkable piece of good fortune that the Commissioners had the courage to adopt it".[13]

Although, up to the eighteen fifties, cast iron was used so effectively, and indeed with indisputable genius, in these great iron and glass structures, the general use of the material did not show a corresponding progress. This was not wholly attributable to the ironfounders. They certainly did not make the same use of brilliant designers, as certain founders had done in the late eighteenth century, but apart from this the standard of design was changing throughout industry. These changes were not caused by indifference to design, but by an increasing incapacity to grasp its fundamental significance. Parliament recognised the seriousness of the situation, and appointed various

Fig. 251 (*Above*). The roof of the Crystal Palace after re-erection at Sydenham.
Reproduced by courtesy of the Architectural Press.

Fig. 252 (*Below*). The Crystal Palace: detail of the cast iron roofing to the barrel vault.
Reproduced by courtesy of F. R. Yerbury.

Committees to investigate the possibilities of Government-sponsored art schools and centres where the necessary study of decoration in relation to products of the new machine age could be made. It was recognised that "our manufacturers were, in all matters connected with machinery, superior to all our foreign competitors" as Sir Robert Peel stated in the House of Commons in 1832, when advocating the building of a National Gallery. But unfortunately design was no longer recognised as a basic operation. The "application" of ornament to a surface was considered the principal function of a designer. Pattern makers and moulders in the foundries had attained such a high standard in the art of casting that they could produce almost any intricate design, and the characteristics of the material enabled this intricate detail to be repeated again and again. So, without proper guidance from trained designers, it was natural that decoration was mistaken for design.

In 1843 that champion of the Gothic revival, A. Welby Pugin, delivered a series of lectures at St. Marie's, Oscott, which were later published, in which he ridiculed the current tendencies in the design of everyday articles, and blamed somewhat unfairly, the ironfounders for the state of affairs. "The fender," he stated, "is a sort of embattled parapet, with a lodge-gate at each end; the end of a poker is a sharp-pointed finial; and at the summit of the tongs is a saint. It is impossible to enumerate half the absurdities of modern metalworkers; but all these proceed from the false notion of disguising instead

FIG. 253. View of the Palm House at Kew Gardens, designed by Decimus Burton and Richard Turner. *Copyright*: *The Times*.

FIG. 254. Interior view of the Palm House at Kew Gardens. This was begun in 1844, and is an example of the use of cast iron and glass units.

Reproduced by courtesy of the Architectural Press.

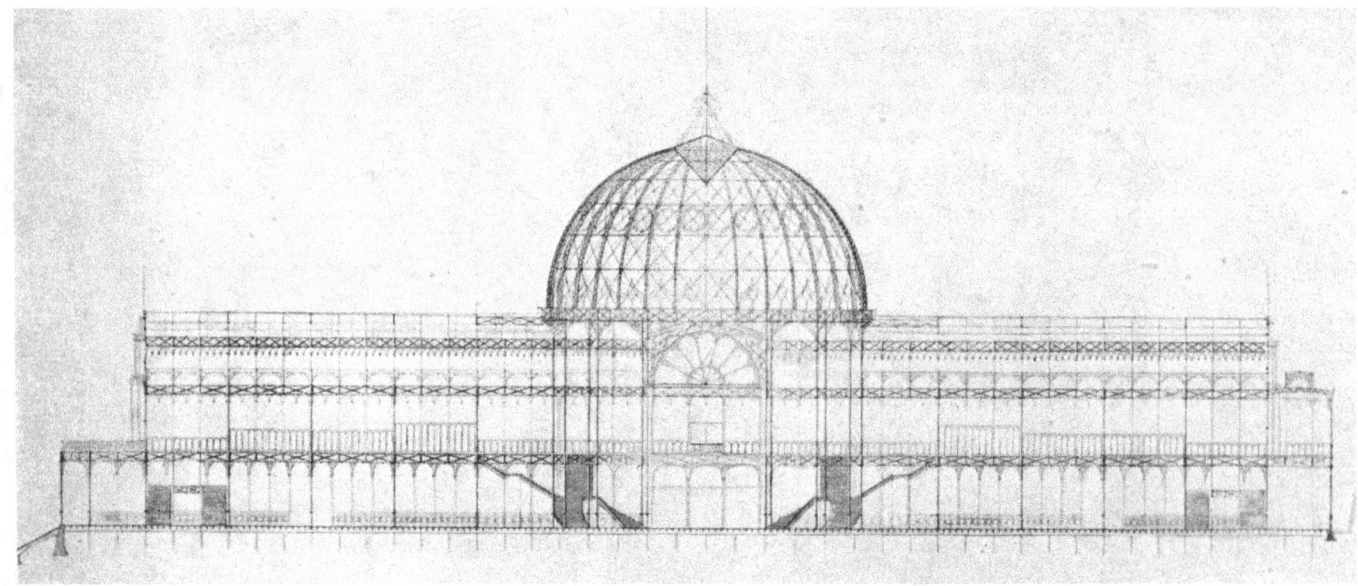

Fig. 255. The building designed to house the New York Exhibition, erected after the example of Paxton's Crystal Palace had started a fashion for such structures, and called the New York Crystal Palace. Designed by G. J. B. Carotensen and G. Gildemeister, Architects, and C. E. Detmold and H. Allen, Engineers. From an engraving in the collection of the Royal Institute of British Architects.
Reproduced by courtesy of the Royal Institute of British Architects.

of beautifying articles of utility. How many objects of ordinary use are rendered monstrous and ridiculous simply because the artist, instead of seeking the most convenient form, and then decorating it, has embodied some extravagance to conceal the real purpose for which the article has been made! If a clock is required, it is not unusual to cast a Roman warrior in a flying chariot, round one of the wheels of which, on close inspection, the hours may be descried; or the whole front of a cathedral church reduced to a few inches in height, with the clock-face occupying the position of a magnificent rose window. Surely the inventor of this patent clock-case could never have reflected that according to the scale on which the edifice was reduced, his clock would be about two hundred feet in circumference; and that such a monster of a dial would crush the proportions of almost any building that could be raised. But this is nothing when compared to what we see continually produced from those inexhaustible mines of bad taste, Birmingham and Sheffield: staircase turrets for inkstands, monumental crosses for light-shades, gable ends hung on handles for door-porters, and four doorways and a cluster of pillars to support a French lamp; while a pair of pinnacles supporting an arch is called a Gothic-pattern scraper, and a wiry compound of quatrefoils and fan tracery an abbey garden-seat. Neither relative scale, form, purpose, nor unity of style, is ever considered by those who design these abominations; if they only introduce a quatrefoil or an acute arch, be the outline and style of the article ever so modern and debased, it is at once denominated and sold as Gothic."

Some years after Pugin's critical explosion about the misuse of cast iron, another less passionate, but far more discerning critic was deploring a new class of ornament which had come into favour and which often forced cast iron into the most incongruous forms. In his *Analysis of Ornament*, Ralph N. Wornum, Secretary and Keeper of the National Gallery, said: "There is a class of ornament which has much increased of late years in England, and, by way of distinction, we may call it the *naturalist* school. The theory appears to be, that as nature is beautiful, ornamental details derived immediately from beautiful natural objects must insure a beautiful design. This, however, can only be true where the original uses of the details chosen have not been obviously violated; and one peculiar feature of the school is, that it often substitutes the *ornament itself* for the thing to be ornamented . . ." In another sentence this critic described exactly the process which led to the widespread misuse for inappropriate decorative purposes, not only of cast iron, but of many other materials, although cast iron, because of its tractable nature, suffered most. Wornum pointed out that "every implement or article of practical utility, as, for instance, a candlestick, that is composed or built up of natural imitations exclusively or as principals, however poetical the idea may be supposed to be, is practically bad as a design.

"There is a very great difference between *ornamenting* a utensil with natural objects, and *substituting* these natural objects for the utensil itself. In the latter case, however true the details, the design is utterly false; in the former, you are in both respects true, and may be also highly suggestive and instructive. Of course, there are many natural objects which at once suggest certain uses; and we can never be wrong if we elaborate these into such implements or vessels as their own very forms or natures may have spontaneously presented to the mind.

"Every article of use has a certain size and character defined for it by the very use it is destined for, and this may never be disregarded by the designer; it is, in fact, the indispensable skeleton of his design, and has nothing to do with ornament. But it is upon this skeleton that the designer must bring all his ornamental knowledge to bear; and he is a poor designer if he can do nothing more than imitate a few sticks and leaves, or other natural objects wherewith to decorate it; he must give it character as well as beauty, and make it suggestive of something more than a display of sprigs and flowers gathered from the fields, or this would be mannerism indeed".[14]

A distinguished architect, Mr. Hartland Thomas, F.R.I.B.A., writing to-day, has been critically retrospective of the inappropriate ornamental enthusiasms that Pugin and Wornum identified in the middle years of the nineteenth century. He has lucidly summarised the history of mid-Victorian misapprehension about architectural and industrial design in relation to the use of cast iron. "Consider what the Victorians did to cast iron," he writes. "They made it masquerade in all kinds of absurd make-believe. Cast iron seats in the park pretend to be logs with the bark on, cast iron railings reproduce the exact modelling of wood treillage or of hempen ropes, cast iron street furniture imitates moulded wood panelling or rusticated stonework, columns apparently of stone and admired as handsome monoliths are revealed by the blitz to be only cast iron. When this kind of subterfuge went out of fashion cast iron went out with it—most unjustly, because it was due to the very virtues of the material that it was used in this way.

"First among its virtues: in each of the examples quoted, iron was chosen because it is more durable than the material supplanted. The iron logs, iron trellis, iron panelling are still there for us to see, whereas timber would long ago have decayed or broken in such exposed situations. Even the iron 'monoliths' can claim the virtue of endurance, for the bomb knocked them down unbroken.

"Secondly, it had the virtue of tractability. The ease with which any desired shape can be approximated by the mould makes cast iron an obvious material for falsification. An intractable material has its own defence against misuse. Cast iron has none.

"Other materials are tractable—putty for instance—but a material cast in moulds adds to tractability convenience for repetition. In the early days of mass production cast iron was the obvious material. The manufacture of a hundred thousand cast iron rustic seats was easier than to make them in real logs. It was cheaper that way.

"Cheapness, adaptability, durability—these were the virtues that contrived cast iron's disgrace and fall from fashion. They are also the virtues that have maintained cast iron in use for humbler indispensable workaday things. Once discovered, we could not do without it for manhole covers, rainwater goods, fire-grates, garden railings, boilers and all sorts of things. And these uses of cast iron persist. We use it since we must, not because we like it. (How tiresome it was when 'The Standard of Wartime Building' insisted upon concrete for manhole covers!) But in its heyday, when the Victorian iron age was carrying all before it, iron was the fashionable material. Not all the applications were suspect to modern ideas of æsthetic truth, like those mentioned above. Much can be learned from Victorian cast iron, from the exuberance of our great-grandfathers' delight in this material, to enable us to give it forms that are characteristic of its qualities.

"Consider things that are unmistakeably of cast iron—those iron tables with marble tops that one finds in public-houses. (Iron is used here, by the way, just because it is heavy, so that the table is hard to overturn.) The shapes into which the iron is cast are flowing, rounded, bulbous shapes. If you want keen arrises and attenuated proportions, tense curves just so, with not an ounce of material wasted; don't think of cast iron. But if you want a fat comfortable richness, ample curves and solidity—real as well as apparent—then cast iron will be your faithful servant. Remember too that cast iron can be as up-to-date as any of them; the most famous of all prefabricated buildings—Paxton's Crystal Palace—was realised in cast iron units for dry erection in Hyde Park, demounting, transportation and re-erection on Sydenham Hill. These advanced techniques were known to cast iron (they could be known again if we wished it) when the more modern materials, such as aluminium, were hardly discovered. But do not let us this time make a fool of cast iron by putting it into shapes that belong to other things. It is a molten material, let it flow comfortably into shape, without distortion".[15]

FIG. 256. The functional simplicity of the Great Exhibition was not matched by the exhibits. Here is an example which shows how Victorian designers, so-called, had departed from the canons of taste and the standards of common sense universally accepted at the beginning of the nineteenth century. This particular exhibit was described in the *Illustrated London News* for August 1851 (page 193) as "One of the most pretentious works in the Building". The comment went on to say that although the casting supported the reputation of the ironfounders, there were many and grave objections to the design. The cupola and vane were described as "very bad". Amid all this fantastic riot was a cast of the Eagle Slayer by J. Bell. The *Illustrated London News* pointed out that "The eagle, transfixed by an arrow at the top, inside, must be considered as an absolutely inexcusable piece of bad taste".

FIG. 257 and FIG. 258 (*Above*). Two cast iron hot air hall stoves exhibited at the 1851 Exhibition. From the *Illustrated London News*.

FIG. 259 and 260 (*Right*). King's gas cooking range, described in the *Illustrated London News* of June 1851, as follows: "A gas cooking range, in the side aisle south of the western part of the Nave, which is constructed on a plan peculiar to the town of Liverpool. It was designed by Mr. King, chief engineer of the gas works of that town. It is divided into three compartments of different sizes for roasting and baking, being furnished with a damper to regulate the flow of air through them. The burner is arranged inside the oven, at bottom, around the sides, back, and front, with a dripping-pan occupying the centre.... Comfort and cleanliness to the cook, and economy to the consumer, are among the qualifications of this useful invention. The gas is lighted with a gas-torch, or portable jet of iron pipe, attached to a flexible pipe".

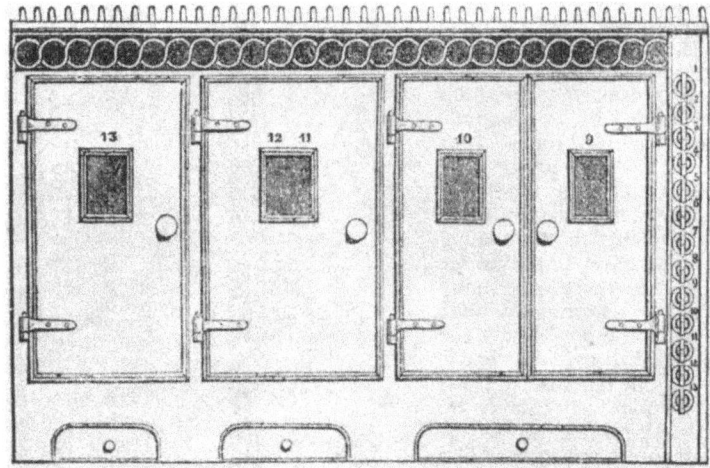

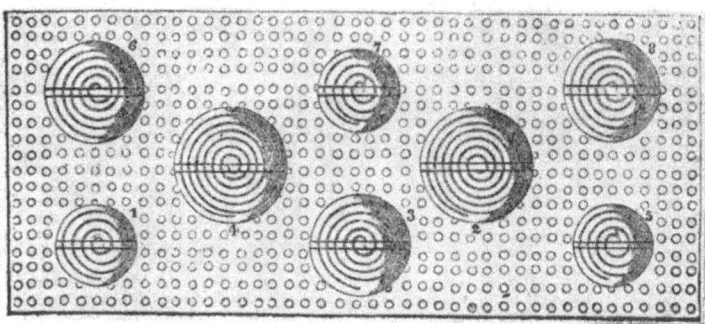

FIG. 261 and FIG. 262. Cast iron fireplace designed for the 1851 Exhibition by the French sculptor, J. P. Vaudre. The engraving is from the *Illustrated London News* of May 1851, which says: "The design of this chimney-piece is essentially French, but at the same time is so admirably adapted to the material, that the more florid characteristics of style are kept down". The photograph below is of the original fireplace, now in the head office of the Falkirk Iron Co.

213

FIG. 263 (*Left*). Cast iron plaque attributed to a Lincolnshire foundry. The detail is exceptionally delicate and is equally good on both sides of the plaque, showing the great technical skill of the foundrymen. The date is unknown. Compare this with Fig. 264.

Original in possession of, and photograph reproduced by courtesy of Frank Pascall.

FIG. 264 (*Right*). German cast iron filigree plate, *circa* 1840, from the Lamprecht Collection. Herr Gustav Lamprecht, a professor of graphic art at the University of Leipzig, spent his life and much of his fortune in amassing what is probably the largest private collection of cast iron art objects in the world. In 1922, he sold this collection, which was ultimately purchased by the American Cast Iron Pipe Company, Birmingham, Alabama, U.S.A. The collection, while it does not illustrate architectural uses, does indicate the astonishing technical accomplishment of the pattern makers moulders and craftsmen. It contains jewellery, statuettes, candlesticks, vases, household utensils, coins, medals, medallions and plaques.

Reproduced by courtesy of the American Cast Iron Pipe Company.

Fig. 265 (*Right*). A cast iron lamp standard in the Market Place at Northampton, erected in 1863. Although this is florid and ornate, it does retain the orderly formal lines of an earlier period; it is wholly different in conception from the chaotic hotch-potch shown in Fig. 256. This represents cast iron under control.

Reproduced by courtesy of the National Buildings Record.

Fig. 266. Cast iron plaque of the first Duke of Wellington. The inscription on the back reads "Moulded by Jobsons patent process, 13th November 1854".

Reproduced by courtesy of J. Billings.

Fig. 267 (*Left*). Railings designed by Alfred Stevens, in the Museum grounds at Leicester.

Photograph: A. Newton & Sons.

Fig. 268 (*Right*). The famous cast iron lion designed by Alfred Stevens for Sir Robert Smirke, for surmounting the low guard rails outside the British Museum, in 1852. (These have since been removed.) Castings of this lion are found in cities all over Great Britain, having, for example, been added to the railings outside Ely House, Dover Street, London, as shown in Fig. 120, on page 118. Another view of the lion is shown in Fig. 271 opposite.

Photograph: A. Newton & Sons.

Fig. 269 (*Left*) and Fig. 270 (*Below*). Railings and entrance gates to the British Museum, London, designed by Sir Robert Smirke, 1823 - 1847. It has been suggested that Alfred Stevens was associated in the design of some of the features, including the vases surmounting the main members. Stevens certainly designed the lion for the low guard rails which have since been removed. The railings are cast hollow, and in Fig. 269, the nineteenth rail from the left has been cut by bomb splinters, and as the rail has dropped, this discloses the fixing rod that runs up the centre.

Fig. 271 (*Left*). Another view of the lion designed by Alfred Stevens. See Figs 267 and 268.

Photograph: *A. Newton & Sons*.

Fig. 272 (*Left*). Gates to Hyde Park, Marble Arch, London, erected in 1841. Design here is still under control; these gates display the same restraint in the use of cast iron which was apparent in the opening decades of the century. The other illustrations on this and the opposite page provide a striking contrast, though the gates shown there are only ten years later than those in this illustration.

Reproduced by courtesy of the National Buildings Record.

Fig. 273 (*Below*). Cast iron gates and railings specially designed for the 1851 Exhibition. These are shown in the interior of the Crystal Palace, in Plate I (frontispiece). They now divide Hyde Park from Kensington Gardens. A detailed elevation of the gates is shown opposite.

FIG. 274. The cast iron gates and railings designed for the 1851 Exhibition. The material is beginning to run away with the designer; but order is still preserved, though chaos is obviously not far off. They are a tribute to the technical ability of ironfounders, and an alarming disclosure of the way designers' minds were working, or rather, were not working, for here, commingled in ornate abundance, are all manner of notions copied from other periods. Rococo extravagance and Italianate gaiety are fighting for the mastery: result, confusion.

Fig. 275. One of the cast iron grilles in the openings over the side porches of the Albert Hall, London. Designed by Captain Fowke and General Scott, in 1868. A new orderliness is apparent in this design, but the detail is coarse.

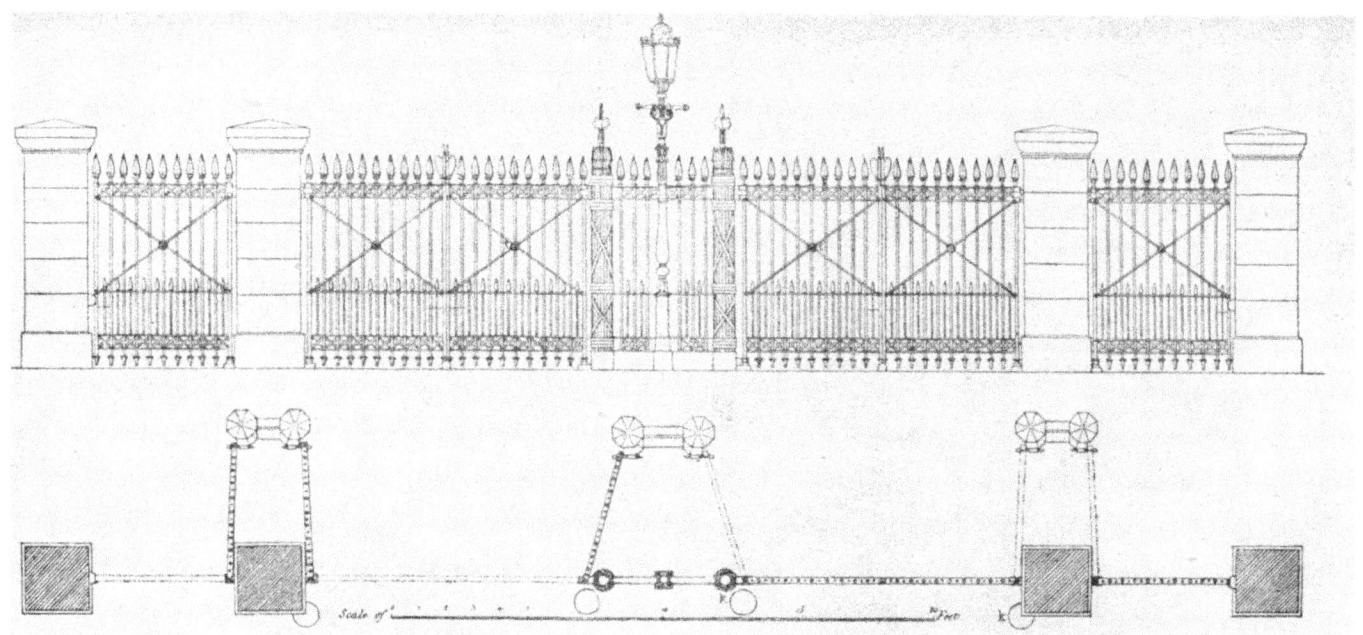

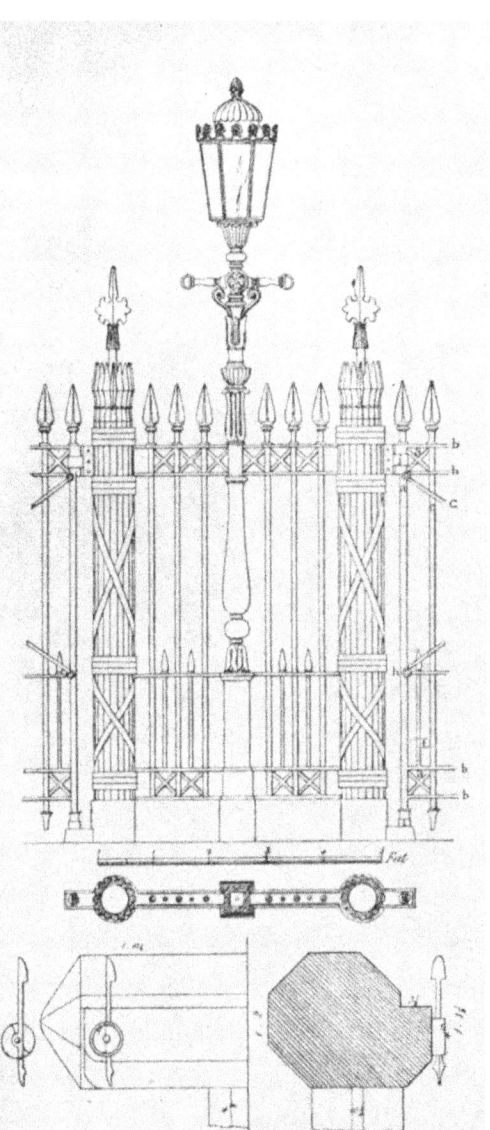

Figs. 276 to 292 inclusive are taken from the edition of L. N. Cottingham's *Smith and Founders Director*, which was published in 1840. The original edition was published in 1823. (See page 116, in Section 2). This book recorded all the best classical sources for cast iron railings, gates, grilles, lamp standards and so forth, that had been produced and used during the first part of the nineteenth century.

Fig. 276 (*Above*). Shows the elevation of the Cumberland Gates at Hyde Park, London.

Fig. 277 (*Left*). Plan and details of the Cumberland Gates.

From L. N. Cottingham's 'Smith and Founders Director'.

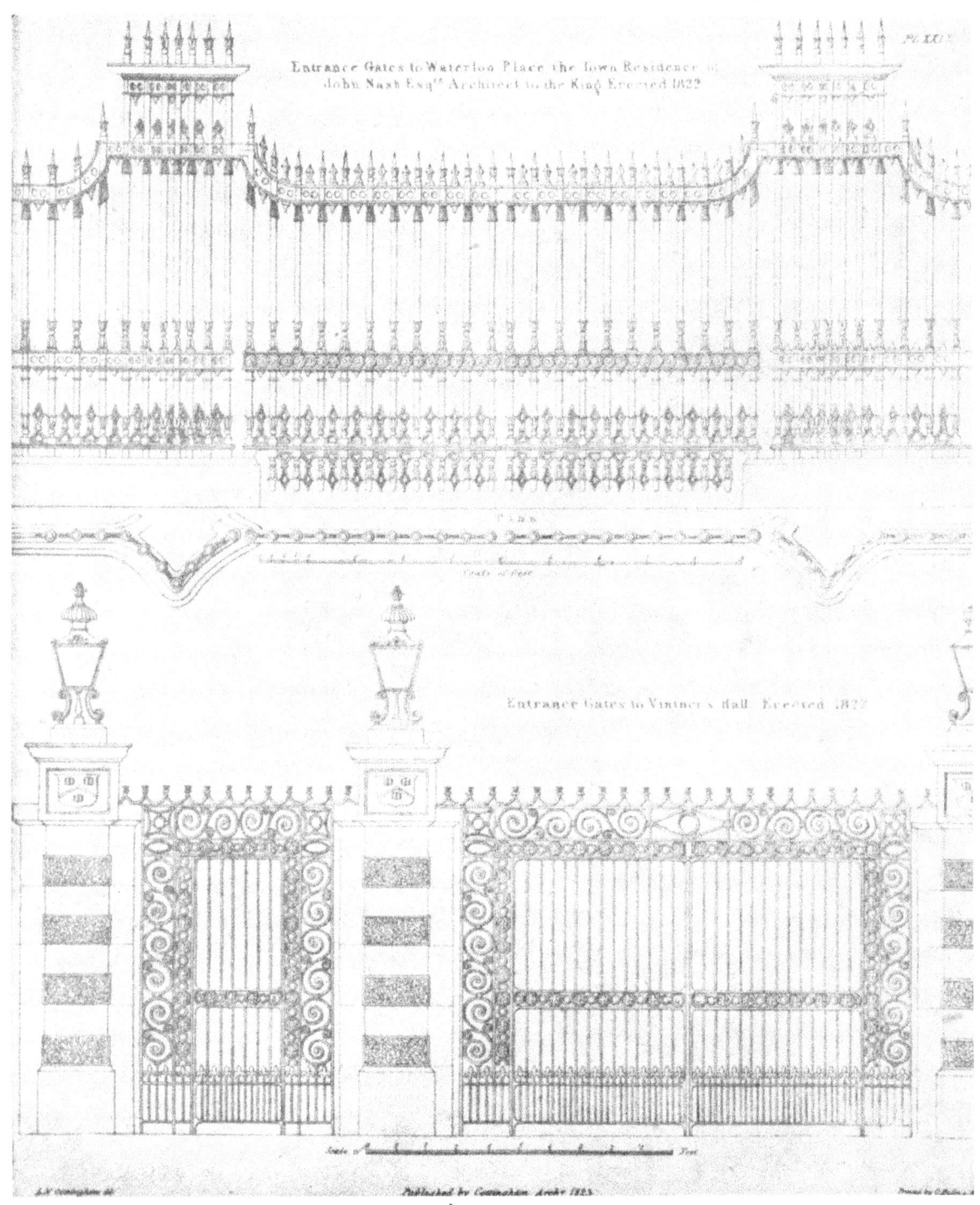

FIG. 278. Above are the entrance gates to Waterloo Place, London, the town house of John Nash, 1822. Below, the entrance gates to Vintners Hall, London.

From L. N. Cottingham's 'Smith and Founders Director'.

Fig. 279. Piers and lamps for gates and palisade fences.
From L. N. Cottingham's 'Smith and Founders Director'.

FIG. 280 (*Above*). Cast iron balusters and spear heads.

FIG. 281 (*Right*). Fanlights, lamps, gate posts and spear heads. Compare these with Figs. 156 to 162 inclusive, on pages 142 to 145.

From L. N. Cottingham's 'Smith and Founders Director'.

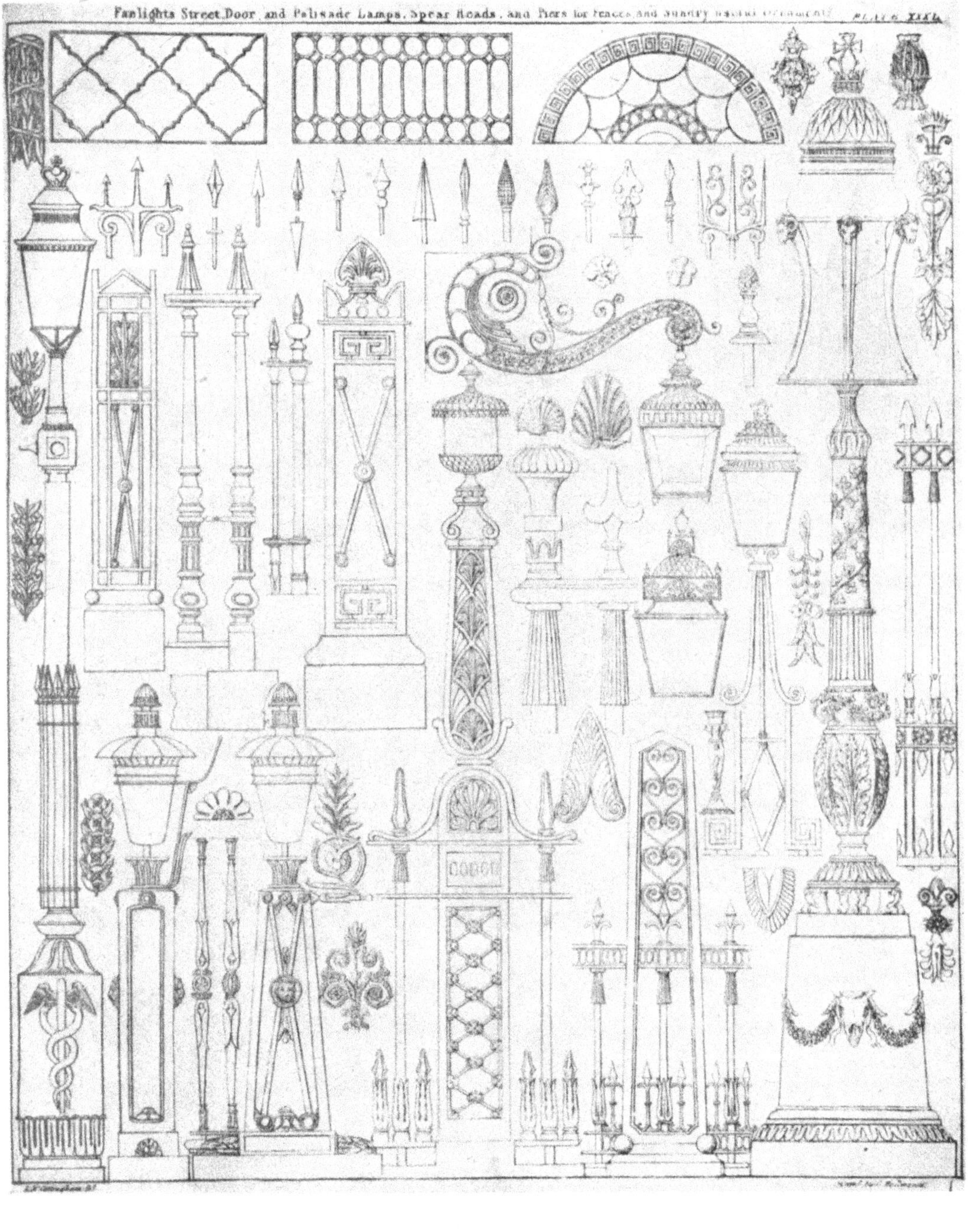

PLATE IV. Some of the hardware exhibits in the Great Exhibition of 1851, showing typical grates of the period. From *Dickinsons Comprehensive Pictures of the Great Exhibition of 1851*, painted by Messrs. Nash, Haghe and Roberts. Published by Dickinson Bros., 114 New Bond Street, 1854.

FIG. 282. Lamp standards and candlesticks.
From L. N. Cottingham's *'Smith and Founders Director'*.

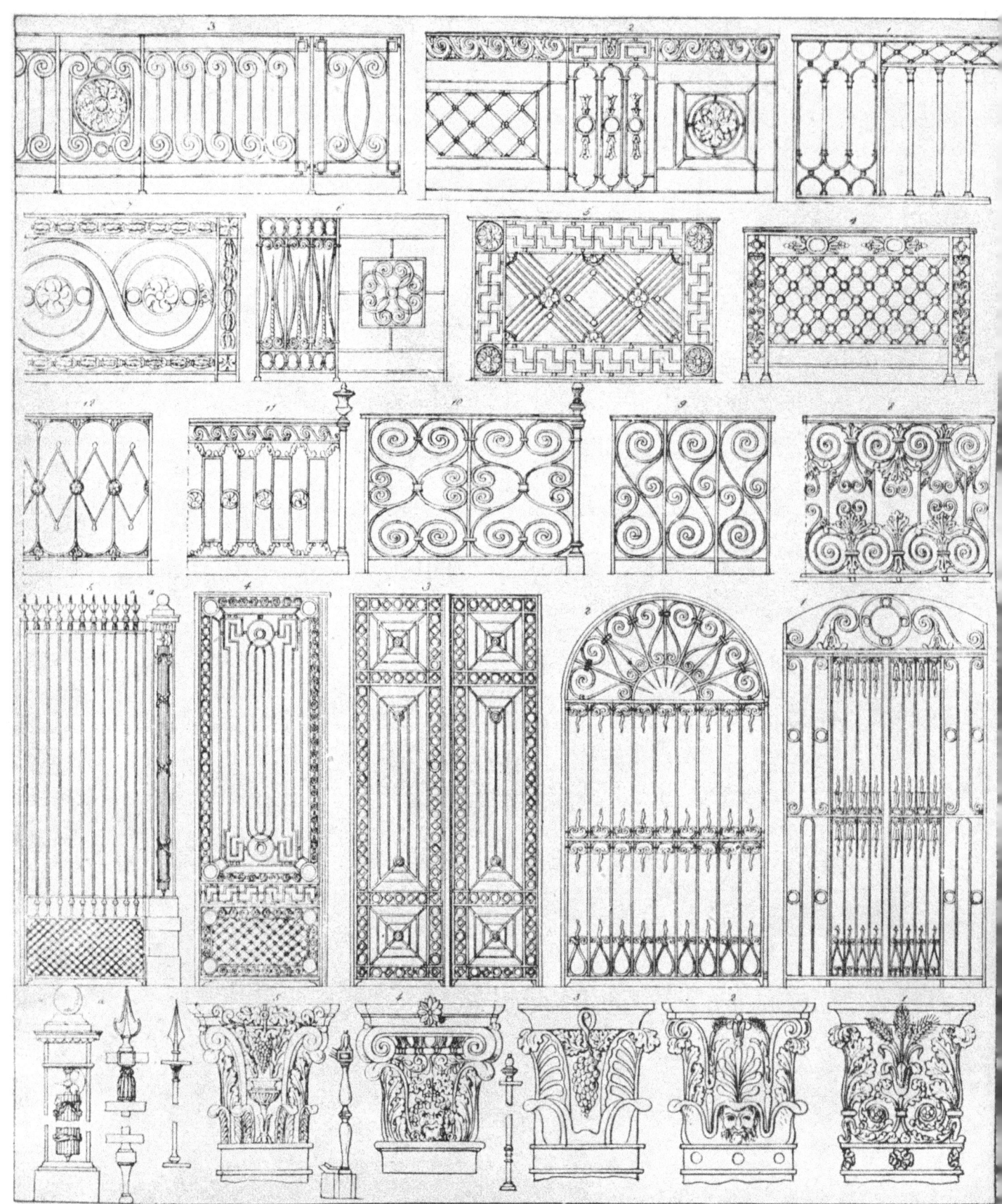

FIG. 283. Balcony railings, gates and capitals.
From L. N. Cottingham's 'Smith and Founders Director'.

Fig. 284. Window guards and balcony railings erected in London.
From L. N. Cottingham's 'Smith and Founders Director'.

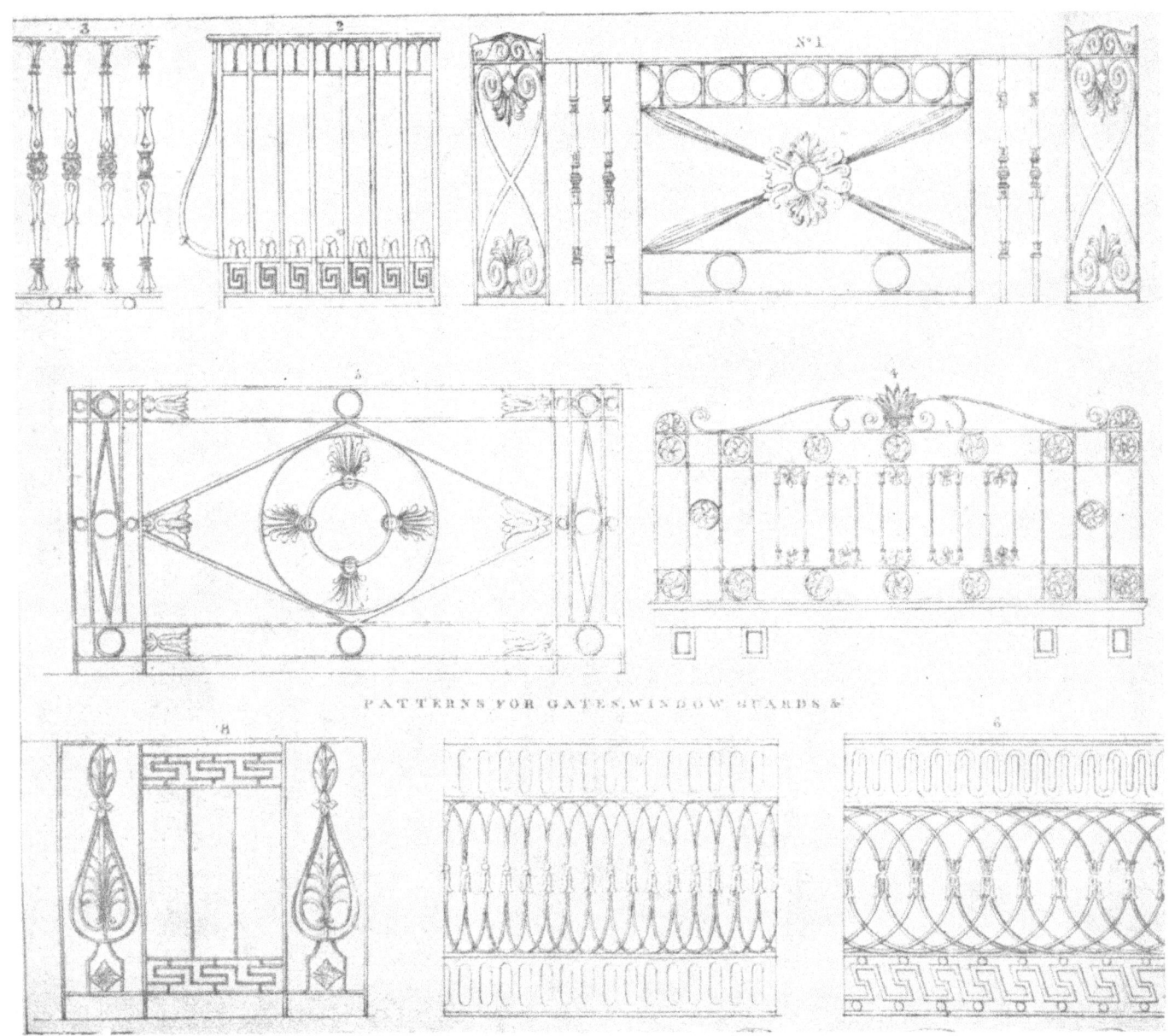

FIG. 285. Window guards and balcony rails.
From L. N. Cottingham's 'Smith and Founders Director'.

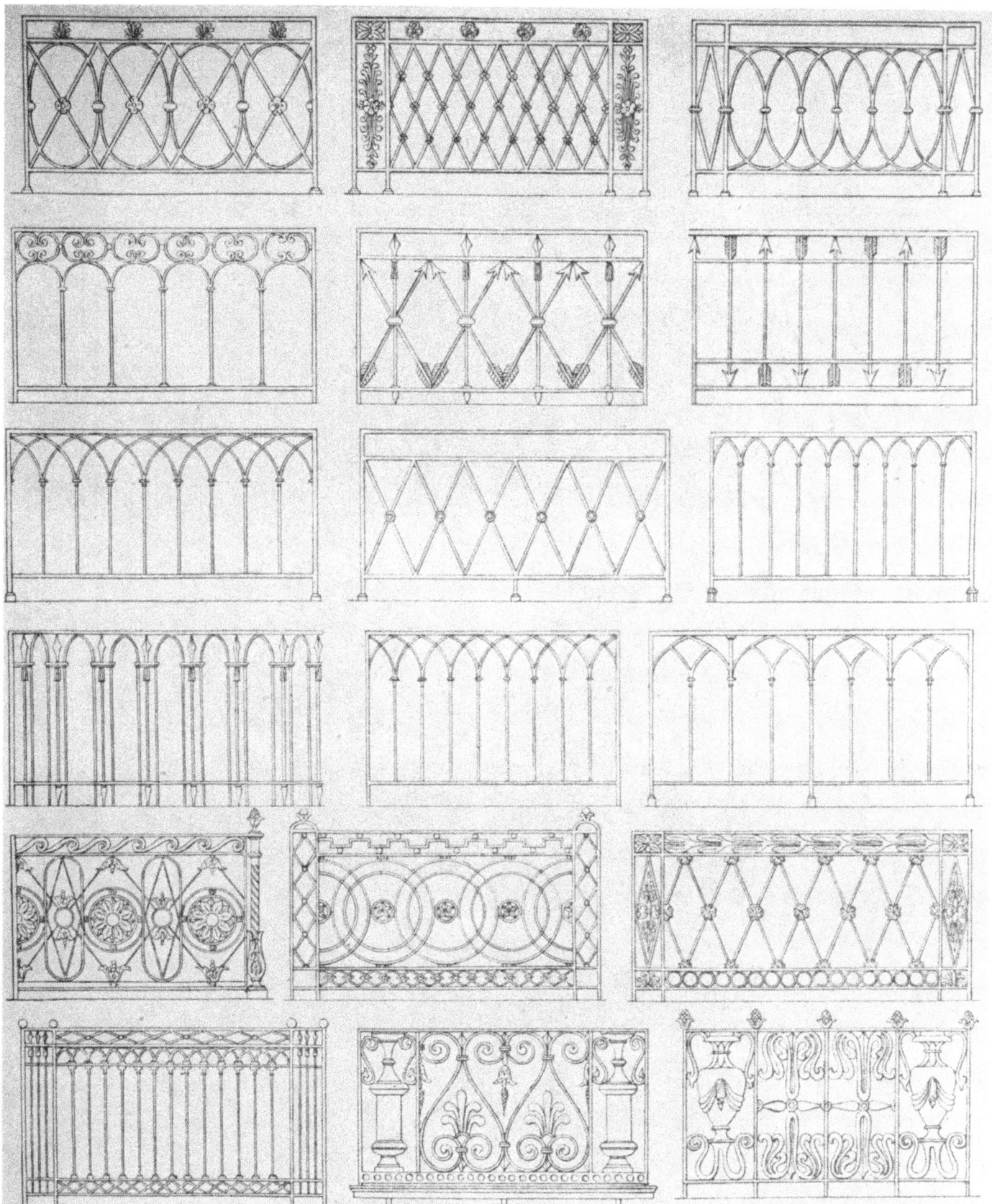

FIG. 286. Balconies and balustrades.
From L. N. Cottingham's 'Smith and Founders Director'.

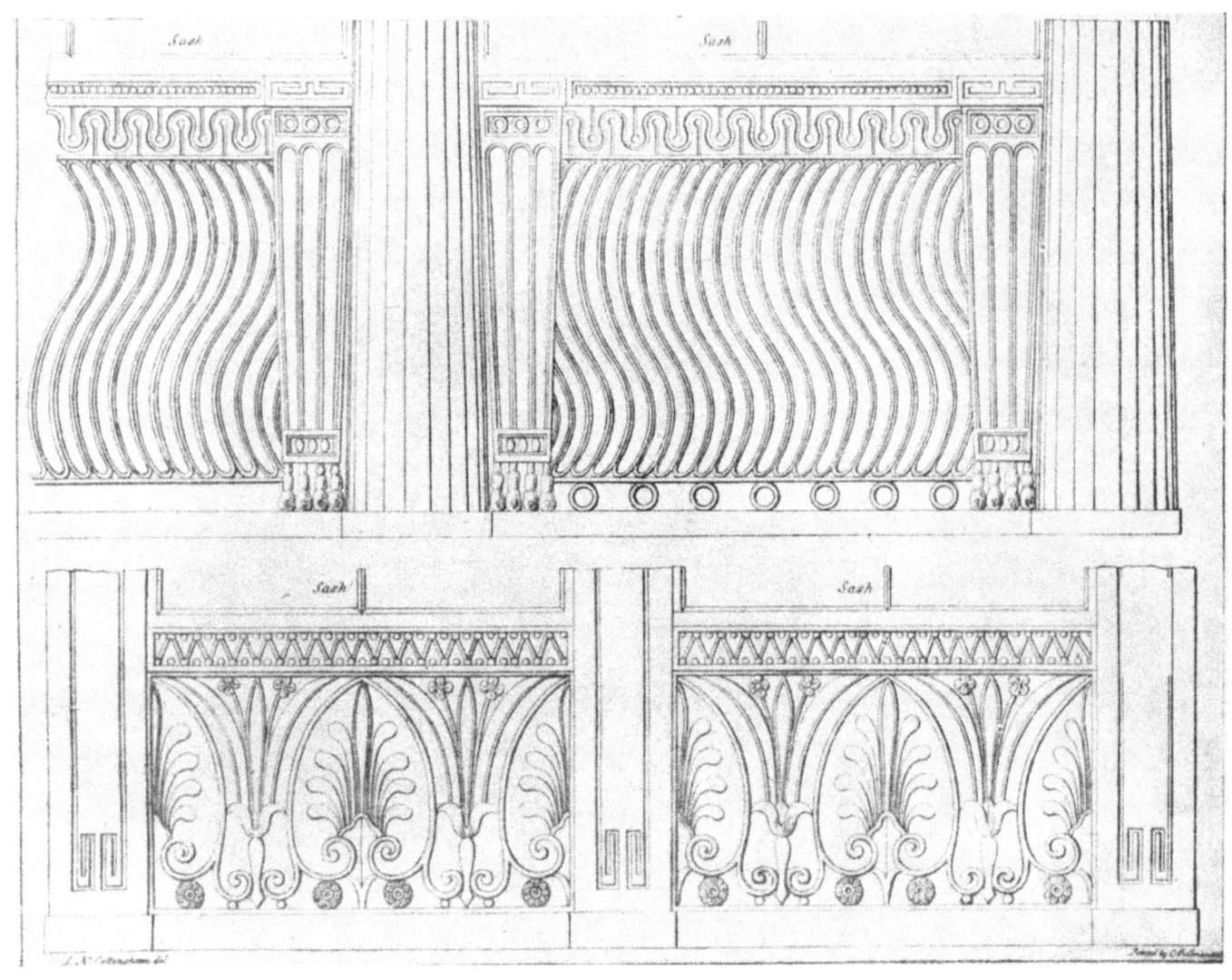

FIG. 287. Guard grilles below shop windows.
From L. N. Cottingham's 'Smith and Founders Director'.

Fig. 288. Guards for shop windows.
From L. N. Cottingham's 'Smith and Founders Director'.

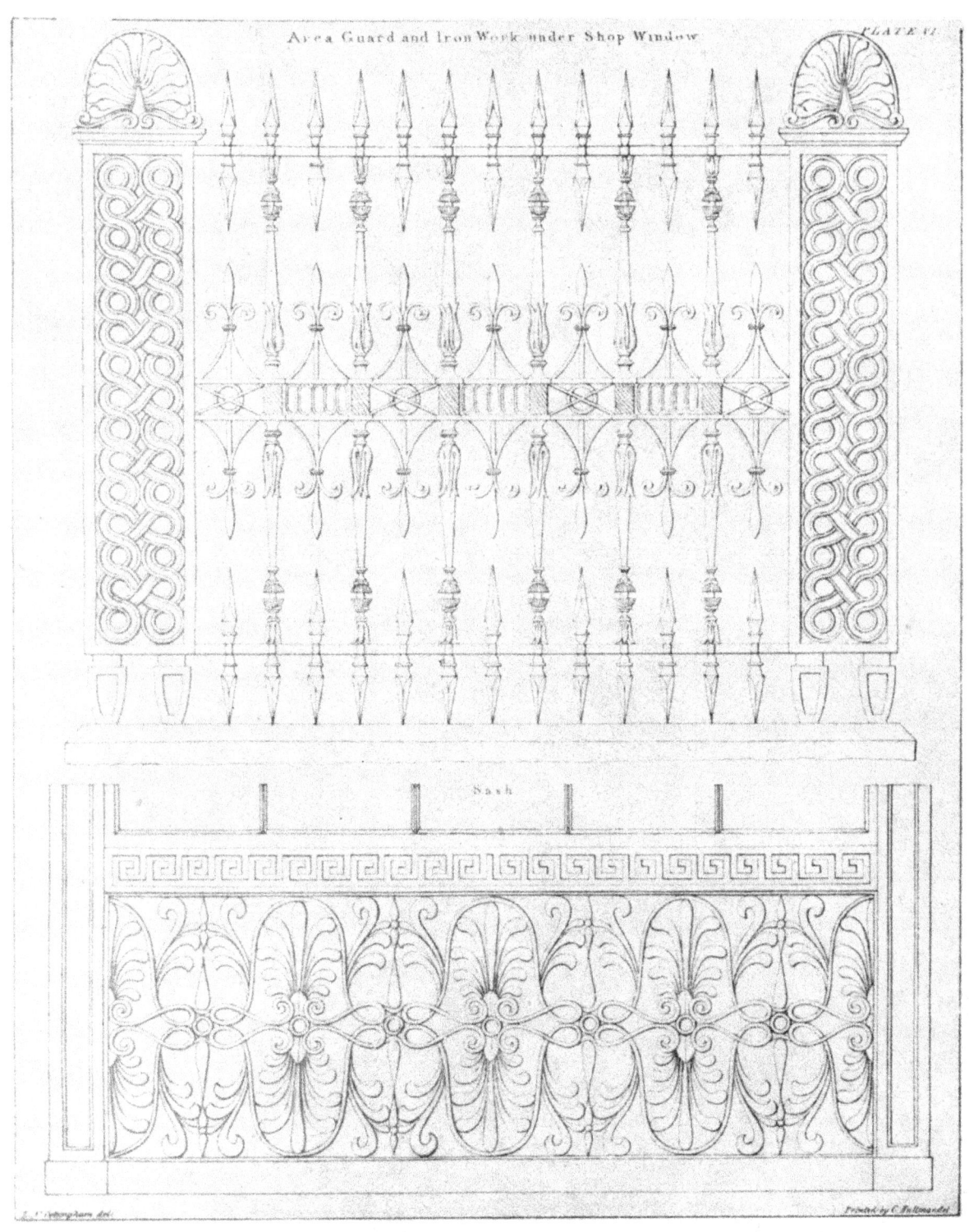

FIG. 289. Area guard and grille beneath shop window.
From L. N. Cottingham's 'Smith and Founders Director'.

FIG. 290. Lamps, newels and balusters for staircases.
From L. N. Cottingham's 'Smith and Founders Director'.

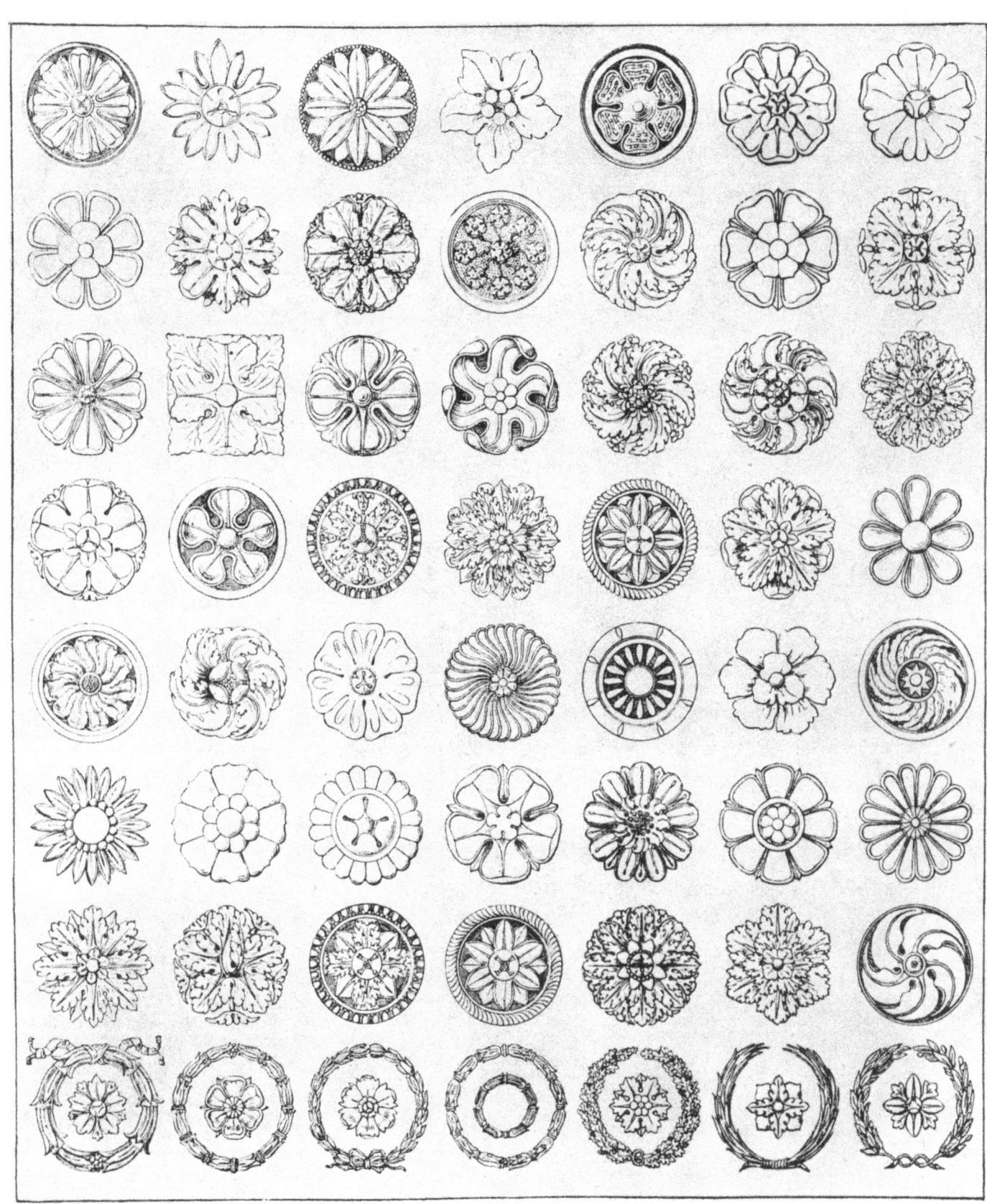

FIG. 291. Rosettes and wreaths.
From L. N. Cottingham's 'Smith and Founders Director'.

FIG. 292. Greek and Roman ornaments.
From L. N. Cottingham's 'Smith and Founders Director'.

SOURCES OF REFERENCE IN SECTION THREE

[1] *Daedalus*, or *Science and the Future*, by J. B. S. Haldane, F.R.S. Kegan Paul, 1924. p. 19.
[2] *Industrial Art Explained*, by John Gloag. George Allen & Unwin Limited. Enlarged and revised edition, 1946. Chapter VI, pp. 108, 109.
[3] *Lives of the Engineers*, by Samuel Smiles. Vol. III. Published by John Murray, Albemarle Street, London, 1863. p. 232.
[4] *Industrial Art Explained*, by John Gloag. George Allen & Unwin Limited. Enlarged and revised edition, 1946, Chapter IV, pp. 67, 68.
[5] *Our Iron Roads, their History, Construction, and Social Influences*, by Frederick S. Williams. Ingram, Cooke & Co., 1852. Chapter X, p. 235.
[6] *Ibid*, Chapter X, pages 221, 222.
[7] "The Strength of Cast-Iron Girder Bridges", by C. S. Chettoe, B.Sc., M.Inst.C.E., N. Davey, B.Sc., M.Inst. C.E., and G. R. Mitchell. A paper given to the Institution of Civil Engineers on May 2nd, 1944.
[8] *Our Iron Roads, their History, Construction, and Social Influences*, by Frederick S. Williams. Ingram, Cooke & Co., 1852. Chapter VIII, p. 178.
[9] *Ibid*, pp. 178, 179.
[10] *On the Application of Cast and Wrought Iron to Building Purposes*, by William Fairbairn, C.E., F.R.S., F.G.S., 3rd edition. Published by Longman, Green, Longman, Roberts & Green, 1864. p. 66.
[11] *Space, Time & Architecture*, by Sigfried Giedion. Published by The Harvard University Press, Cambridge, U.S.A. p. 134.
[12] *Glass in Architecture and Decoration*, by Raymond McGrath and A. C. Frost. Architectural Press, 1937. p. 123.
[13] *History of the Modern Styles of Architecture*, by James Fergusson, D.C.L., F.R.S., etc. 3rd edition, Vol. II. Published by John Murray, Albemarle Street, London, 1902. p. 419.
[14] *Analysis of Ornament*, by Ralph N. Wornum. Chapman & Hall Ltd., 1882. 7th edition. Chapter II, pp. 8-11.
[15] "What is wrong with Cast Iron?", by M. Hartland Thomas, M.A., F.R.I.B.A. *The Official Architect*. Volume VIII, No. 5, p. 238.

SECTION FOUR

SECTION FOUR

CHANGES IN THE INDUSTRY FOLLOWING THE DEVELOPMENT OF NEW USES: 1860 - 1900

UNTIL the end of the eighteenth century in this country all vertical supports in building had been made of wood, brick or masonry, and all beams had been of wood. The work of Boulton and Watt resulted, as we have seen, in the use of cast iron columns as vertical supports and specially designed cast iron beams to take the floor loads. This development was greatly stimulated by the researches of Tredgold, Fairbairn and Hodgkinson, until, by the middle of the nineteenth century, this form of construction was common for factories, warehouses and public buildings of every description. Fairbairn, after his tests and experiments, maintained his faith in cast iron as a structural member when in compression, but advocated the use of wrought iron for members in tension. By 1850, cast and wrought iron were thus not only in common use but were the only ferrous structural materials available for the engineer and architect.

Between 1850 and 1856 Henry Bessemer conducted a series of important experiments in seeking to rid wrought iron of the slag and other impurities which had been responsible for inequalities in the material and unreliability in use. By blowing air through molten pig iron he produced iron in the liquid state and free of impurities, but on solidification it was full of blowholes. By the addition of a pig iron containing a high percentage of manganese—an idea which originated from David Mushet—the blowholes were eliminated, but as this pig iron also contained about five per cent carbon, Bessemer unintentionally re-carburised the iron and produced steel. This alloy of iron and carbon, the latter being in a percentage of about 0·65, was produced in 1856 and was at once recognised as an improvement on wrought iron as a structural material, being harder and stronger, although less ductile. Previously steel had been produced in small special crucibles and of a high carbon quality—in the nature of 0·65 to 1·7 per cent—which was suitable only for cutting-tools and springs. Bessemer's discovery made possible the simple commercial production of steel of a low carbon quality for structural purposes.

It might be thought that such material would have been immediately used in building, but this was not so. The usual troubles attending all new inventions were experienced. The majority of the British ores could not be used in the process, and special ores were imported from Spain. The original Bessemer process—the so-called acid process—could not remove phosphorous from phosphoric ores and its use was thus confined to hematite and low phosphoric ores. Various efforts were made to revise the process, and in 1877 Sydney Thomas and his cousin, Percy Gilchrist, succeeded in developing the Bessemer process so that phosphoric ores—which occurred widely in this country—could be used, and by 1880 steel was produced commercially by this so-called basic process, and though, in building, stanchions were still for many years to be of cast iron, the superiority of the new material for all forms of beams and members in tension was gradually established.

The change-over was surprisingly gradual. The technical text books of the period mostly indicate some doubt about steel and show that it was not in any great use in building, though it was considered to be a new and promising material. These books reflect the general trend of Fairbairn's earlier work (see page 192) in suggesting that if greater scientific control was exercised in the making of cast iron, the material had qualities not then fully realised. In spite of the many purely technical books on the subject, this lack of research and metallurgical knowledge and the need for better control of the foundry processes seem often to be stressed in the various engineering and structural text books of the period. For example, Ewing Matheson says "There are very wide differences of quality in the various kinds of cast iron which are made in Great Britain; and although the experiments of the last thirty years have done much to make these differences better known, they are not so generally appreciated as are those in wrought iron. . . Cast iron is used to a large extent for articles (gas-pipes, short columns, etc.) whose dimensions are determined, not so much by an investigation into the forces they may have to resist, as by purely practical considerations, such as architectural outline, intended long duration under unfavourable conditions, the impossibility of casting large pieces with a small thickness of iron, etc., most of which favour a consumption of metal often greatly in excess of what would be needed if only the capability (which is very great in cast iron) of resting [? resisting] compressive strains were considered. For such purposes the most rudimentary criticism of the quality of cast iron will be sufficient, *i.e.* that it is stronger than stone or clay, that it can assume almost every imaginable shape, and that it lasts very long under a coat of oil paint. These qualities are not wanting in any substance which is sold under the name of cast iron. But there are many articles or structures made of cast iron which have to support great loads, and to resist very considerable forces, under circumstances where no excess or waste of metal is required. Such are columns of great height in proportion to their diameter (almost all columns are here comprised) where the quiescent central pressure produces a transverse strain besides the compression; such, also, are girders where the transverse strains are direct; and such, again, are articles subjected to powerful vibrations and concussions which altogether evade calculations of an ordinary kind. In these cases the intensity of molecular cohesion and the elasticity possessed by good cast iron become of the greatest importance".[1]

These are general remarks on the material, but in dealing with the building uses of cast iron the author indicates the general lack of control and the fact not one, but a series of materials with different characteristics according to its composition, was not indisputably demonstrated and generally recognised until the results of the research work conducted in the last few decades of the present century had become available.

Matheson's views on the value of steel and the likelihood of its increased uses compared with wrought and cast iron are interesting. Writing in 1877, he says: "Up to the present time steel has not been rolled into so many different sections as iron; and, as the same rolls are not suitable for the two metals, the choice of sections is limited, unless the quantity of a new kind which is required is sufficient to induce the manufacturer to make a set of rolls costing from £40 to £100. Both L and T sections of all sizes up to 4 in. by 4 in. are now made, but beyond this limit the sections for which rolls exist are few in number, and, as in the case with iron, cost more per ton than the smaller sizes... With regard to the strength of steel, it may be stated generally that it is in most respects twice as strong as wrought iron, and that the price for

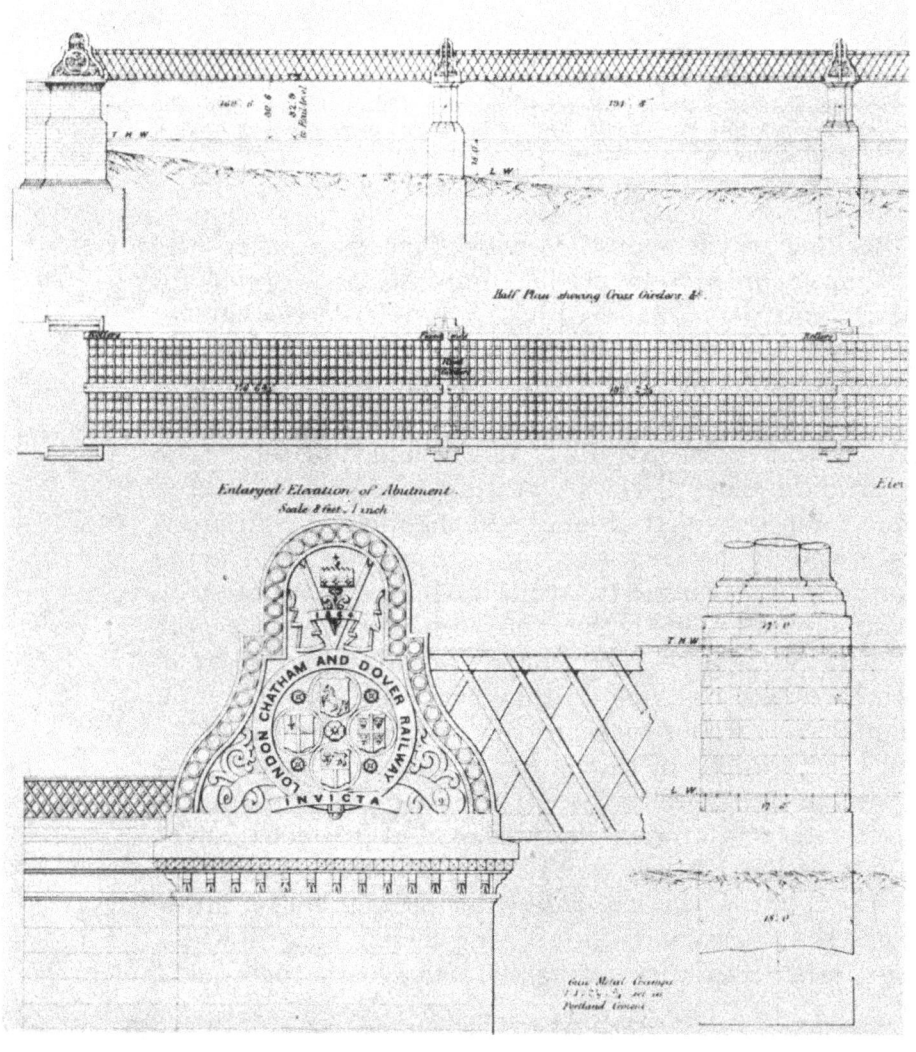

Fig. 294. Railway bridge over the Thames at Blackfriars, London, built originally for the London, Chatham and Dover Railway, and designed by Joseph Cubitt and F. T. Turner, Engineers. Each pier consists of a group of four massive cast iron columns, filled solid, with a total diameter of 14 ft. Ornamental castings mask the junction of the main girders and the column supports. This illustration, and the details shown in Fig. 295 opposite, are from *A Record of the Progress of Modern Engineering*, edited by William Humber, 1863-1864.

Reproduced by courtesy of the Royal Institute of British Architects.

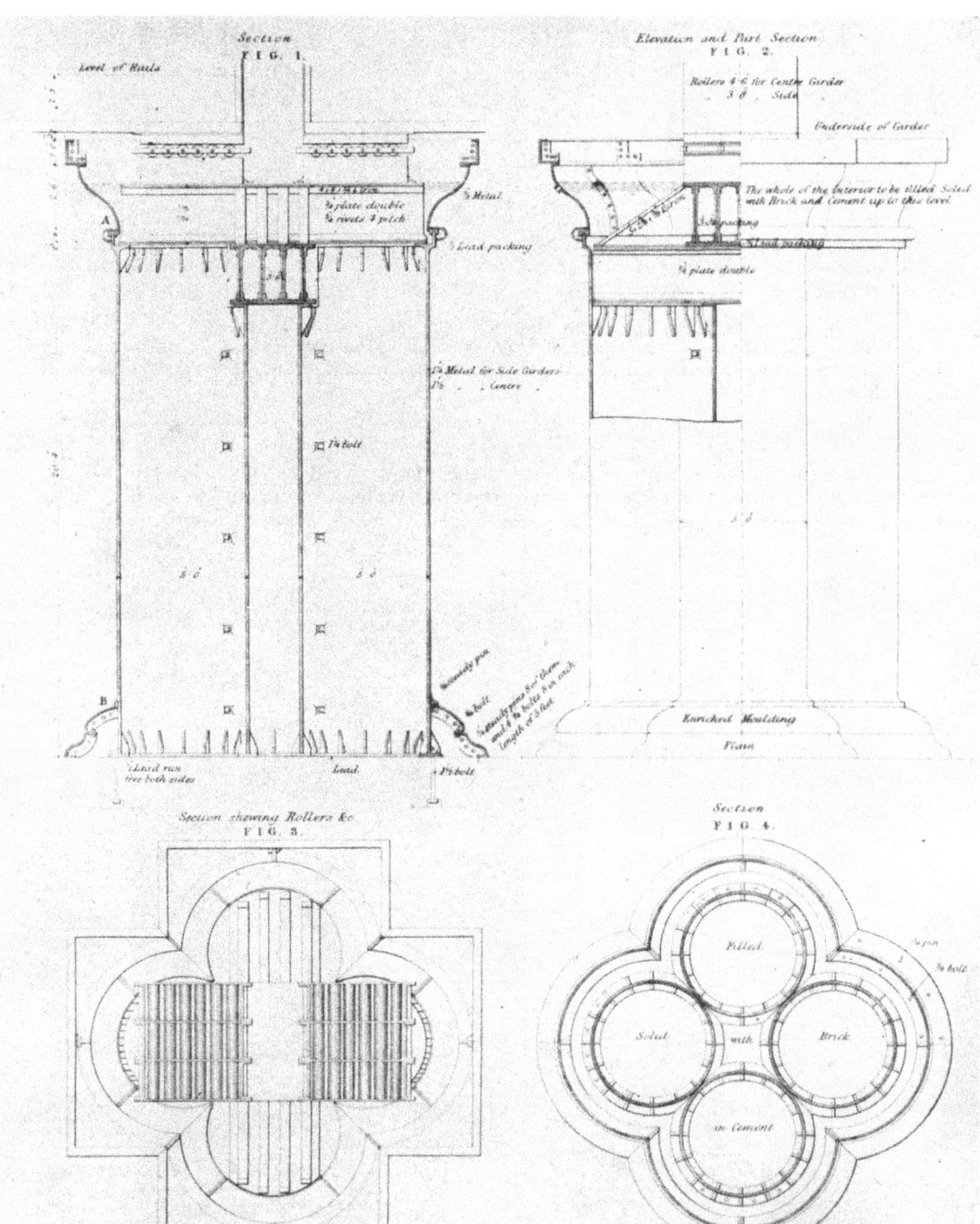

FIG. 295. Details of the columns, capital and base of the Blackfriars Railway Bridge shown in Fig. 294.

Reproduced by courtesy of the Royal Institute of British Architects.

ordinary bars and plates has been in about the same ratio. There would appear, therefore, to have been no economy in using steel in cases where strength is the only consideration, and it must depend on the importance which attaches to other qualities whether the application of steel is preferable to that of wrought iron. . . The conclusion may be drawn that steel can successfully replace wrought iron only in parts of the most simple shape, and that even then its application is expedient only in very large structures. As, however, there is a growing disposition in this and in other countries to construct bridges and roofs of very large span, it cannot be doubted that the real capabilities of steel will be tested in actual structures on a greater scale than has hitherto been attempted. The importance of the subject is fully appreciated, and during the last few years engineers and steel manufacturers

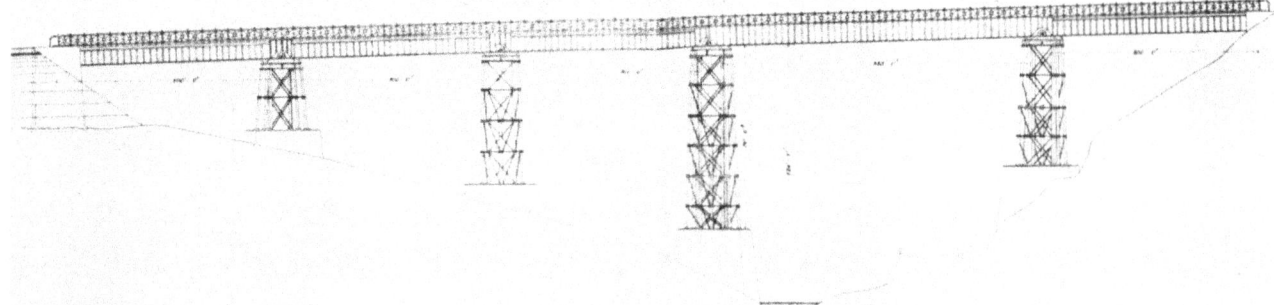

FIG. 296 (*Above*). The Maquis Viaduct on the Santiago and Valparaiso Railway, Chile; made in Britain and designed by W. Lloyd, Engineer. The piers consist of cast iron columns 8, 10 and 12 ins. in diameter, the material varying from ⅝ in. to 1 in. in thickness.

FIG. 297 (*Below*). Plans and elevation of one pier of the Maquis Viaduct, with details of the cast iron columns and joists From *A Record of the Progress of Modern Engineering*, edited by William Humber, 1863 - 1864.
Reproduced by courtesy of the Royal Institute of British Architects.

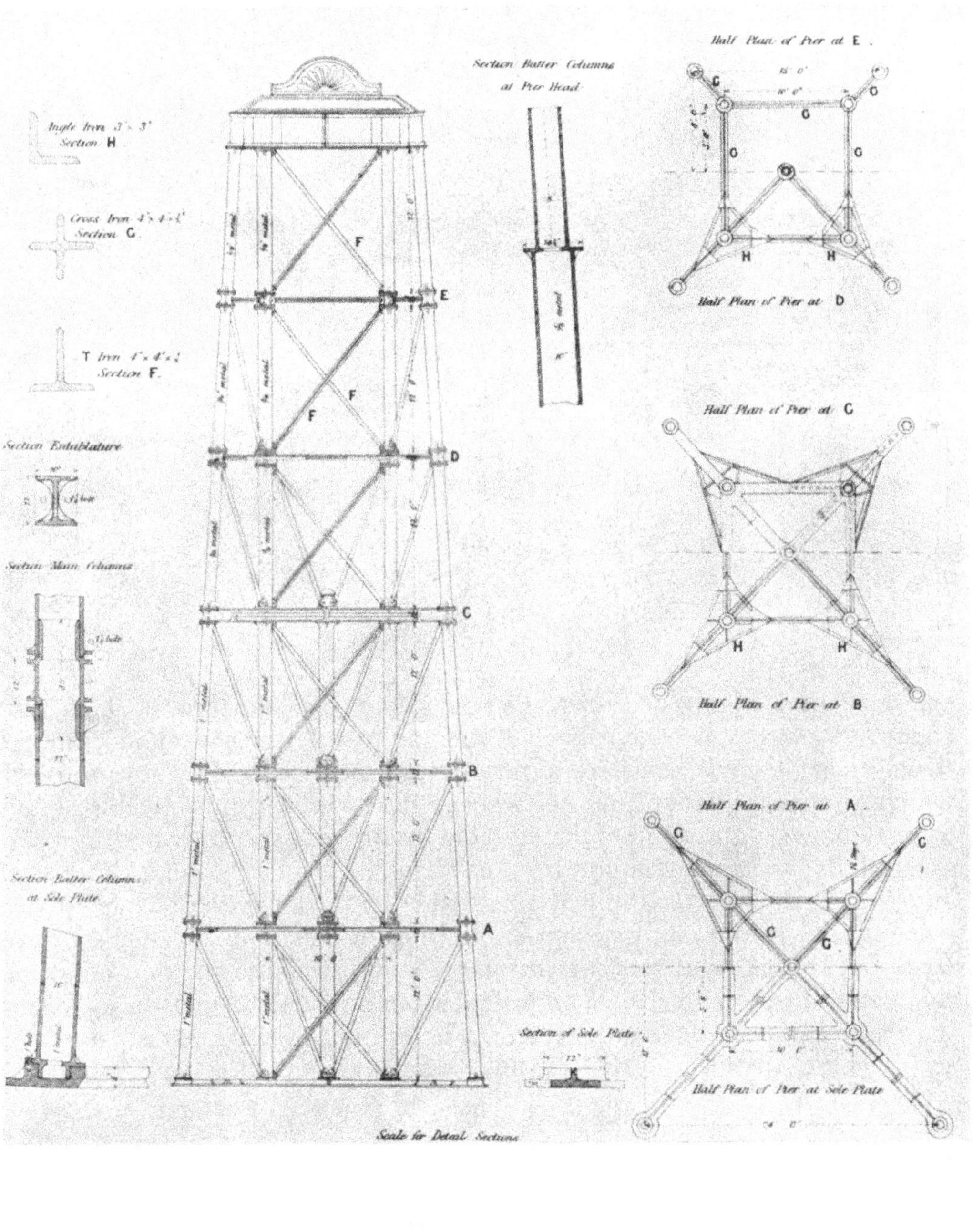

244

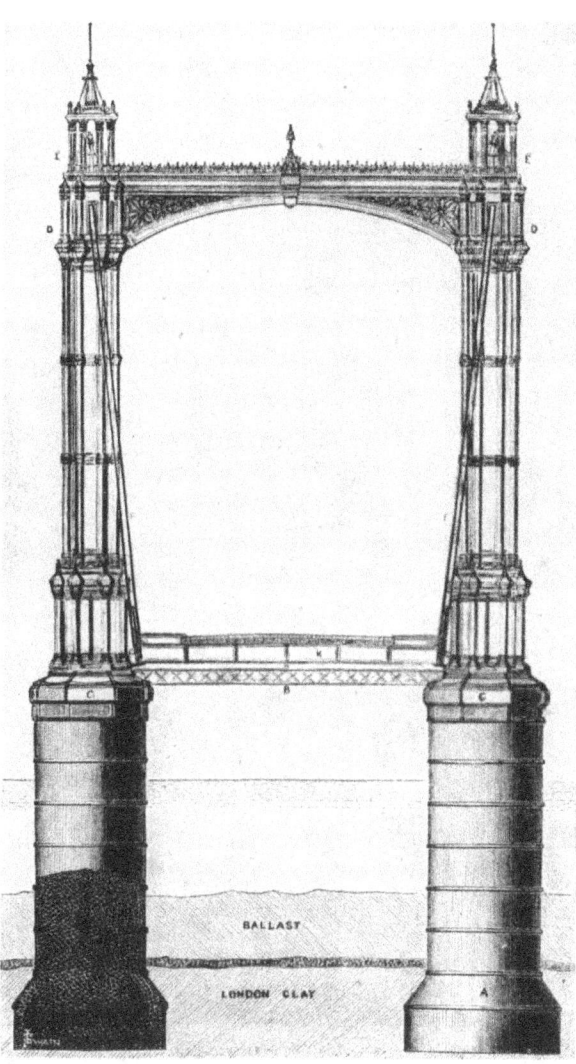

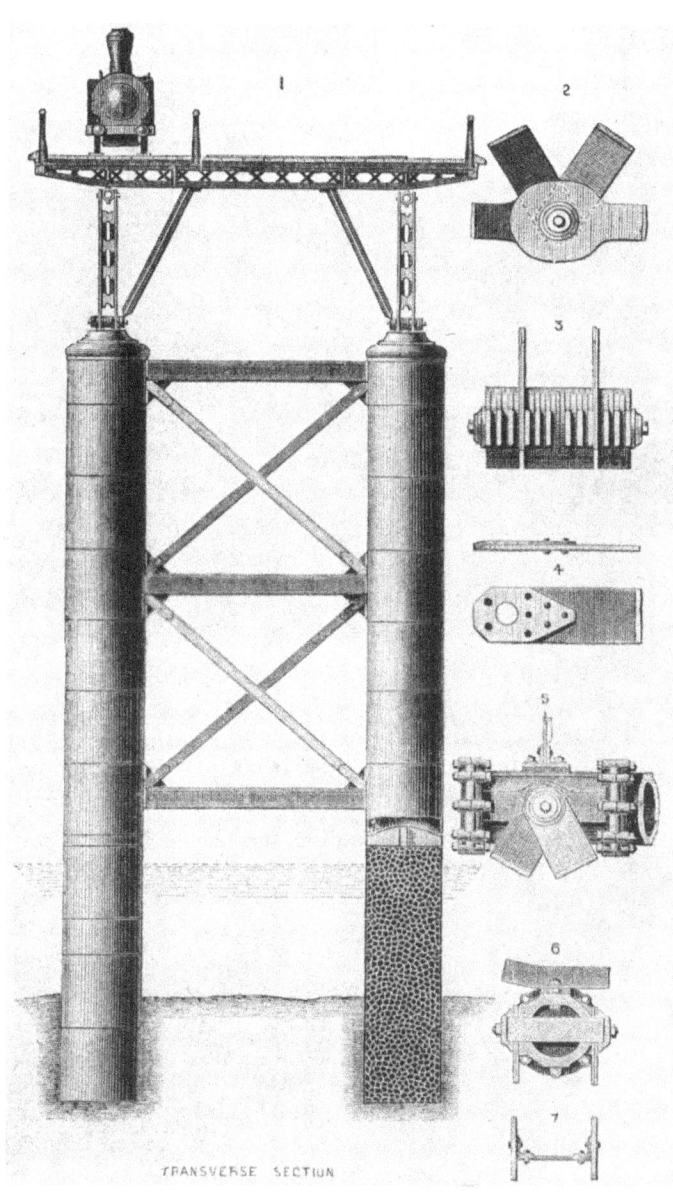

FIG. 298 (*Above*). The Albert Suspension Bridge over the Thames at Chelsea, London. Designed by R. M. Ordish in 1872. Length 710 ft., centre span 400 ft., 41 ft. between the parapets. Each river pier consists of two cast iron columns 21 ft. in diameter and filled with concrete. The columns are capped with an octagonal base for the towers, and the towers are made entirely of cast iron. Total weight of the material in the piers is 274 tons, and in the towers and toll houses 547 tons.

FIG. 299 (*Right*). Road and railway bridge over the River Bremer, Queensland, Australia. Designed by Sir Charles Fox & Sons, *circa* 1870. The river piers are cast iron columns, 6 ft. 6 ins. in diameter, filled with concrete and carrying a structure mainly of wrought iron.

have conducted a series of experiments with a view of obtaining such reliable information as will show how steel may be best applied in the many situations where iron has hitherto been the only material available. The marked progress that has been made since 1870 in the processes of steel-making, renders it extremely probable that the remarks made here as to the present position which steel occupies may be rendered obsolete if, as is hoped, and as appears probable, some of the objections to its adoption are removed and its use extended".[3]

245

Fig. 300 (*Above*). A contemporary drawing of the Trent Bridge, Nottingham, designed by M. C. Tarbotton, M.INST.C.E., 1871. The bridge consists of three 100 ft. spans, the width between the parapets is 40 ft., and each arch has eight cast iron ribs, 3 ft. deep at the springing point, and 2 ft. 6 ins. deep at the crown. The face ribs are ornamented; and there is a cast iron cornice and balustrade.

Fig. 301 (*Below*). Details of the cast iron work in the Trent bridge. See Fig. 300 above and Fig. 302 opposite.

BRIDGES, PIERS AND HARBOURS.

Just as cast iron introduced a new era into bridge building in the late eighteenth century, so the same material was later of the utmost importance in solving many of the foundation problems for the late nineteenth-century bridges, allowing these to be constructed in otherwise impossible or extremely difficult positions. Iron screw piles were first used in the early eighteen thirties, and later cast iron columns, cylinders and caissons, either independently or filled with concrete or masonry, were to become the normal type of foundation for underwater piers, carrying the bridge girders. Cast iron was still often used for the complete structure of small bridges, but in larger works a considerably increased proportion of wrought iron was intro-

FIG. 302. The Trent Bridge, Nottingham, photographed in November 1932. See Figs. 300 and 301 opposite.
Reproduced by courtesy of the Public Works, Roads and Transport Congress.

duced. Matheson again indicates the trend of thought and the doubts of the period when he wrote: "In the modern construction of girders and bridges, cast iron has, to a great extent, been superseded by wrought iron, because of the superiority which the latter possesses over the former against every kind of strain (tensile, transverse, shearing, vibrating strains, and those caused by the effects of temperature), with the exception of the quiescent compressive strain. Cast iron is much stronger than wrought iron in resistance to compression; and, owing to the existence of a natural crust or skin on its surface, it is not so liable as wrought iron to deterioration by rust, and, if kept painted, with ordinary care will prove the more enduring of the two. In the facility which it possesses of taking all possible shapes—whether required by a

Fig. 293. Blackfriars Bridge, over the Thames, London, designed by Joseph Cubitt, 1863 - 1869. The arches are of cast and wrought iron, with a cast iron cornice and parapet. The piers are of brickwork and granite. Central arch 185 ft. and the intermediate arches 175 ft. and 155 ft.
Reproduced by courtesy of the Public Works, Road and Transport Congress.

prejudice that had by this time grown up in the minds of many engineers and architects. He says: "A large proportion of the iron castings supplied for building purposes in this country are made without any reference whatever to the quality of the iron. It is often the case that an increase in the dimensions is supposed to compensate for inferiority of quality; but it is well known to engineers that certain shapes should be avoided, and that beyond a moderate thickness cast iron becomes spongy or open. . . In the minds of those who are not aware of the great differences of quality that have been here referred to, the numerous accidents and risks that often attend the use of cheap metal have created a prejudice against the use of cast iron altogether; and many people, considering it to be without elasticity, avoid it wherever wrought iron can by any means be used instead. This is the more to be regretted, because cast iron allows infinite variety of shapes, more nearly approaching the exact forms and sizes required by design or strength, than is possible with wrought iron; and for many situations cast iron is even stronger and more enduring". [2]

It seems obvious that specifications of iron work, if they were given, were rarely enforced, and perhaps this careless use of the material without the necessary foundry control and metallurgical knowledge, and the fact that the material could be and was easily cast into any desired shape in hundreds of small foundries all over the country, was responsible for the general but quite unjustified prejudice against its use, and the growing feeling that it was a cheap and unreliable material. Fairbairn's suggestion that cast iron was in minute regard to the forces in a structure or for purposes of ornament—cast iron is also superior to wrought iron; while, weight for weight, it is much cheaper.

"Considering the advantages and disadvantages which thus exist with regard to either material, it is obvious that any arbitrary rule against the use of cast iron in bridge construction would be injudicious; but, no doubt, the general impression of the superiority of wrought iron over cast iron has produced a prejudice against the use of the latter which is shared by some engineers. No inconsiderable skill is required to discriminate between the relative advantages of each material, and these depend upon the circumstances of every case which comes in question. It cannot be wondered at that failures arising from the improper application of cast iron, or the belief that all cast iron is of uniform quality, should have deterred many people from using it at all in bridge and girder construction".[4]

It is clear from the catalogues, technical books and periodicals, that there was at this period a large export of parts of bridges and even whole buildings to all parts of the world. Matheson's book alone, in illustrating some of the

FIG. 303. A conservatory in Glasgow, constructed of cast iron. The straightforward simplicity of forerunners of this type of structure, such as the Palm House at Kew Gardens, has been lost, though some attempt has been made to preserve good proportions.

Fig. 304. The Arcade, Johannesburg, South Africa. Cast iron is used for the shop fronts, balcony, top lights and general structure.

bridges constructed by Handyside & Company of Derby, refers to bridges in South America, Japan, Sweden, Russia, Norway, Australia and India. In all, he gives full descriptions and details of cost of over thirty bridges. Apart from a few small bridges made entirely of cast iron, most of the works show a considerable increase in the use of wrought iron compared with the bridges of the previous half-century. Perhaps the three most interesting examples are the road bridge over the river Trent, at Nottingham, designed by M. O. Tarbotton, in 1871, and containing 696 tons of cast iron and 173 tons of wrought iron; the road and railway bridge over the river Bremer in Eastern Australia, designed by Sir Charles Fox and containing 221 tons of cast iron and 288 tons of wrought iron; and the Albert Suspension Bridge over the river Thames at Chelsea, designed by R. M. Ordish, in 1872, where cast iron was used for the cylindrical piers, which were filled with concrete, and for the Gothic towers and toll houses. The cast iron used totalled 821 tons and the wrought iron 619 tons. (Fig. 298, page 245.)

BUILDINGS.

Just as wrought iron had replaced cast iron beams and girders in bridge work, so a similar substitution took place in building work generally, and for the same reasons. The cast iron stanchion or column still survived, and the same material was used more than previously in this country for the infilling or apron pieces between the main masonry piers of a façade. But the chief use for cast iron was now for pavilions, conservatories, palm-houses, shelters, bandstands and arcades, and these, like the bridges, were often for the export market. It was a new and considerable development of the architectural use of the material. The simplicity of the work of Paxton and Decimus Burton was unfortunately forgotten, and engineers and architects frequently produced designs that were really conceived for carved marble or terra cotta, although they were carried out in cast iron. Ornament, as Wornum had recorded in his *Analysis*, was becoming an end in itself, except in some outstanding engineering works.

A typical conservatory of this period was erected near London for Mr. Henry Bessemer and designed by Messrs. Banks & Barry, in 1868. It was cruciform in plan with a central dome of 21 feet span, and, with the exception of the outer walls and the wrought iron ribs in the dome, was entirely of cast iron. In 1860 Sir George Gilbert Scott, R.A., designed a winter garden for the Infirmary at Leeds, the engineer being R. M. Ordish. (Fig. 309.) The building was 151 feet long, 63 feet wide and 60 feet high, and, save for the clerestory lattice girders, was wholly of cast iron. A similar building,

FIG. 305. Winter Garden, designed for the gardens of the Royal Horticultural Society, London, by Captain Fowke, R.E., 1860. Length 210 ft., central aisle 45 ft. wide, height 71 ft., columns 8 ins. in diameter. The total weight of cast iron used in the structure is 175 tons, with 70 tons of wrought iron.

Fig. 306. A kiosk for erection in India made entirely of cast iron from the design of Owen Jones, and R. M. Ordish, *circa* 1870. The structure was 80 ft. long, 40 ft. wide and 42 ft. high.

a kiosk in India, was designed by Owen Jones, author of the *Grammar of Ornament*, with R. M. Ordish as engineer. A much simpler type of winter garden or conservatory, uninfluenced in design by any excessive regard for traditional styles, had been built a little earlier, in 1860, for the Royal Horticultural Society in London, by Captain Fowke. It was 210 feet long, with a centre aisle 45 feet wide and a total height of nearly 72 feet.

Nearly all the new railway stations of this period used cast iron columns as main supports, but generally the material was used only for small roof members in compression. An exception was the Dutch-Rhenish Railway Station at Amsterdam, erected by Messrs. Handyside of Derby in 1863, in which much of the roof, as well as the piers, were in cast iron. Cast iron louvres, acting as ridge purlins, provided ventilation. (Figs. 313 to 316.)

One of the last of the great buildings in which a large amount of cast iron was used for structural purposes, was the terminus of the London and North Western and Great Western Joint Railways, at Woodside, Birkenhead. Here the cast iron columns are 21 feet high and the height from the top of the columns to the springing of the roof is 12 feet, 9 inches. Below the track level the columns descend to a depth of 2 feet $3\frac{1}{4}$ inches, and they are bedded on to plates with concrete filling between them, this plated filling occupying a further depth of 5 feet. The original drawings, dated 1876, show all the details of these cast columns. Pseudo-classical capitals are used, showing the contemporary respect for Gothic ornament, and these support cast Gothic cantilevers, representing really considerable sections of cast iron work. The decorative character of these cantilevers has a genuine functional significance, because the castings are so devised that the cantilevers furnish all the necessary support and acquire all the requisite strength without solidity. The decoration, which accords with a series of formalised Gothic perforations in the middle, is really practical. (Figs. 317 to 319.)

The bold, realistic use of a convenient material in Woodside Station, bears out the critical appreciation of these Victorian buildings made by Howard Robertson in his book *Architecture Arising*. He suggests that "some of the older railway stations in England might restore to present-day engineers a not unwholesome humility, for their great steel and iron vaults are often things of beautiful shape and extraordinary daring, unnoticed to-day because they are both begrimed and familiar". He believes that "one thing that makes them satisfactory is that they were built as it were in the stride of industry of the day, to serve a purpose in the building programme, and if sometimes it happens that they were unwisely decorated, it was all in a good cause. No one took them as much more than what in fact they represented, a very competent effort to solve a problem. We are not doing very much better to-day than the Victorians, though we have more alternatives, some of which have still to stand the test of time".[5] The decoration that we have outgrown was as typically Victorian as the boldness that so often distinguished the design of large scale buildings in the middle years of the last century.

An interesting development in America at this time, which appears to have received little consideration in this country, was the use of cast iron plates as a roofing material. As early as the middle of the fifteenth century, cast iron pantiles were used to roof the buildings of the famous Chinese temple at Tai Shan, and are still in service. The great dome of the Capitol at Washington has cast iron plates which have been in position since 1870; and three years later the Hinkley Foundry of San Francisco, California, cast all the parts for a complete cast iron building, including the roof covering. This was the only building satisfactorily to withstand the disastrous fire that followed the earthquake of 1906. About the same time, 1872, Messrs. H. B. Smith built a factory at Smithville, New Jersey, which was roofed with cast iron plates, which required no expenditure on maintenance for over sixty years.

In 1935, after an investigation of these old roofs, the United States Pipe & Foundry Company of Burlington, New Jersey, patented a new type of cast iron plate as a roofing material for industrial buildings. The plate and ridge pieces were ingeniously designed with flanges and lugs to facilitate erection, and large areas have been successfully roofed at the works of the Otis Elevator Company, Harrison, N.J., and at other industrial plants.

Another large scale use of the material, which gave legitimate scope to ornate decoration, was the seaside pier. Richard Sheppard, in his instructive work, *Cast Iron in Building*, has expressed the view that "Most of our piers are highly characteristic and have a genius of design all their own; they were the Victorian equivalent of the folk festival, and the quality of their decoration

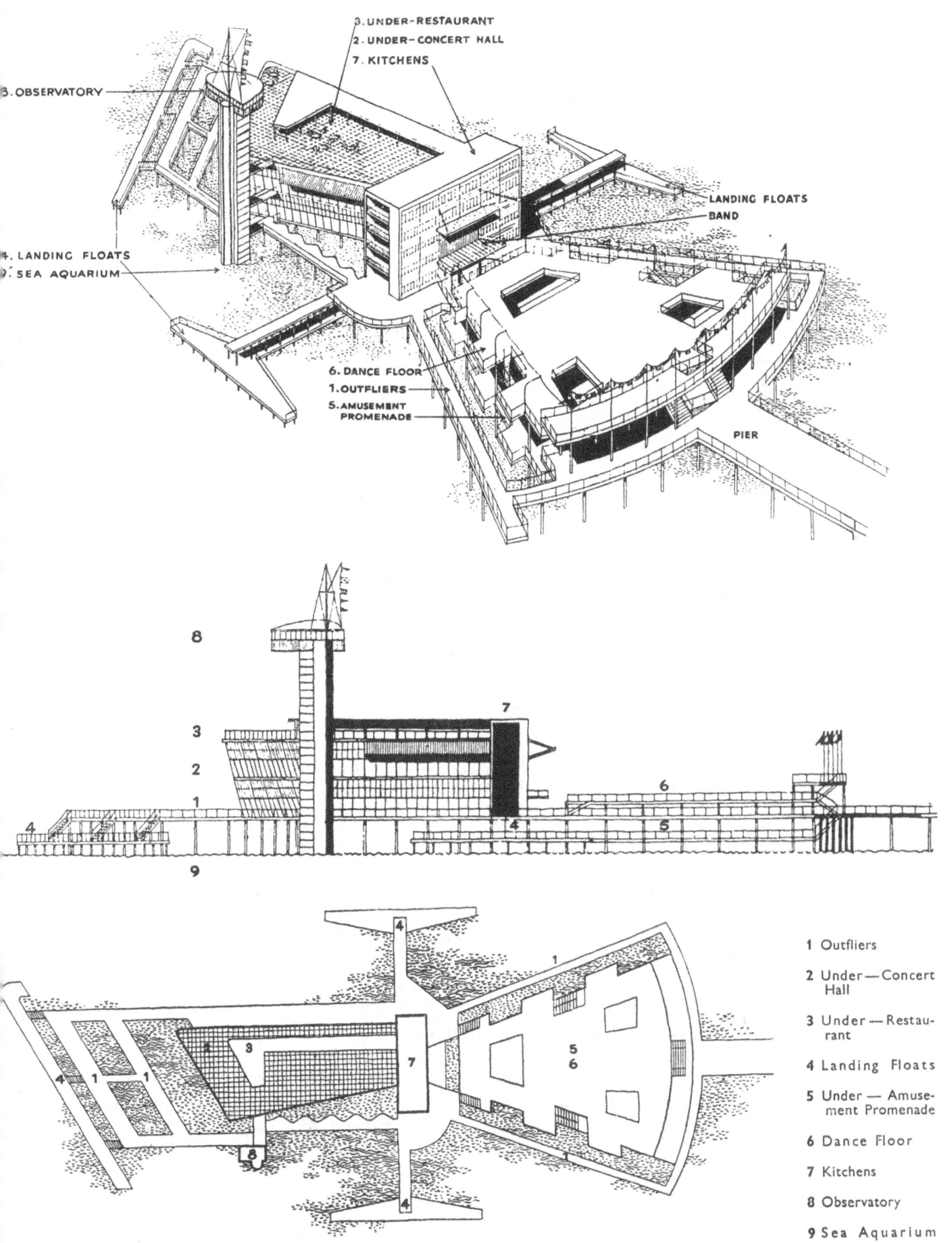

FIG. 307. A design for a pier to be executed in cast iron, by Richard Sheppard, F.R.I.B.A.
Reproduced from 'Cast Iron in Building', by Richard Sheppard, by courtesy of the author.

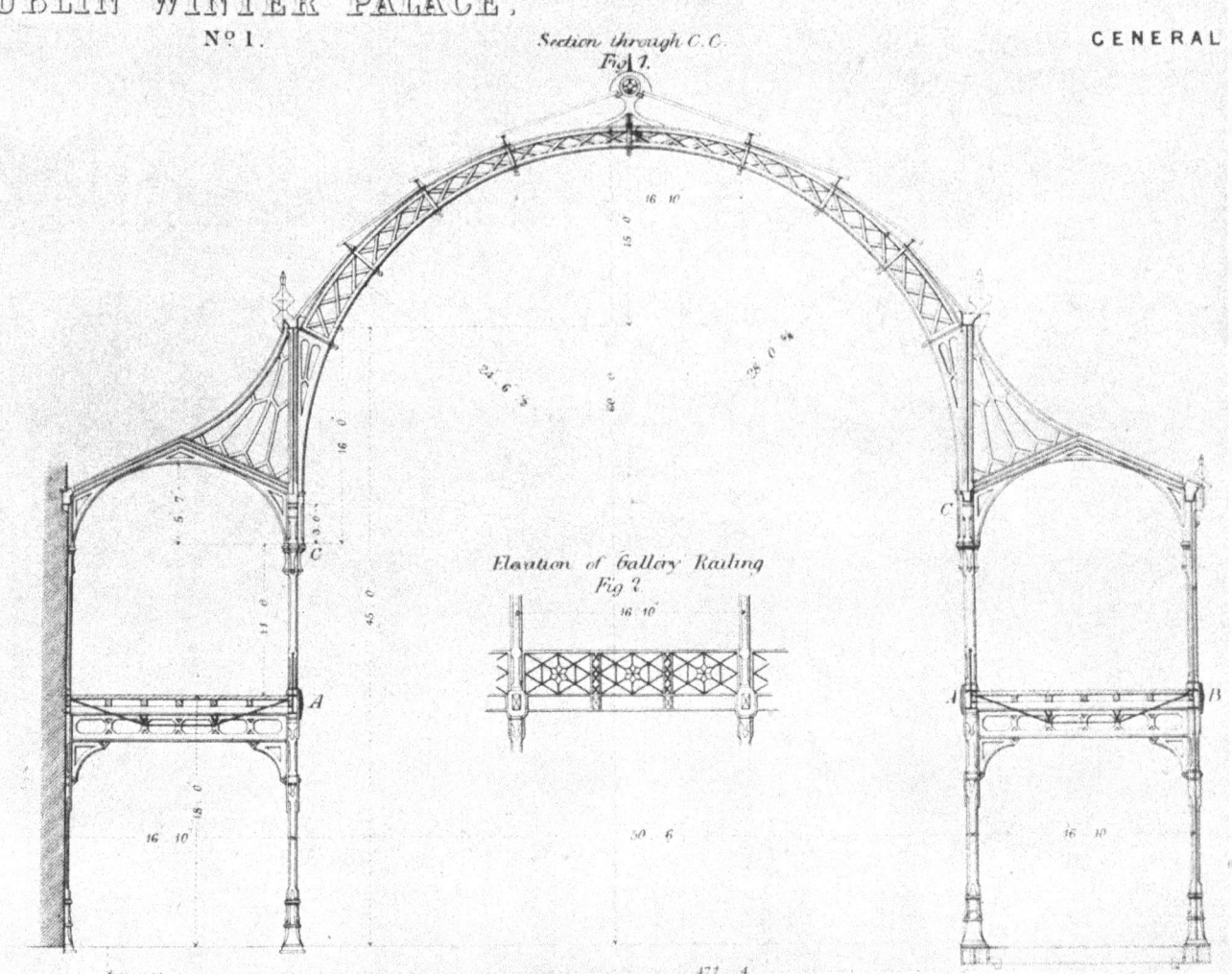

Fig. 308. Winter Palace at Dublin, designed by A. G. Jones, Architect, R. M. Ordish, and Le Feuvre, Engineers, 1865. Cast iron is used for the columns, and for the girders to the side aisles and open-work buttresses and purlins. From *A Record of the Progress of Modern Engineering*, second edition, edited by William Humber, 1864 - 1865. *Reproduced by courtesy of the Royal Institute of British Architects.*

is expressive; it is bold, coarse and vigorous. The iron is used flamboyantly in panels and columns and crockets and gables, for windows and doors, for turnstiles and slot machines. Buried under the paint and obscured by later accretions the social investigator can discover a record of the habits, characteristics and amusements of mid-Victorian England".[6]

Towns on the coast grew enormously, and piers, pavilions, bandstands and shelters were erected in their hundreds, rarely designed by an architect, unless the local surveyor took some small part in giving the foundry an idea of what was wanted. Usually the foundry had a stock line, and such prefabricated components of the architecture of pleasure were often "bought off the peg". To-day they are almost museum pieces with their mixture of ornate styles and wealth of intricate detail. But they still exist, their originators having been, at any rate, practical men who realised that cast iron was less affected by the salt-laden atmosphere of the seaside and by salt water than any other structural material.

Fig. 309. Winter Garden, the Infirmary, Leeds. Designed by Sir George Gilbert Scott, R.A., 1866: Engineer, R. M. Ordish. Length 151 ft., central aisle 37 ft. wide by 60 ft. high, side aisles 13 ft. wide, columns 10 ins. in diameter. It is constructed entirely of cast iron save for the lattice girders over the columns. Total weight of cast iron 108 tons; total weight of wrought iron 130 tons.

Fig. 310. Structural columns, typical of those used in railway architecture and such buildings as mills, warehouses, hotels and office blocks. The classical purity of the early nineteenth century has almost passed away; but some degree of elegance is retained in the elongated versions of Tuscan and Doric columns, particularly in the example on the extreme right and in the fluted column third from right, and the plain column fifth from right. The Greek Doric specimen would never have been tolerated by the architects who used cast iron in the opening decades of the century; and the specimens on the extreme left would have horrified them. These examples, and those shown on the opposite page in Figs. 311 and 312, are from the catalogue of the Coalbrookdale Company, published in 1875.

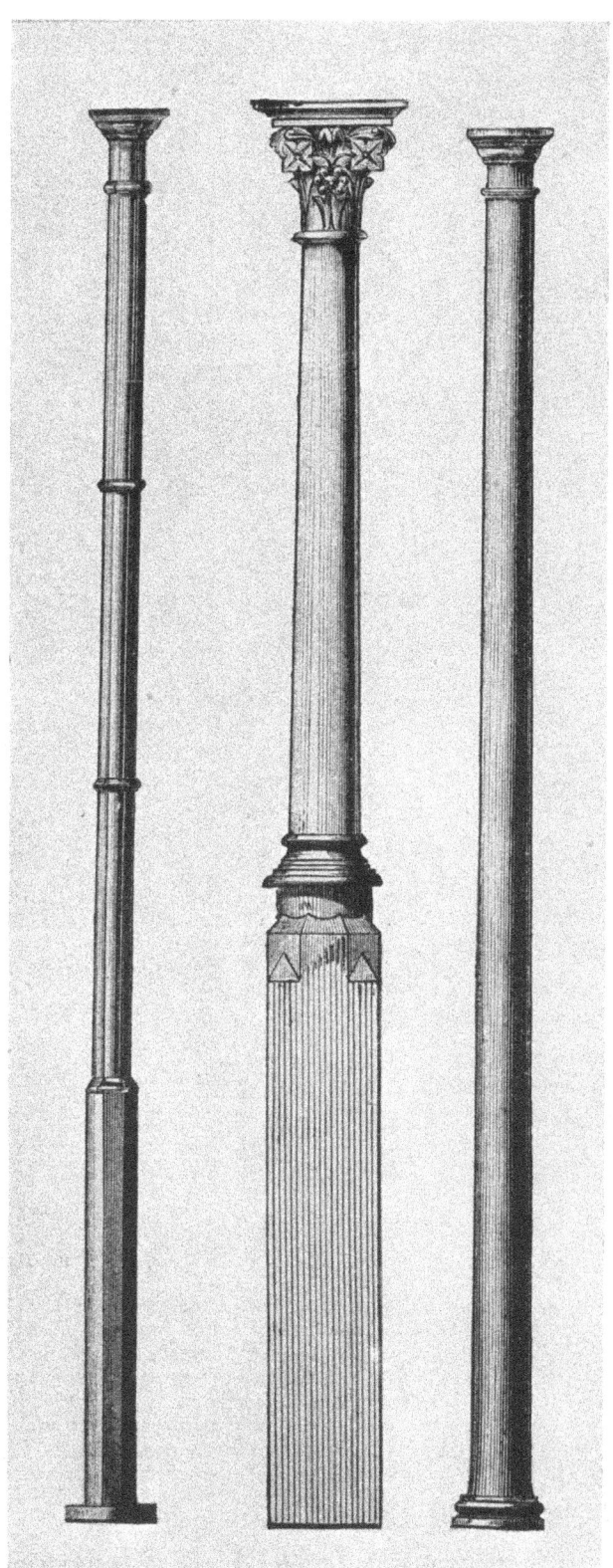

FIG. 311 (*Above*). Further specimens of structural columns. See Fig. 310 opposite.

FIG. 312 (*Left*). Two structural columns in position. See Fig. 310 opposite, and Fig. 311 above.

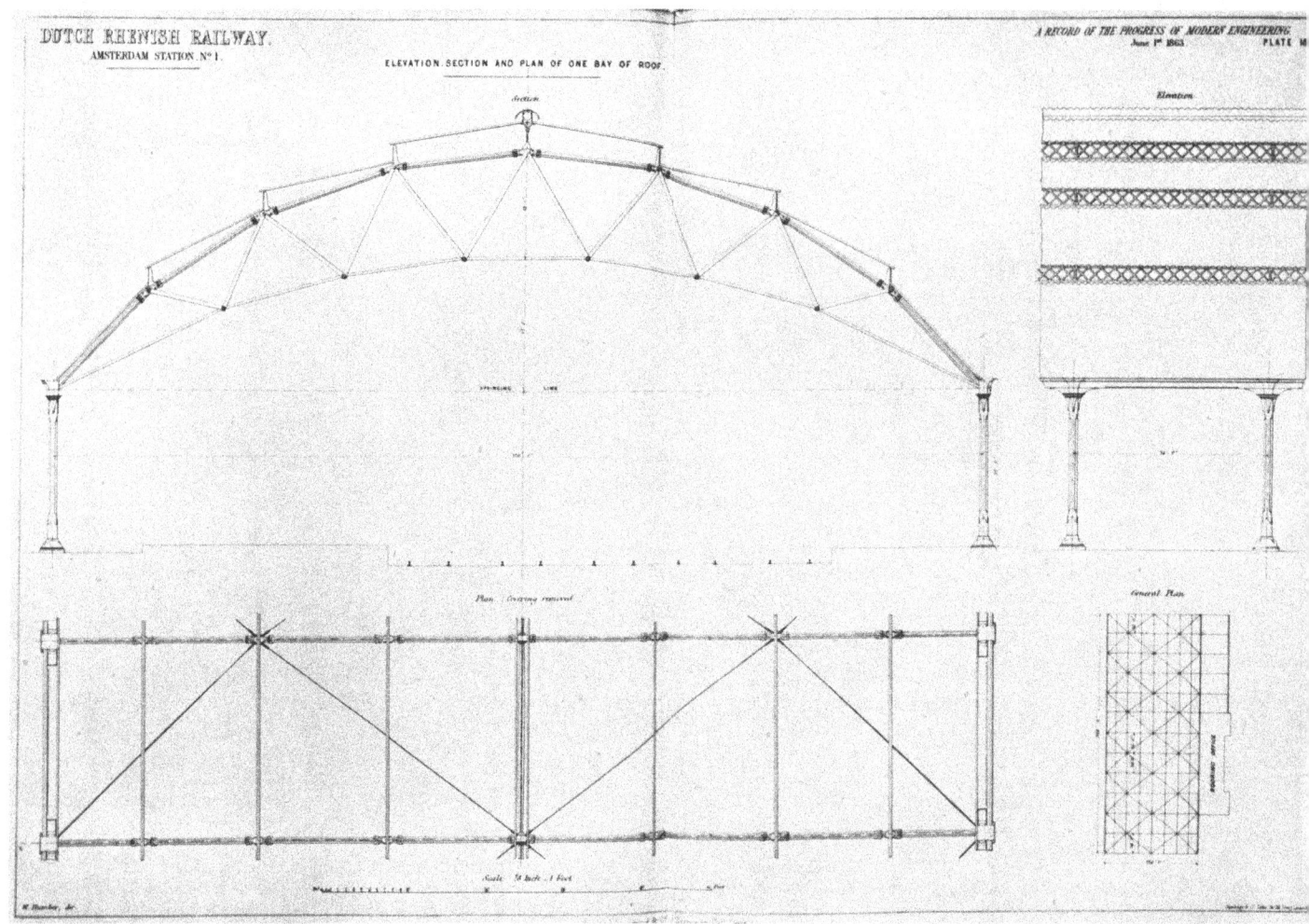

FIG. 313. Amsterdam Station, Dutch Rhenish Railway, designed by R. M. Ordish, Engineer, in 1863, and made in Britain. The elevation, section, and plan of one bay of the roof are shown above. Cast iron octagonal columns are used as down pipes for rainwater, and cast iron gutters in one length between each pair of columns, act as girders for supporting the roof covering. Trusses are formed from straight lengths of cast iron tubes, 8 ins. in diameter. Cast iron purlins are glazed with sheet glass, and cast iron open-work ridge purlins are pierced for ventilation. The length of the station is 300 ft., divided into twelve bays of 25 ft. each. The span is 120 ft. with a rise of 30 ft. Further details are given in Fig. 314 opposite, and Figs. 315 and 316 on the two pages that follow. From *A Record of the Progress of Modern Engineering*, edited by William Humber, 1863 - 1865.

Figs. 313 to 316 inclusive, are reproduced by courtesy of the Royal Institute of British Architects.

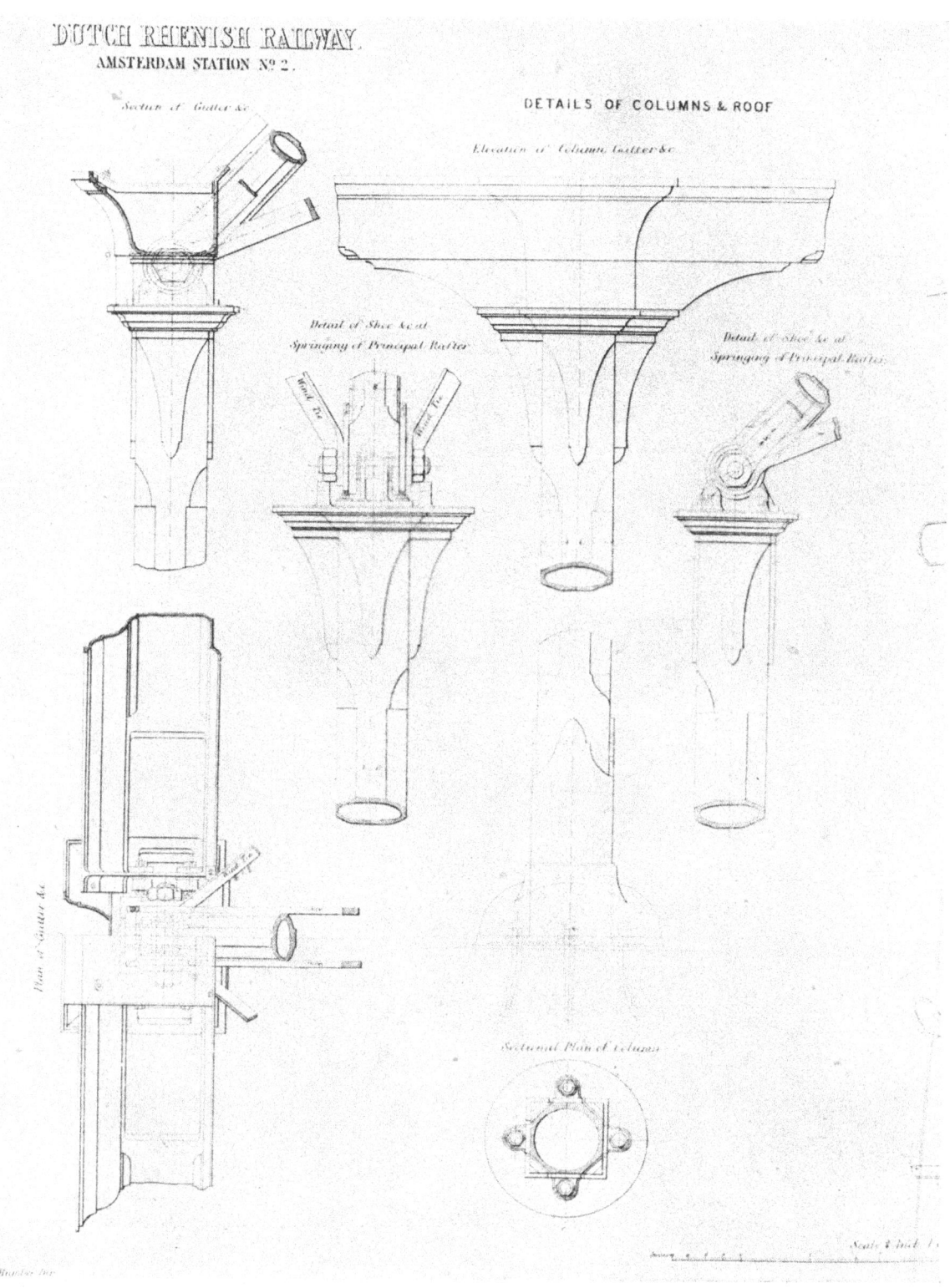

FIG. 314. Amsterdam Station. Details of cast iron columns, roof and structural gutters. See Fig. 313 opposite, and Figs. 315 and 316 on the pages that follow.

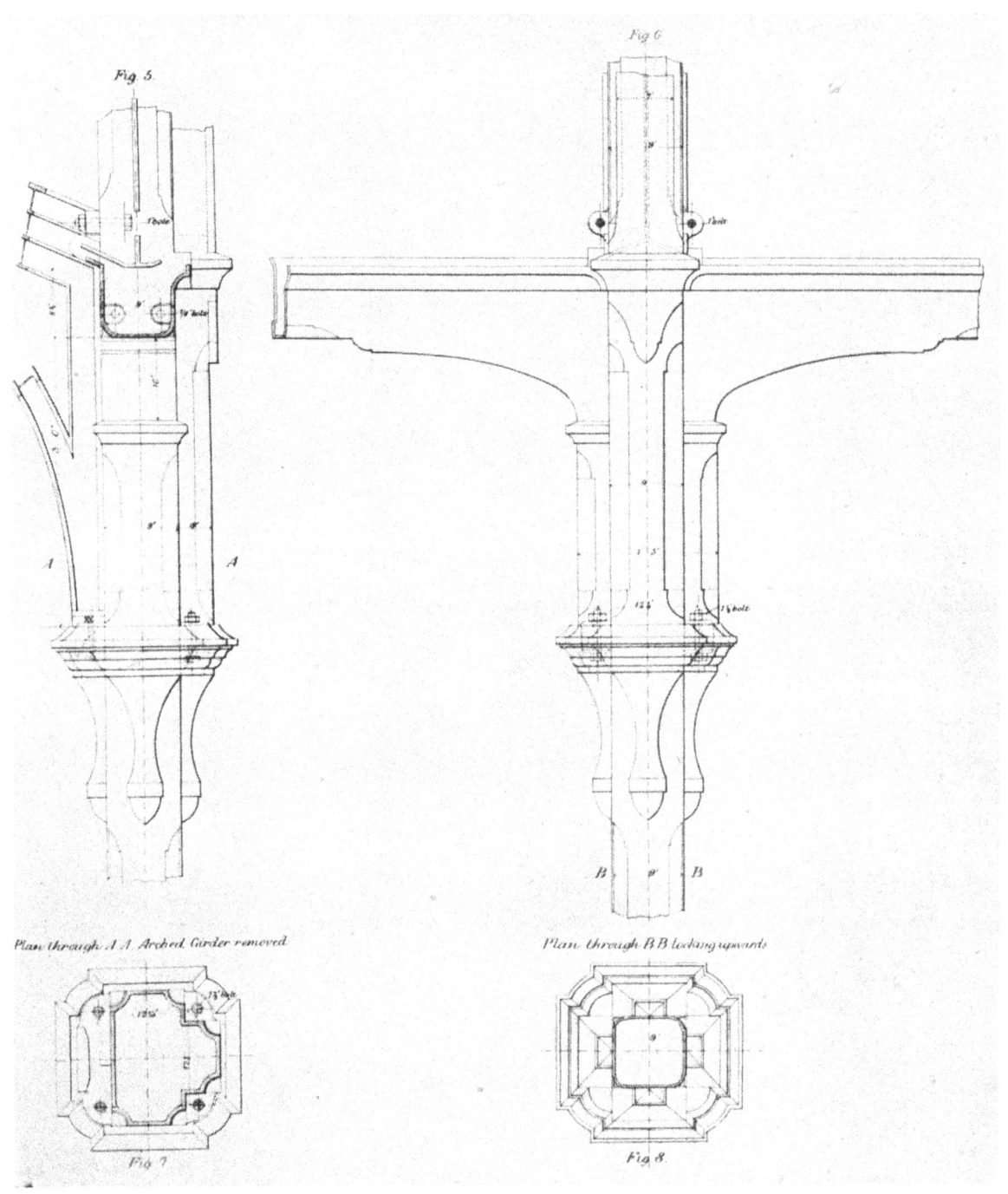

FIG. 315. Amsterdam Station. Details of cast iron columns and structural gutters. See Figs. 313, 314 and 316.

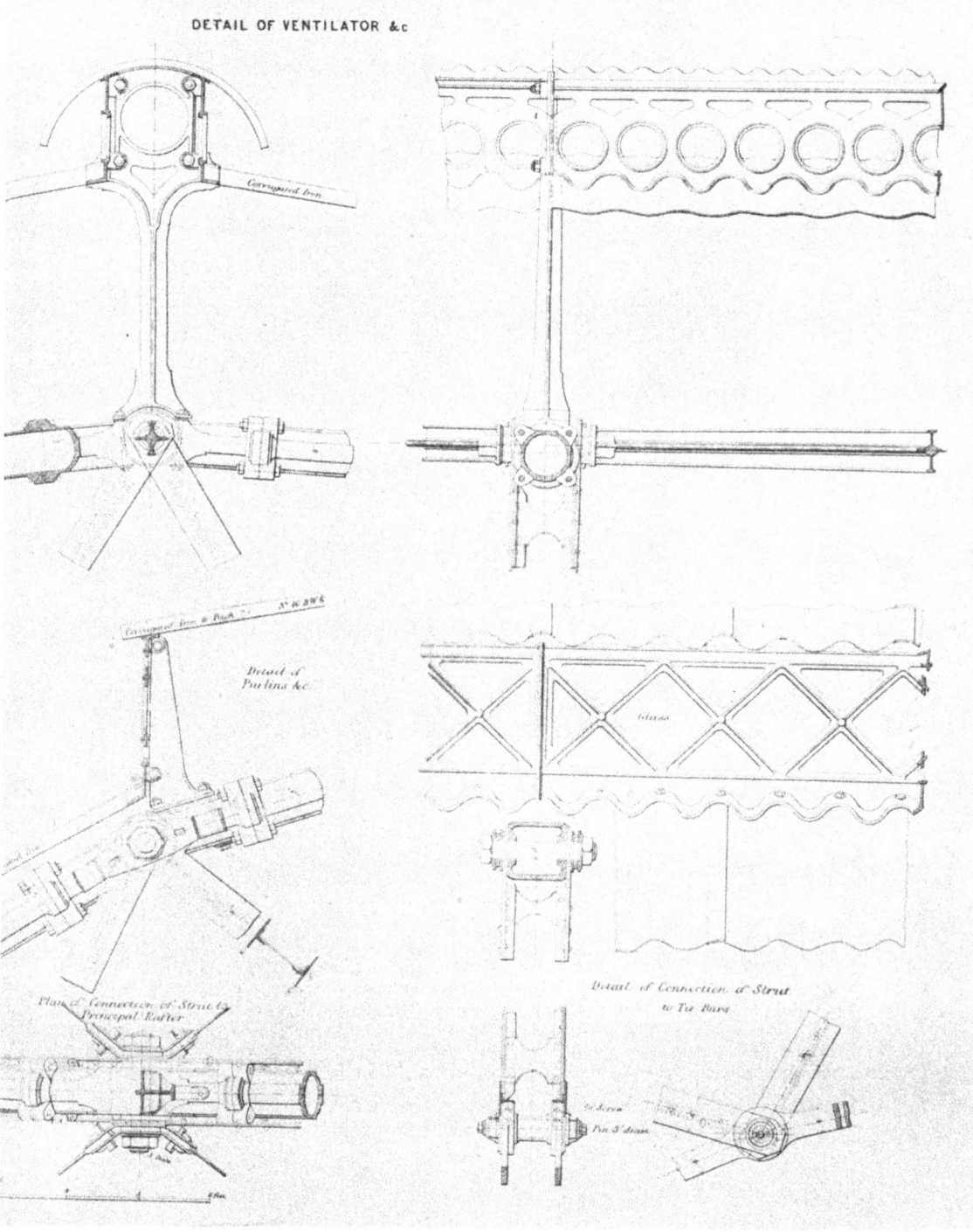

Fig. 316. Amsterdam Station. Details of cast iron roof truss, ventilating ridge purlin, and glazed purlins. See Figs. 313, 314 and 315.

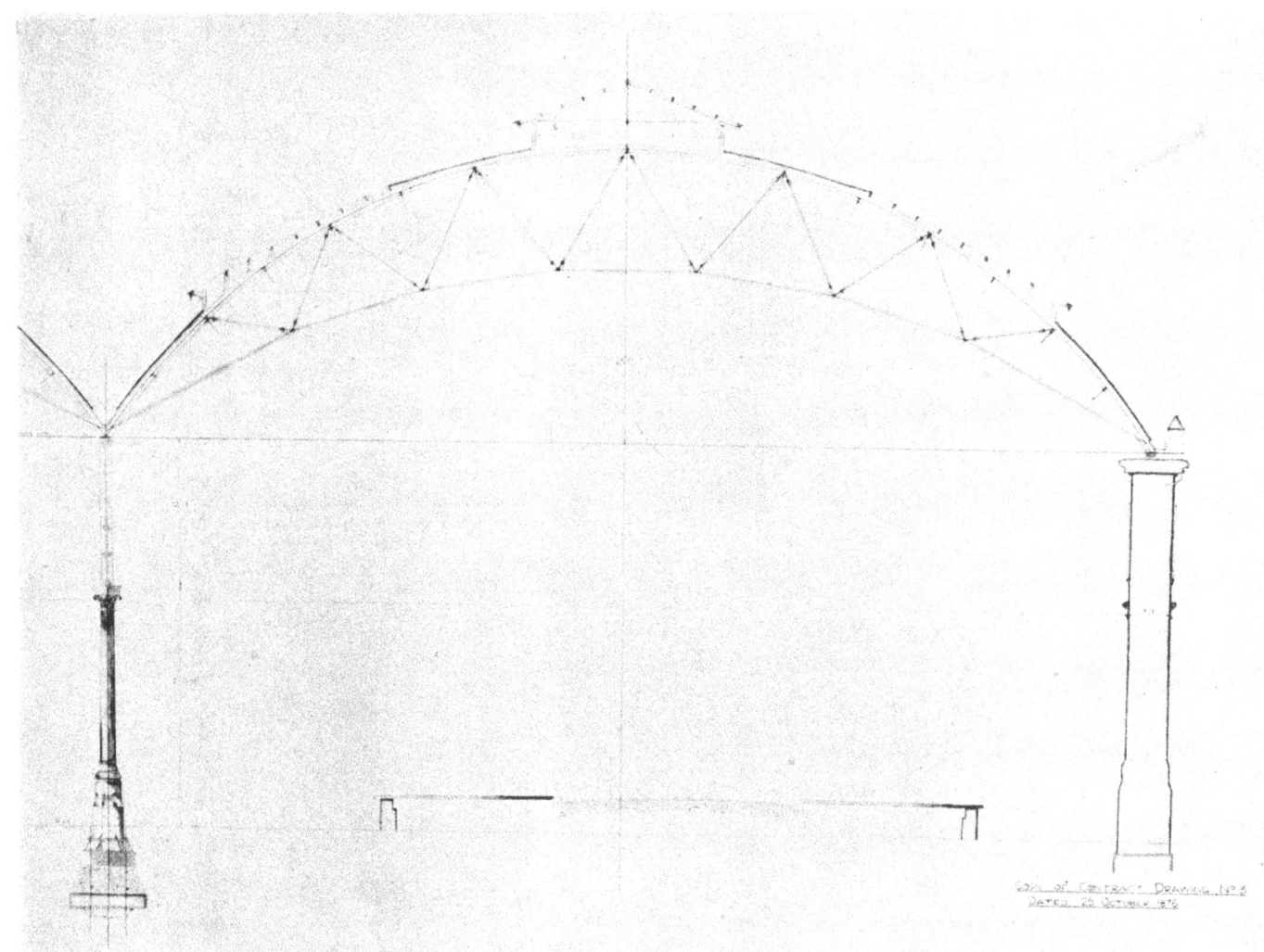

Fig. 317. Woodside Station, Birkenhead, the terminus of the London and North Western and the Great Western Joint Railways. This was one of the last large-scale buildings in which a considerable amount of cast iron was used structurally. See pages 251 - 252. The illustration above, and the details given in Figs. 318 and 319 on the opposite page, are copies of the drawings, made in 1876, which were put at the disposal of the authors by the Engineering Department of the London, Midland & Scottish Railway.

Figs. 317 to 319 inclusive are reproduced by the courtesy of the London, Midland & Scottish Railway.

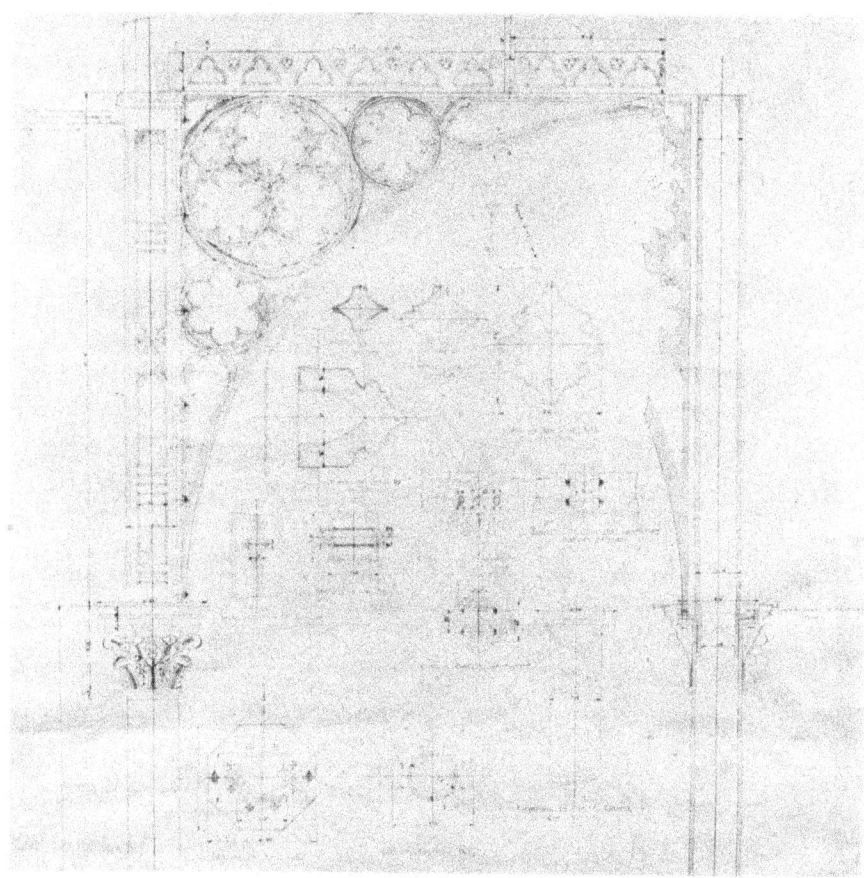

FIG. 318 (*Above*). Pierced cast iron Gothic spandrels in Woodside Station, Birkenhead.

FIG. 319 (*Below*). Details of the bases of the columns supporting the roof of Woodside Station. See Fig. 317.

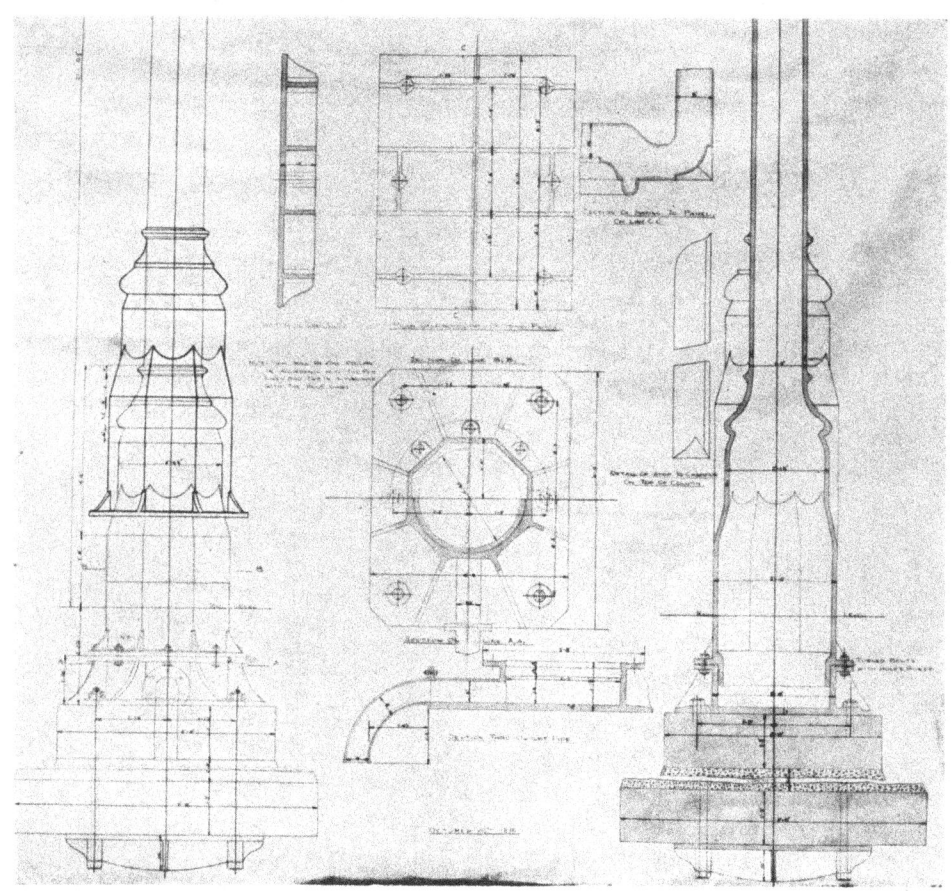

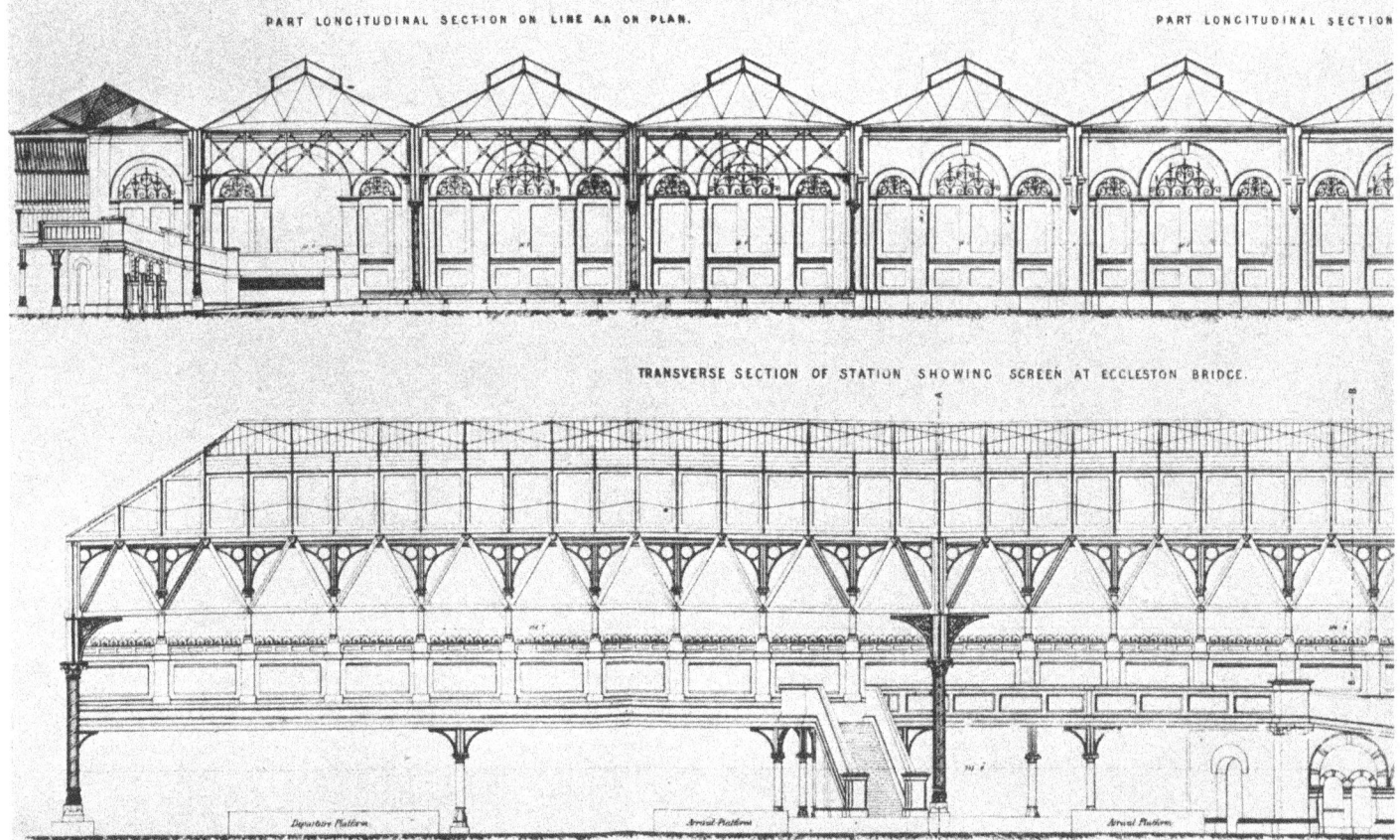

Fig. 320. Victoria Station, London, erected in 1861. Designed by Jacomb Hood, Engineer, for the London, Brighton & South Coast Railway. Cast iron columns are used, 1 ft. 6 ins. in diameter, and 30 ft. high, the metal being 1½ ins. thick at the thickest part. Each column is fixed at the base in a cast iron shoe, and each weighs 6 tons, 2 cwt. and carries a 78 ton load. Details of the columns are given in Fig. 324 on page 268. Structural use is made of the cast iron gutters. Further structural details are given on the opposite page and on the three pages that follow, in Figs. 321 to 325 inclusive. All these drawings are reproduced from *A Record of the Progress of Modern Engineering*, edited by William Humber, 1863 - 1865.

Figs. 320 to 325 inclusive are reproduced by courtesy of the Royal Institute of British Architects.

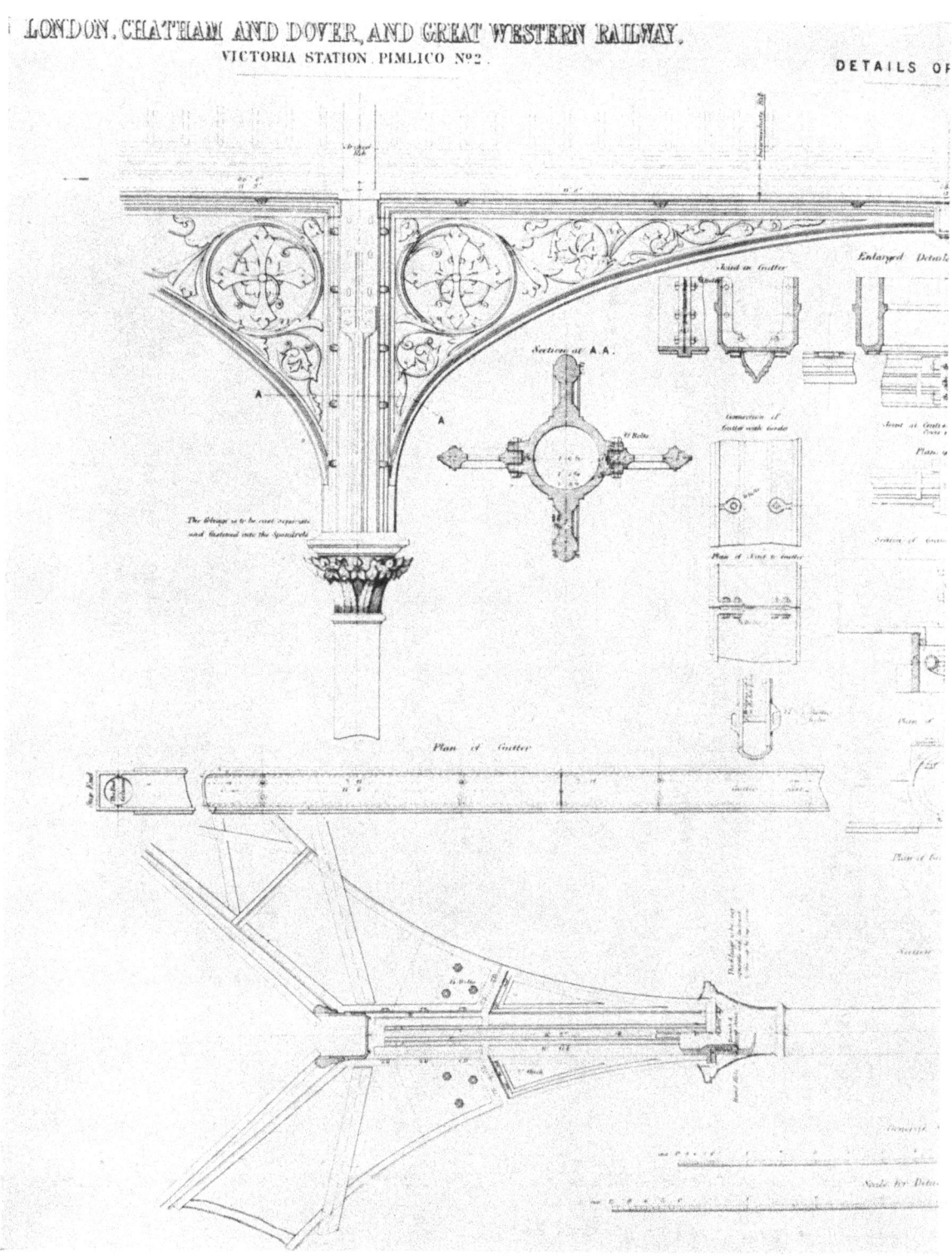

Fig. 321. Details of the cast iron columns, capitals and spandrels of Victoria Station (Dover Section), London. See Fig. 320 opposite, and Figs. 322, 323 and 324.

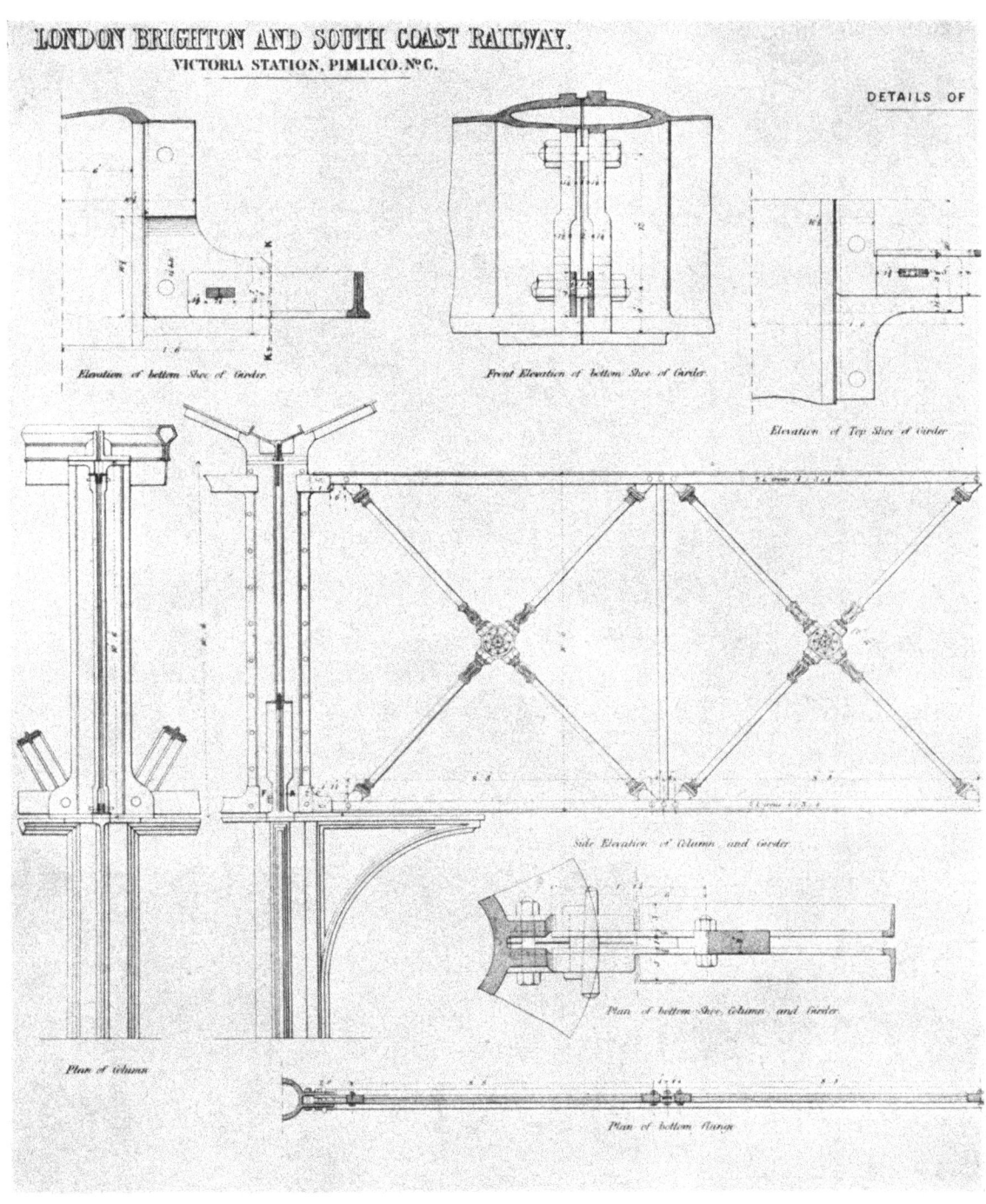

FIG. 322. Victoria Station, London. Details of cast iron longitudinal girders and shoes. See Figs. 320 and 321 on the previous pages, and Figs. 323 and 324.

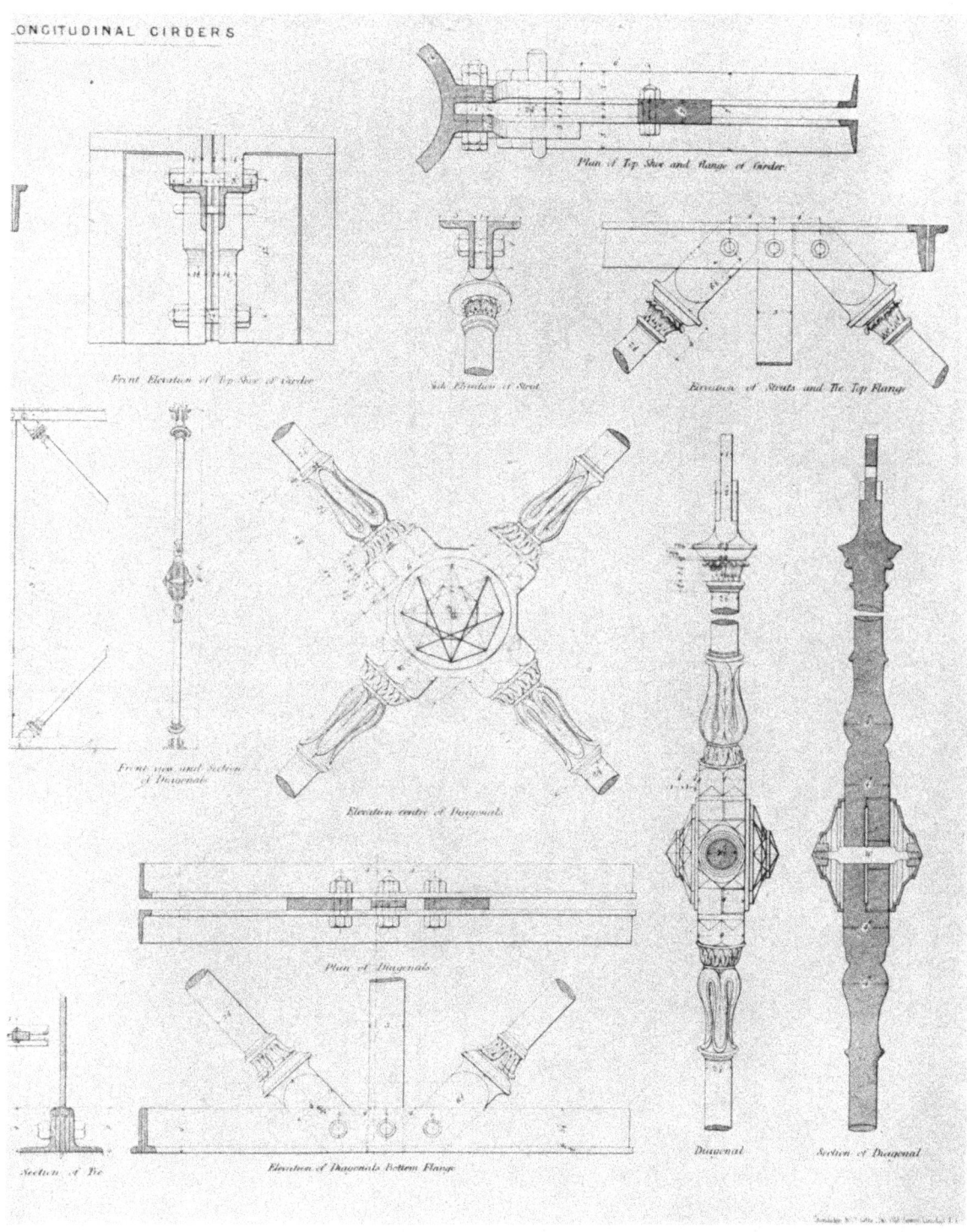

FIG. 323. Victoria Station, London. Details of cast iron longitudinal girders. See Figs. 320, 321, 322 and 324.

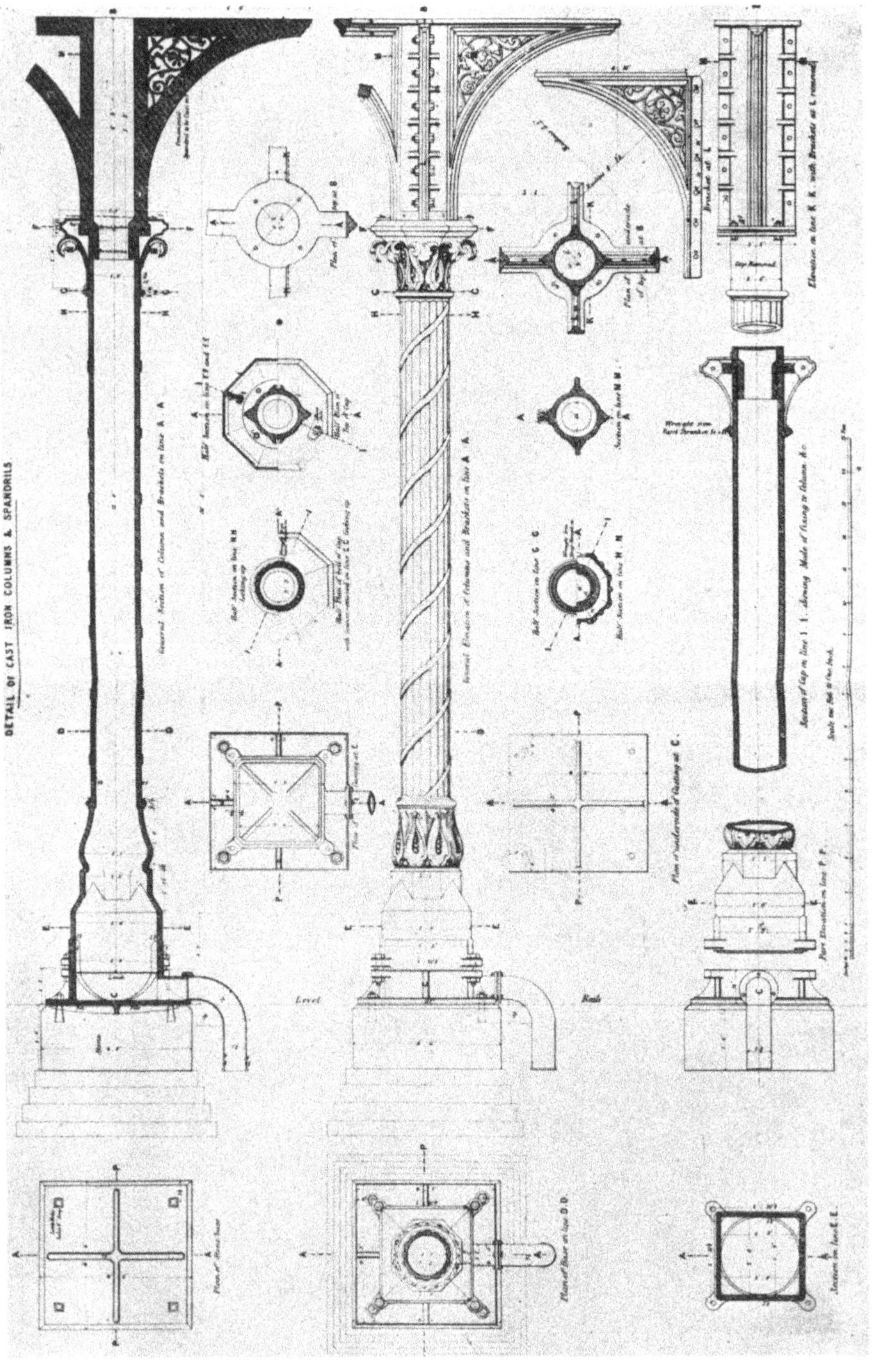

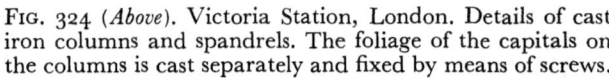

Fig. 324 (*Above*). Victoria Station, London. Details of cast iron columns and spandrels. The foliage of the capitals on the columns is cast separately and fixed by means of screws.

Fig. 325 (*Right*). Detail of platform lamp standard at Victoria Station, London. See Figs. 320 to 324 inclusive.

Fig. 326. Cast iron bandstand at Southend-on-Sea. This highly decorative structure displays not only the enthusiasm of Victorian designers for ornament, but the high degree of skill attained by ironfounders of the period. Of such seaside structures, Richard Sheppard has said: "Most of our piers are highly characteristic and have a genius of design all their own; they were the Victorian equivalent of the folk festivals; the quality of their decoration is expressive; it is bold, coarse and vigorous. The iron is used flamboyantly in panels and columns and crockets and gables..." (From *Cast Iron in Building*, by Richard Sheppard, Section 5, page 63).

Fig. 327. Brackets and street lamps of the period 1860 to 1875. These are extravagant versions of some of Cottingham's designs. Compare these examples, taken from the catalogue of the Coalbrookdale Company, published in 1875, with those shown in Fig. 281 on page 224.

FIG. 328. Three designs for street lamps from the catalogue of the Coalbrookdale Company, 1875. Compare these designs with Cottingham's examples shown in Fig. 282 on page 225.

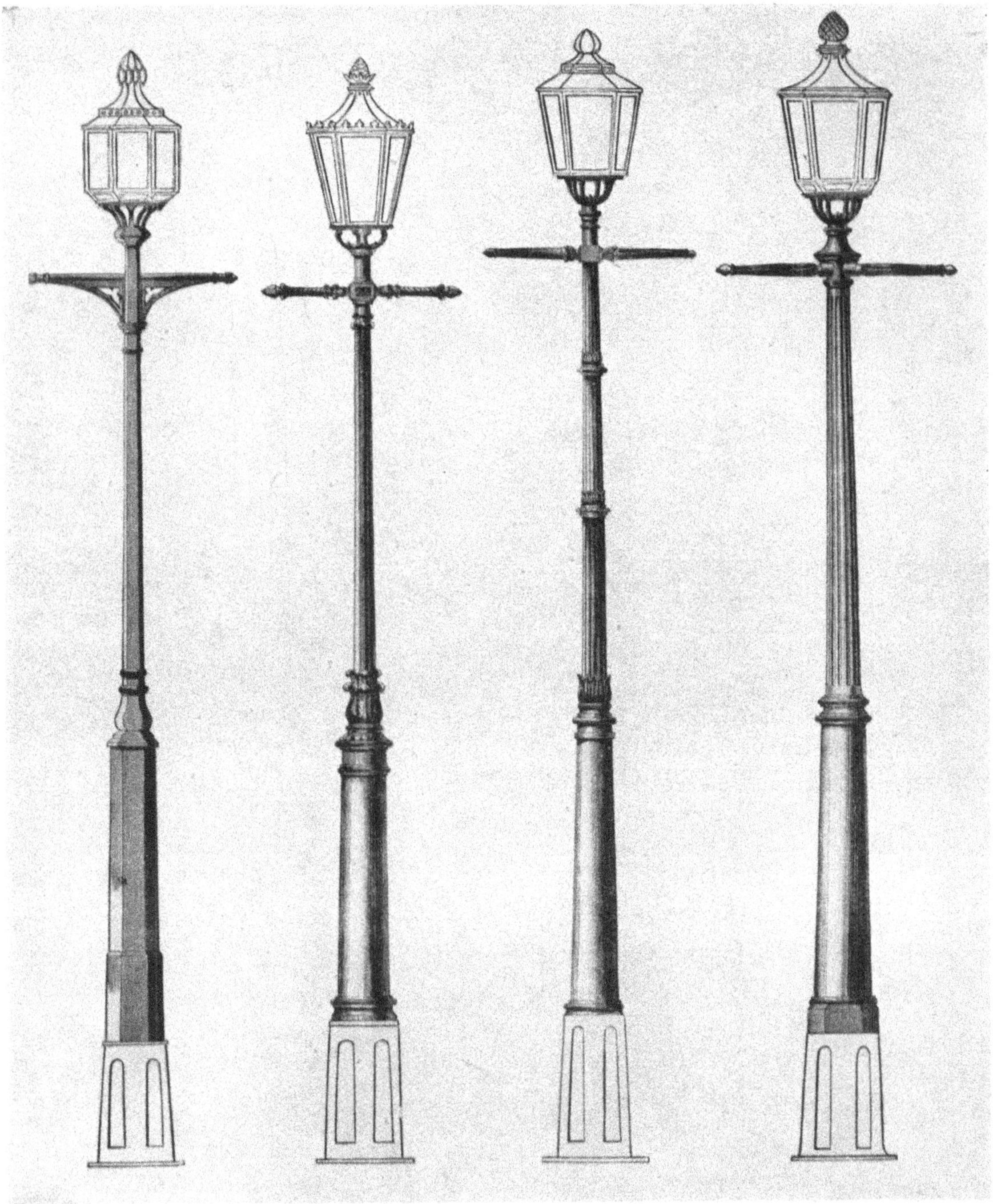

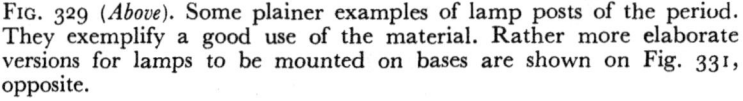

Fig. 329 (*Above*). Some plainer examples of lamp posts of the period. They exemplify a good use of the material. Rather more elaborate versions for lamps to be mounted on bases are shown on Fig. 331, opposite.

Fig. 330 (*Right*). A design in cast iron, made famous by its use on the Victoria Embankment, London. Ornate but orderly in conception, it has undoubted decorative quality.

Fig. 331 (*Above*). A range of lamp standards in cast iron. Compare these with the more restrained designs for street lamps shown opposite in Fig. 329.

Fig. 332 (*Left*). The "Sunflower" lamp standard in the New Walk, Leicester. A mid-nineteenth century design.

Photograph: *A. Newton & Sons.*

Fig. 333. The Royal Coat of Arms in cast iron. This is typical of the ornamental treatment of mid-nineteenth century heraldic work. Although the handling of small scale detail discloses a mastery of technique, the whole design suffers from the rigidity that is apparent in the fireback of the same period, shown in Fig, 37 on page 33. Compare both these heraldic compositions with the monogram of George III (Fig. 51 on page 65). The confidence and ease of the eighteenth century pattern makers have been lost: skill expressed in terms of design has gone—only executive skill remains.

Fig. 334 (*Above*) and Fig. 335 (*Below*). Dogs in cast iron, attributed to Landseer. Originally, they stood outside a tailor's premises in Conduit Street, London; but the building was destroyed in an air raid during the second world war, and the dogs were removed to their present position, outside St. George's Church, Hanover Square, London. The only serious damage they suffered was the partial loss of one tail.

Fig. 336. A pair of gates in cast iron showing the potent influence of the Gothic Revival, and the rigidity acquired by Gothic forms when they were mechanically reproduced in cast metal. (Compare this with the cast iron capital in Fig. 354, page 291, and the tracery panel in Fig. 355 on the same page.)

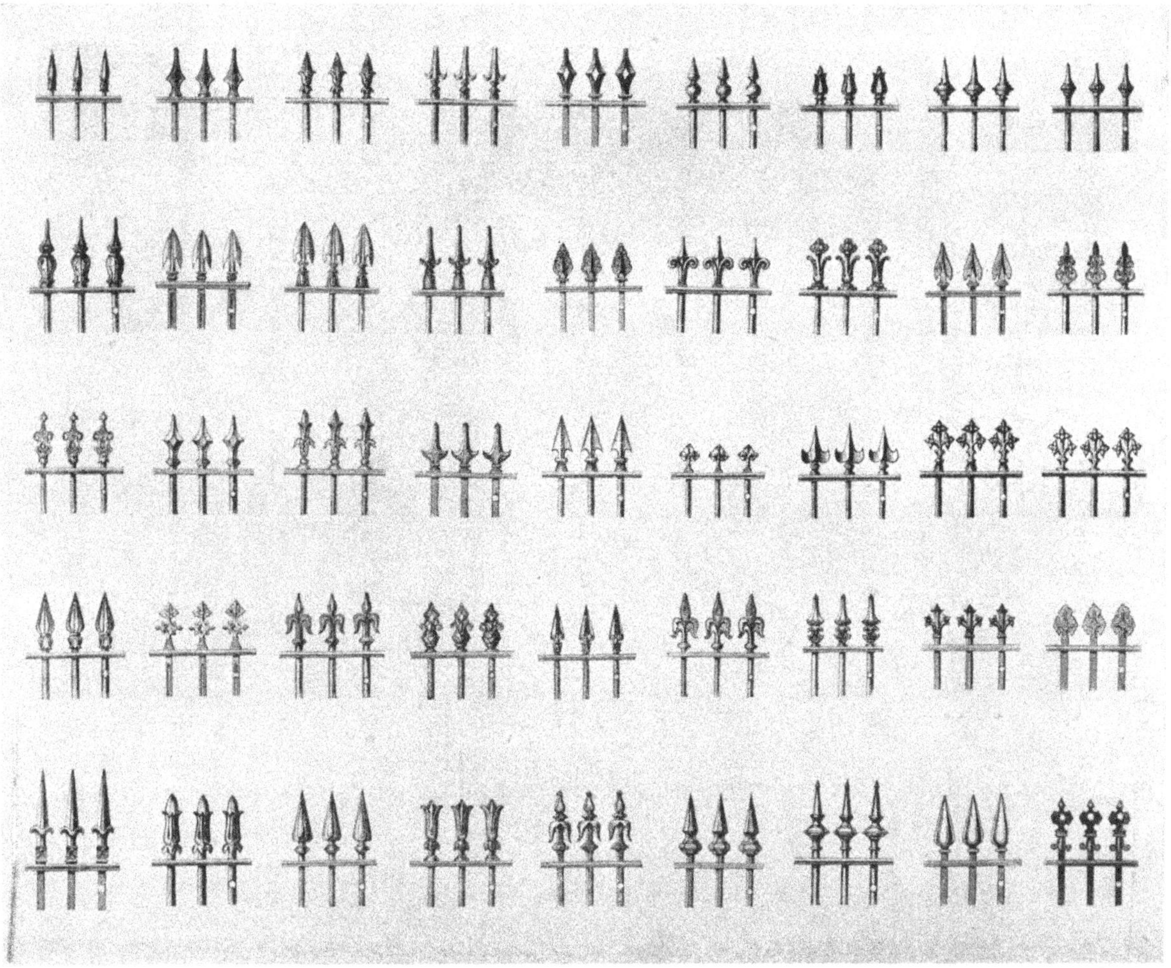

FIG. 337 (*Above*). A range of spear heads for railing terminals. Compare these with the Cottingham designs in Fig. 280, page 224.

FIG. 338 (*Below*). Cast iron railings. Compare the posts on the right with the bollards in Fig. 166 on page 147. The archaic cannon form persists.

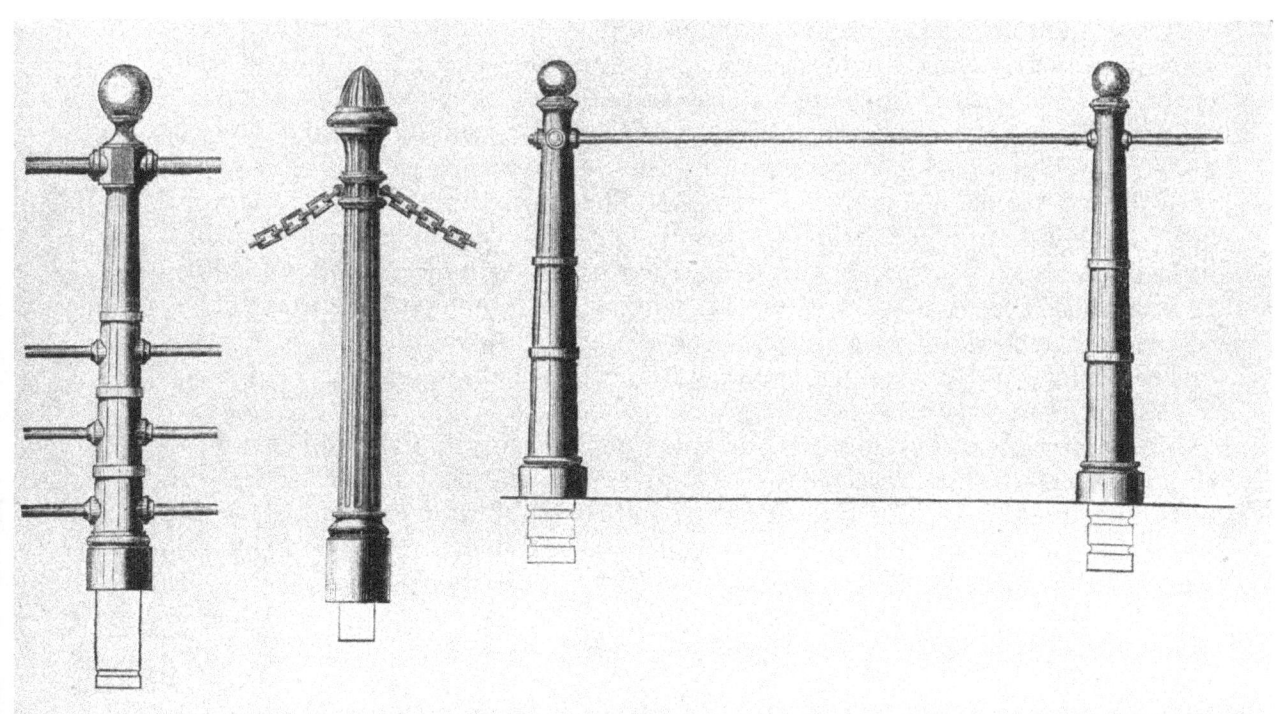

Fig. 339. A range of stair balusters united at the base.

Design.

Every conceivable piece of equipment for buildings was produced in cast iron: gates, railings, verandahs, grilles, treads and risers and complete stairways, columns, windows, pipes, rainwater goods, manhole covers, heating and cooking appliances, baths, mantels, and, in addition, tables, chairs, umbrella stands, vases, clocks, lamps and ornaments of every kind and almost of any size. Fountains, monuments and sculpture were the fashion in the growing towns, and in cast iron they became lasting memorials of Victorian taste.

With the huge increase in the number of industrial and commercial buildings, churches and schools, it was an obvious convenience that cast iron windows should become almost standardised in this period. They had been used in domestic work for many years, and anyone who has had experience of them will vouch for the snug and satisfactory way they fit after much more than a century of use. Some of the late Georgian Gothic designs had a delicacy and distinction to which the crude Gothic windows of the later Victorian church and mansion could not pretend. The industrial window, a simple, light and efficient type, survives in use to-day all over the country, needing little of the upkeep or protection demanded by modern steel windows. Many thousands of such windows were cast during the nineteenth century. They represented a straightforward use of the material, uncomplicated by excesses of decoration, and it is unlikely that the merit of their design received much if any contemporary appreciation.

The confusion of design with the invention, or rather the proliferation of ornament, was characteristic of the nineteenth century. The effect of this misunderstanding of the operation of industrial and architectural design upon the use of a material like cast iron was described by Ewing Matheson who, writing in 1877 of the value of cast iron in ornamental work, said: "In

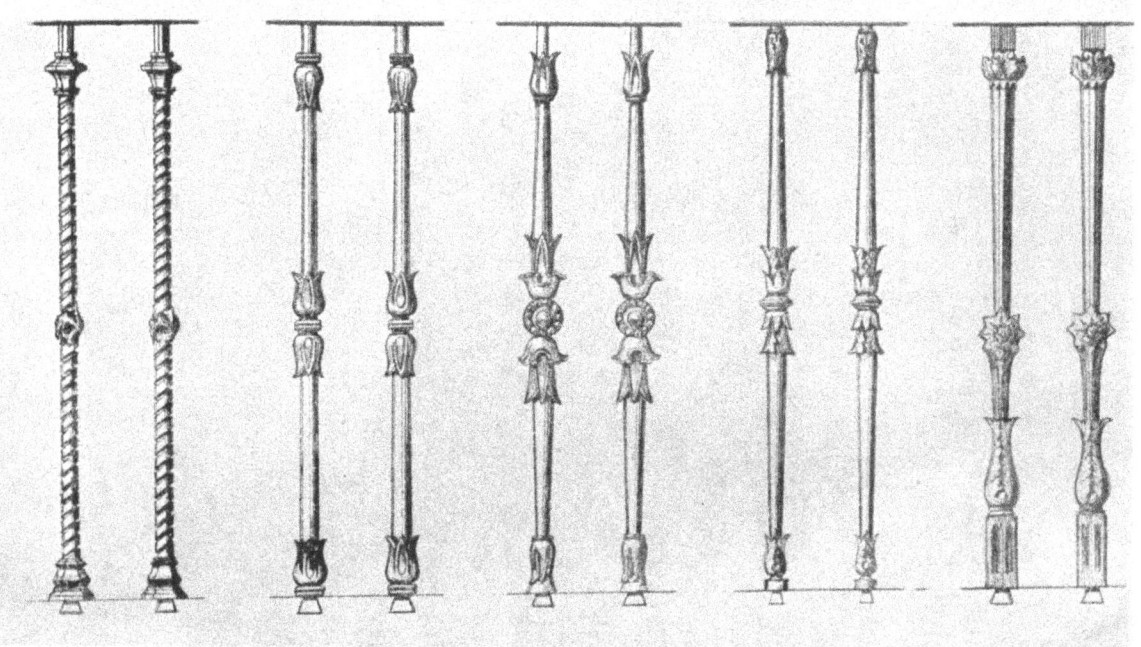

Fig. 340. Stair baluster patterns.

short, while cast iron may be made to any shape that plastic clay will take, or to any which can be carved in stone, marble, mahogany, or oak, it is nearly equal to hammered iron in giving delicacy of form. Moreover, some designs which are too delicate to be utilised in such materials as clay, stone and wood, are of sufficient strength if made in cast iron. Herein lies its superiority over all other materials used in architecture; and this advantage is not lessened by the fact that another material—such as clay or wood—has to be temporarily employed in its production. Indeed, an art-workman has greater scope for design in the earlier stages of cast iron manufacture (*i.e.* in the clay model) than in wrought iron. It would, therefore, be unjust to deny to cast iron the important place in the order of building materials which is now given to it by engineers and builders, notwithstanding the æsthetical objections continually made to it by many architects. But although, as stated above, cast iron can assume any form usual in wood, stone, or clay, its peculiar qualities—hardness, sharpness, strength and durability—demand a new style of ornamentation. The attempts hitherto made at this, though constantly improving on the past, cannot be said to have perfected such a style; and as cast iron is likely to maintain its position of utility, increasing success may be hoped for in the future. But while the merits of cast iron are stated as above, it cannot be doubted that the quality and the rude appearance of many of the cheap castings supplied to builders in this country have justified considerable prejudice among architects, and have prevented the use of iron in many kinds of work where artistic ornament is desired, and where, if properly manufactured, cast iron should be adopted to great advantage".[7]

A few young architects presently attempted to protest against the general misuse of materials, men like C. F. A. Voysey, George Walton, Charles Rennie Mackintosh and W. R. Lethaby, but they were the "modernists" of their day: a little too "advanced" in their views to be taken seriously in the

Fig. 341. Cast iron balcony railings, supports, verandah and area railings of mid-nineteenth century design, at Lansdowne Place, Bristol. The area railings resemble those designed by Stephenson for use on the line outside Euston Station, and were probably produced at the same period. (See Fig. 185, page 165.)
Photograph by Gerald Sanville, F.R.I.B.A., by whose courtesy it is reproduced.

closing decades of the Victorian period. The effects of such misuse were more noticeable in cast iron work because of the ease with which large numbers of articles could be produced and the permanent nature of the material. Catalogues of the period show a remarkable range of goods, an infinite number of styles, but an almost complete absence of æsthetic appreciation. This deficiency was typical of the period: the forms then popular were not typical of cast iron as a material. It has been pointed out in another book, specifically concerned with problems of industrial design, that "An immense amount of skill was available in foundries; some of the early mould-makers were superb craftsmen; but owing largely to the absence of the designer from industrial production, this skill was undirected. Cast iron was used to imitate ornamental forms that had been evolved with other materials; it became a cheap dodge; and it remained in this substitute phase for a long time, with the most damaging results to its reputation. Architects, who were once the most ardent advocates of the use of cast iron, are now inclined to associate the material with the bandstands, public conveniences, and lamp-standards, smothered with 'Gothic' ornament, which were produced in the latter part of the nineteenth century when manufacturers were doing their own designing, and advancing year by year into a deeper and darker jungle of complicated decoration.

FIG. 342. Cast iron balcony railings still performing their function of giving horizontal unity to a façade. The design of these first floor railings is still based on a Greek ornamental motif.
Photograph: Bedford Lemere & Co.

Fig. 343. Mid-nineteenth century balcony and area railings at Craven Hill Gardens, Paddington, London. This type of railing was typical of the West End building developments in London during the middle years of the century.

Reproduced by courtesy of the National Buildings Record.

"There is always a tendency to carry over old forms into situations that demand new forms; and this tendency has not only hampered the development of new materials by forcing them to pass through a substitute phase, but has relegated materials like cast iron to a wholly undeserved obsolescence, from which they can only be rescued by a reassessment of their properties by enterprising manufacturers and designers".[8]

Richard Sheppard, in *Cast Iron in Building*, includes in his classification of building components "such things as window frames and porches, gutters and gates, drain pipes and manhole covers, which are fixed or built into a structure, and all components necessary to complete a structure and to ensure its maintenance in good order". He goes on to say that "for many of these purposes the use of iron has become traditional and for drain pipes and gutters it has long been the only material which is both cheap and durable. For others the changing basis of building technique makes it likely that its use

Fig. 344 (*Right*). Compare the grace of the early nineteenth century balcony and area railings with those shown opposite in Fig. 343. This example is from Munster Square, Regents Park, London. The balcony railing is shown in Cottingham's *Director*. Fig. 284, page 227.

Reproduced by courtesy of F. R. Yerbury.

Fig. 345 (*Left*). Another example of early nineteenth century railings, before the character of design had become debased and coarsened. These are at Silwood Place, Brighton. This is based on one of the designs shown in Cottingham's *Director*. Fig. 285, centre right, page 228.

Reproduced by courtesy of the National Buildings Record.

Fig. 346. Mid-nineteenth century balcony railings and brackets, at a house in St. Paul's Road, Clifton, Bristol.

Reproduced by courtesy of the National Buildings Record.

Fig. 347. Another typical pattern of cast iron balcony railings of the period, at Alexander Place, Kensington, London.

Reproduced by courtesy of the National Buildings Record.

will be extended, and such things as window frames and porches are likely to be made in cast iron to an increasing extent owing to the fact that the building craftsman is becoming an expensive luxury, that costs have risen rapidly since the war, and mass-production represents the only hope of getting homes in sufficient quantities and at a low cost. Many of these products, moreover, were first fabricated in the last century, and although the designs have been modified in course of time, some of them require to be re-designed to bring them into line with modern requirements. The industry is very much alive to the importance of overcoming the prejudices which arose in the latter part of the nineteenth century".[9]

Although the true character of industrial design was not identified or appreciated during the last half of the nineteenth century, sound workmanship, especially in anything mechanical, was taken for granted in all British industrial products and the need for maintaining our leadership in this direction was sharpened by foreign competition. An example of this acute awareness of the growing significance of foreign industrial art, occurred in 1878, when the Society of Arts inaugurated a scheme to send men from various industries to the International Exhibition in Paris. In the summer of 1889 a further exhibition was held in Paris, and the Lord Mayor of London sponsored a similar scheme among the industries of the metropolis; an action imitated by other centres of industry. In London a small Committee of employers and workers was formed; an adequate sum was easily raised to meet all expenses; and artisans from some seventy different trades, chosen by the workmen in those trades, were sent to Paris for a fortnight. Eventually their reports were published in book form.[10]

The building that housed this Paris exhibition was typical of the period with its exuberant cast iron ornament. Glass and iron alone, in an orderly, simple arrangement, were deemed insufficient; everything had to be highly decorated and oppressively rich in effect, to accord with that "atmosphere of banknotes and gold" of which du Maurier wrote in *Trilby*. Vierendeel, the Belgian engineer, criticising this exhibition building and its ornate columns, said: "The enormous danger of this sort of support lies in its revolting vulgarity".

The general impression on reading the reports of the British artisans sent by the Mansion House Committee is that the visitors came away feeling fairly satisfied with what their own country was capable of doing in their own particular trades and with the conditions under which British workpeople lived. The terms of reference are of interest. The men were "to note the quality of stock; differences in make, materials and tools employed; new ideas and improvements; prices and cost as compared with England; wages, modes of payment and hours of work, special attention being directed to the use of machinery and general labour-saving apparatus; the proportion of male, female and juvenile labour; the manner of life of foreign artisans in each particular trade; the cost of living and house rent, etc." There was no emphasis on design or general appearance; perhaps a wise omission as the members were not really qualified to judge this aspect.

The iron foundries were represented by a Mr. S. Masterson, who gave due credit to certain exhibits, but seems to have held the opinion that British ironfounders had little to learn. He describes a large casting in the Belgian section consisting of cylinders, condenser, bed, stand, and so forth, for marine engines, all constructed in one piece, and though admitting that it was an excellent casting, he was unable to see that it had any special mechanical merit for practical purposes. Though it was shown as an example of difficult moulding and casting, he says: "Any average English loam moulder could make the same kind of castings if time were given him to do it; apart from this, it is the cleanest and best unfinished casting shown in the Exhibition, and if it is a fair specimen of the work turned out by the firm that sent it, they need not fear comparison with any English firm that I know of. The castings in the different machines and engines that are shown here by the French and Belgian firms are, as far as can be seen where it is possible to judge of them, of fair, average quality. But in many cases I noticed, where castings are bright and finished, that the metal does not seem so close and

Fig. 348. A heavy type of cast iron railing used outside the Prison at Leicester.

Photograph: A. Newton & Sons.

good as that usually seen in England." Again, of the architectural work he reports: "Several French firms here show unfinished castings (or rather castings as they have left the foundry); most of these are ornamental in design, and as such are of great interest to the trade I represent, as we are often told that we are surpassed by the Frenchmen in the beauty of their designs for ornamental cast iron work. The panels, ornamental parts, columns, designs for fountains, and cast iron staircase with balcony, etc., exhibited in this department, do not bear out this assertion, and with regard to the quality of the work there is much of it shown here that would hardly be allowed to pass as an ordinary casting at a good English firm... English foundry masters do not need to send to France for their designs, as they are quite competent to hold their own with anything that I saw". Later he states:

FIG. 349 (*Above*). Cast iron balcony railings and verandah at No. 14 Albert Road, Regents Park, London.

Photograph: Herbert Felton.

FIG. 350 (*Left*). Garden gate and piers at the entrance to the Insurance Committee offices, New Walk, Leicester. Although these are of mid-nineteenth century design, they preserve the virtues of an earlier period. Compare this example with Fig. 160, page 144.

Photograph: A. Newton & Sons.

FIG. 351. Cast iron railings and gates are used here to emphasise the horizontal lines of an architectural composition, typical of the Greek Revival. The High School at Edinburgh.

Photograph: E. R. Jarrett.

"I was informed that many of the so-called bronze figures shown in the French bronze department are made of cast iron, but as no English firm that I know of execute this class of work, it has nothing to do with my mission here".

Mr. Masterson was evidently ignorant of the work of his fellow founders in Scotland and Shropshire who, at this period, could and did turn out any amount of sculpture in cast iron.

From the exhibition, he toured the foundries of the Paris district, and, after carefully examining many castings, he wrote: "They are certainly not up to the average standard of English firms, the workmanship and finish being deficient; but, more especially, the quality of the metal is far inferior to that generally used here. In several firms I visited they use French and Spanish pig iron; but the castings that are made from this metal are, as a rule, very dirty, and one foreman informed me that he could not get clean castings unless he mixed Scotch pig iron with this other metal".

He reported that nothing in the general working appliances would be of any use in English foundries, and that most of the moulds which would be made in loam in England were made of sand, which affected the quality and cost of the work. After describing the life and habits of the foreign worker, he ended on this critical note: "The French workman will not exert himself; he moves steadily along, smokes his cigarette or pipe whilst at work, and appears to think that it is not necessary that he should work himself to death for the sake of earning a living".

FIG. 352. The gateway to the Sailors' Home at Liverpool, which has been described by John Summerson as "an astonishing tour-de-force . . . a wonderful flourish of arabesques and trophies quite miraculously interpreted by the founder from the drawings of the fulsome neo-Elizabethan architect, Cunningham".

Reproduced by courtesy of the National Buildings Record.

Fig. 353. The Lecture Theatre in the Medical Institution, Liverpool. Cast iron railings of a repeating pattern are used on the front of the balcony.

Reproduced by courtesy of the National Buildings Record.

Fig. 354 (*Right*). Again, the Gothic Revival raises its florid head: these cast iron capitals are reproduced from *Works in Iron* by Ewing Matheson (1877). That author says: "These are shown, not as specimens of design but merely to indicate the kind of forms which may be introduced in cast iron".

Fig. 355 (*Left*). Cast iron tracery panel in the oak door of a pew at Holy Trinity Church, Coalbrookdale, Shropshire.

Fig. 356

Fig. 357

Fig. 358

Fig. 359

Fig. 360

Figs. 356 to 360 inclusive show cast iron grates of the period 1860 to 1875. They have been selected from the catalogue of the Coalbrookdale Company, published in 1875. They present considerable variety of form, and further examples of the semi-circular headed type are shown in Figs. 361 to 366 inclusive, on the opposite page.

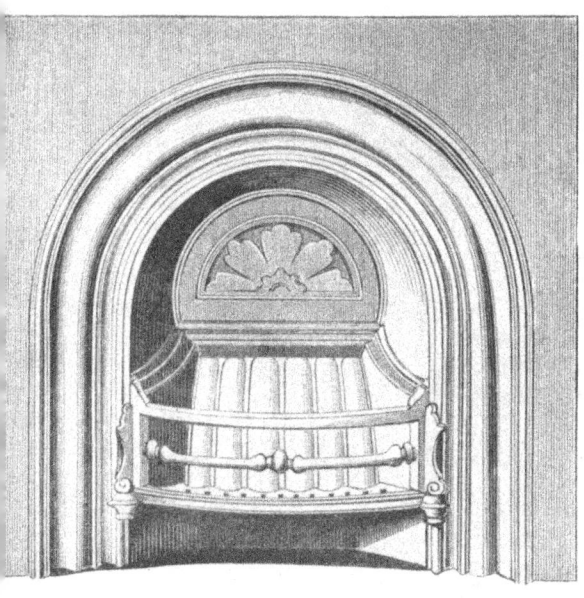
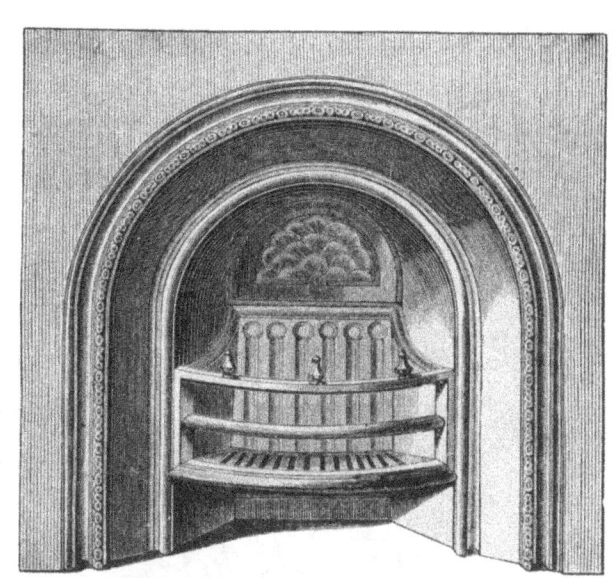

Fig. 361 (Left).
Fig. 362 (Right).

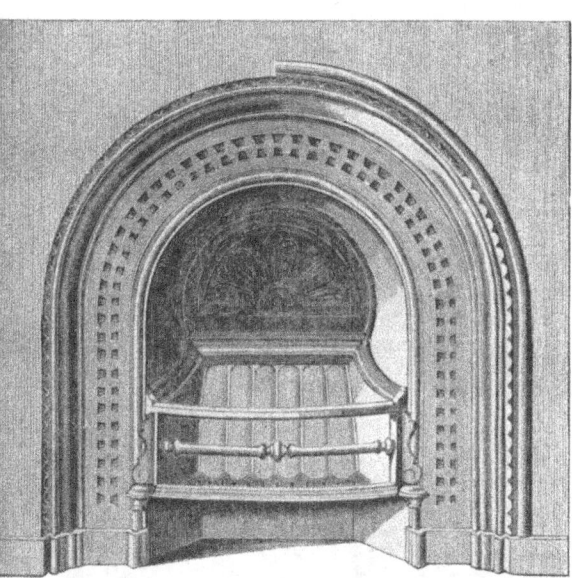

Fig. 363 (Left).
Fig. 364 (Right).

Fig. 365 (Left).
Fig. 366 (Right).

FIGS. 361 to 366 inclusive. With the exception of Fig. 365 which is from a photograph, these designs are from the catalogue of the Coalbrookdale Company, published in 1875.

Fig. 367 (*Above*) and Fig. 368 (*Below*). Both these ornate examples are elaborations of the designs shown on the previous page, Figs. 361 to 366. To the semi-circular headed grate, a vast cast iron mantelpiece has been added, reproducing forms which were originally conceived for execution in marble, or compositions of wood and plaster.

Fig. 369 (*Above*) and Fig. 370 (*Left*). Above, is a highly ornate example of a cast iron mantelpiece. Compare this with the fireplace on the left, designed by Alfred Stevens, *circa* 1860, with cast iron reeded panels and a vivid decorative feature introduced above the grate opening.

Figs. 367 and 368 opposite, and Fig. 369 on this page, are from the catalogue of the Coalbrookdale Company, published in 1875.

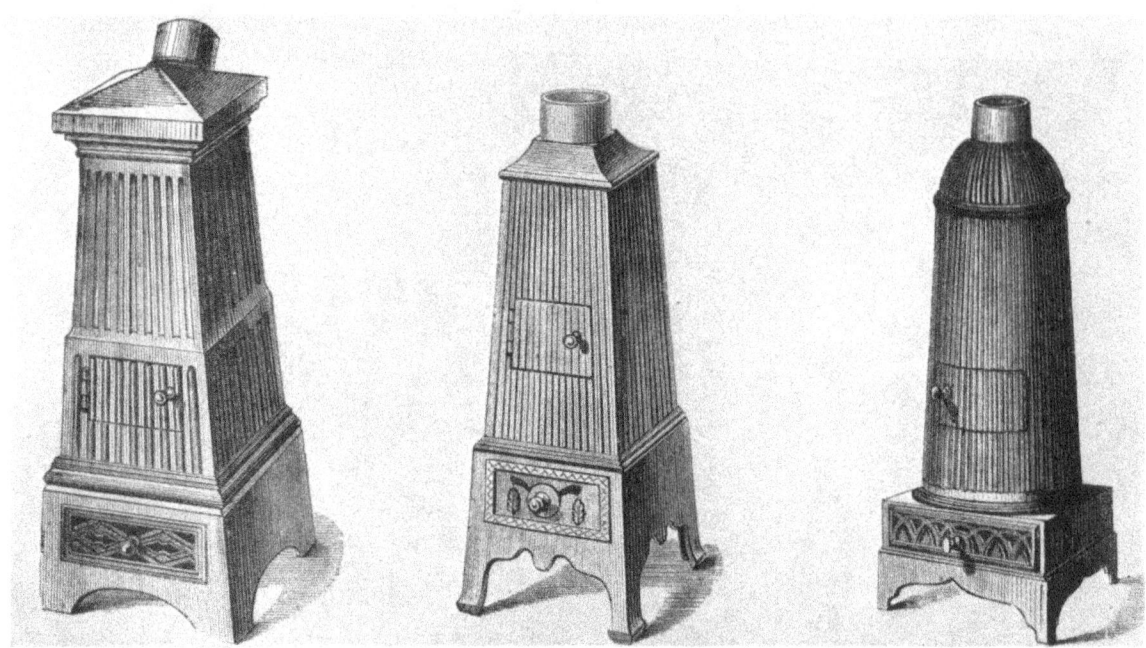

Fig. 371 (*Above*). Three examples of cast iron heating stoves of the eighteen-sixties and eighteen-seventies.

Fig. 372 (*Below*). A more ambitious type of heating stove which reproduces the characteristics of the fireplaces shown in Figs. 361 to 366 on page 293.

Fig. 373 (*Above*). A cast iron hot air stove of the period 1860 to 1875.
All the illustrations on this page, and opposite, are from the catalogue of the Coalbrookdale Company, published in 1875.

FIG. 374 (*Above*). Cottage range and boiler in cast iron, *circa* 1850.

FIG. 375 (*Below*). Kitchen range and boiler in cast iron, *circa* 1870.

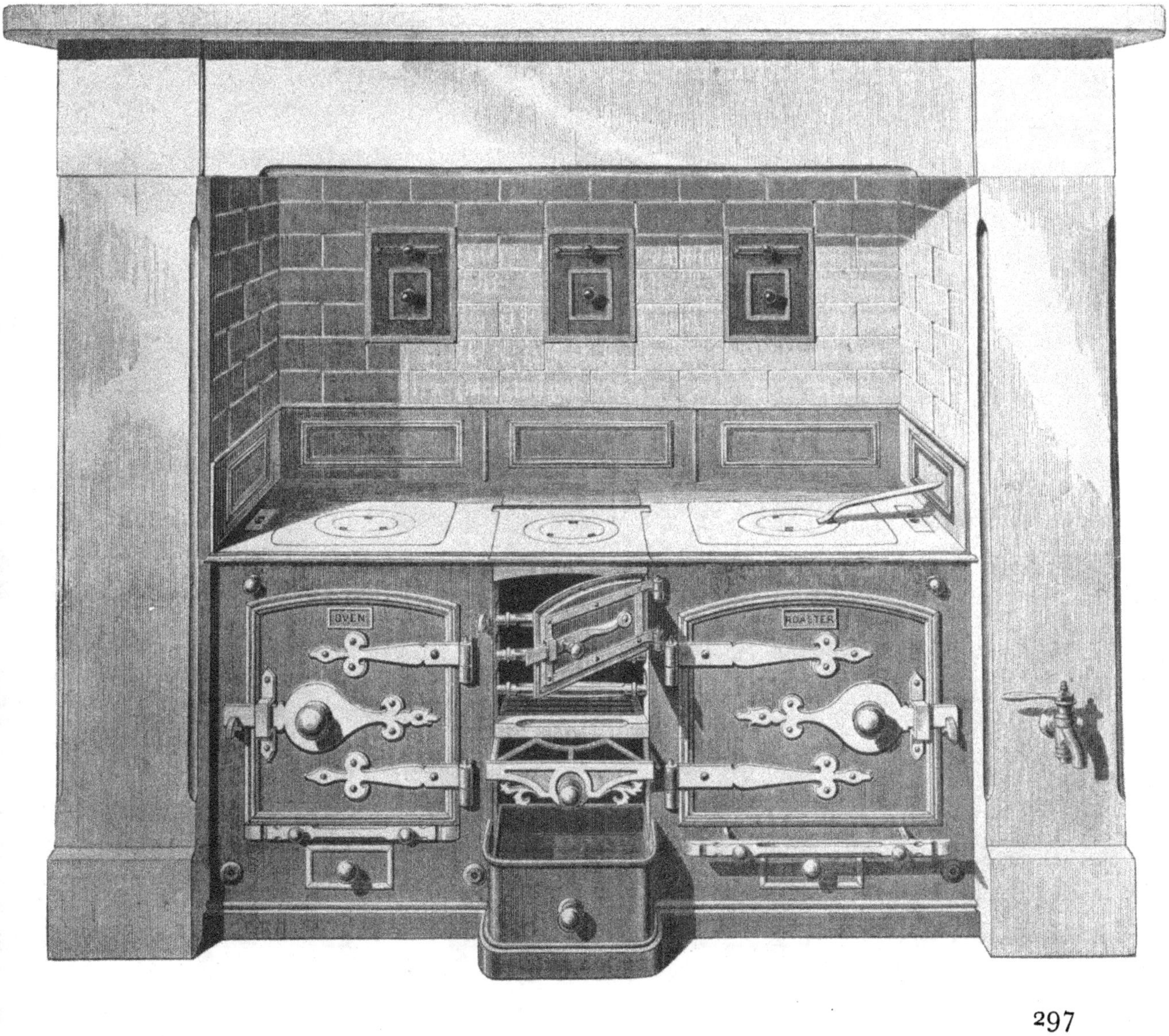

FIG. 376 (*Above*). Cast iron gratings of the period 1860 - 1875. Further examples are shown on the opposite page in Fig. 378.

FIG. 377 (*Right*). Edging for garden paths in cast iron, period 1860 - 1875.

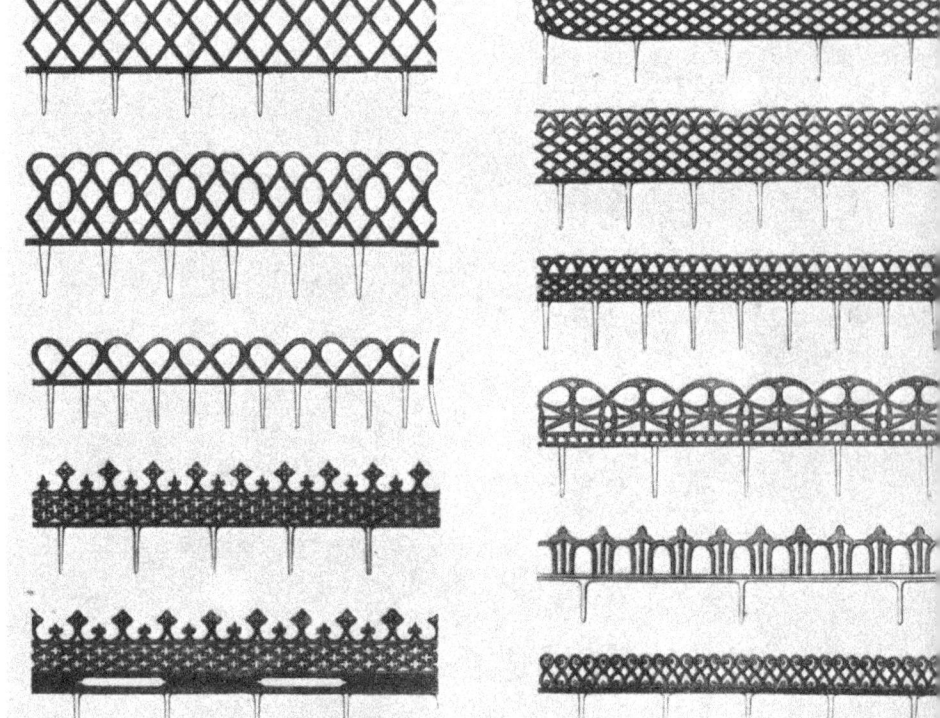

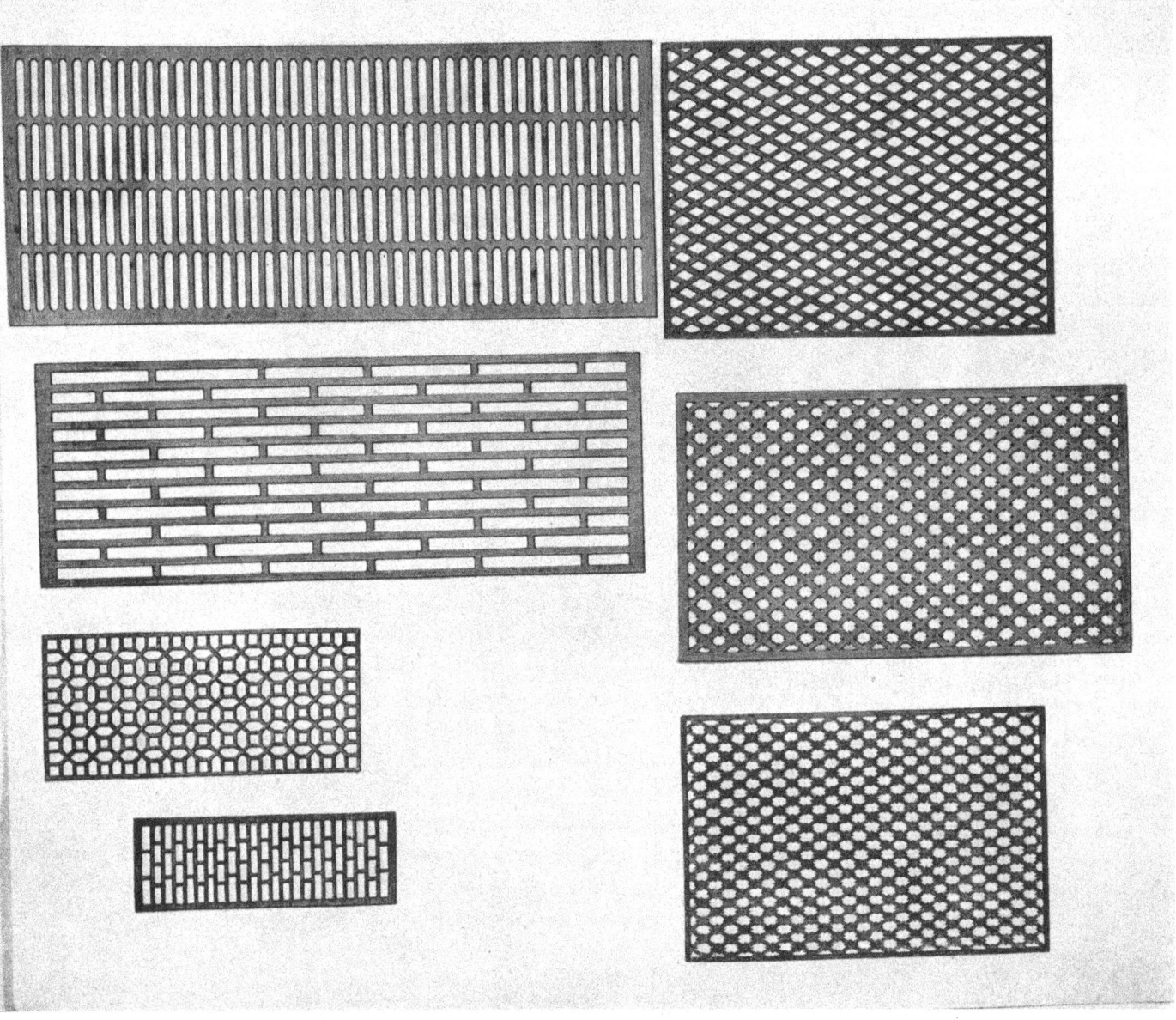

Fig. 378. The cast iron gratings shown above and in Fig. 376 opposite, demonstrate what is probably the chief characteristic to be expressed in the design of cast iron articles: the idea of a *plate*, pierced or moulded, instead of the *bar*, which is the basic theme of wrought ironwork. The coal plates shown on the following page, in Fig. 379, are intrinsically decorative, as indeed are the gratings shown above.

The illustrations on these two pages are from the catalogue of the Coalbrookdale Company, published in 1875.

Fig. 379. Coal cellar plates in cast iron. These and the other forms of gratings shown on the two previous pages in Figs. 376 and 378, represent some of the best achievements in cast iron, in terms of design, of the mid-nineteenth century.

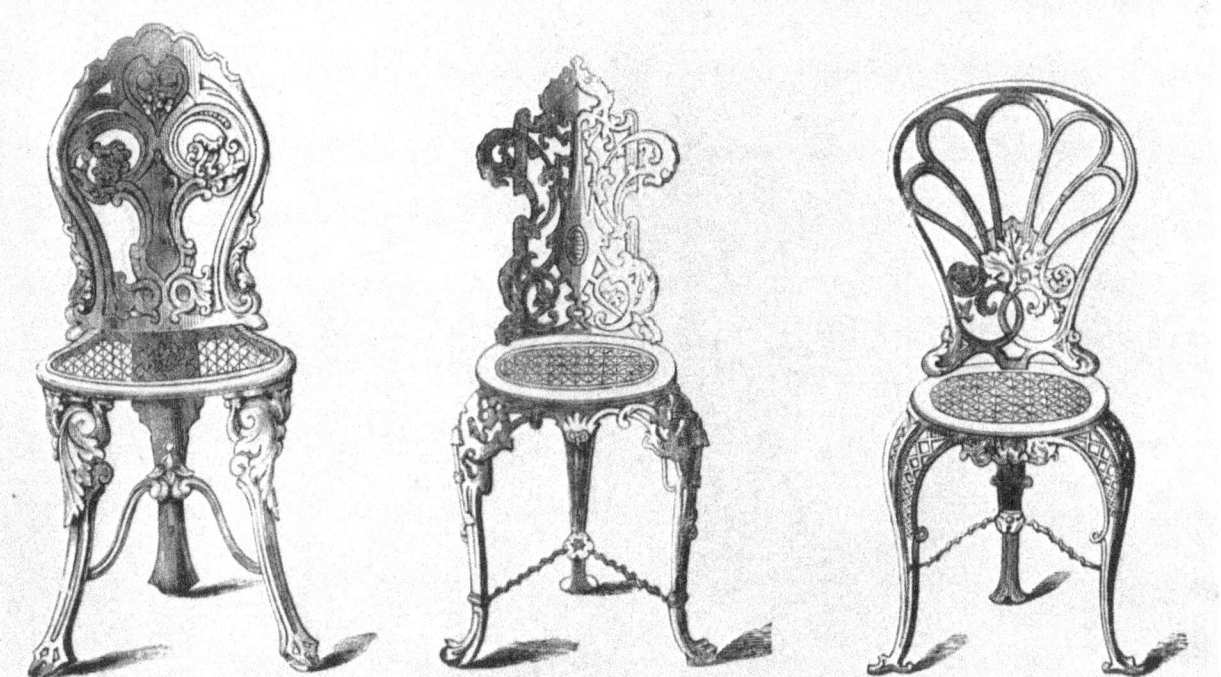

Fig. 380 (*Above*). Three garden chairs in cast iron, which show the founders' skill and the designers' incompetence.

Fig. 381 (*Right*). This is a still more complicated example of "drawing board" design, typical of the nineteenth century, and the sort of thing that gave cast iron a bad name when taste improved and standards of design were raised.

Fig. 382 (*Below*). Photograph of a cast iron garden seat. Complexity and bad proportion are united in this example.

All the illustrations on this page indicate the great technical competence of ironfounders of the period. It was not the fault of the ironfounding industry that mid-Victorian design was at such a low level.

Fig. 383. The cast iron hall stand for hats, coats and umbrellas was as much a part of Victorian house furnishing as the heavy red velvet hangings, the thick, over-patterned oriental carpets, and the bulbous mahogany furniture. Overcrowded as this example is with ornament, some semblance of control has still been maintained by the designer. Compare this with Figs. 384 and 386 on the opposite page. In these hall stands the designer has lost control of the material completely.

Fig. 385 (*Above*). Two umbrella stands of the "ultra naturalistic" school of design, displaying a fatuous concurrence of competing natural objects.

Fig. 386 (*Left*). Another complicated example of the Victorian hall stand.

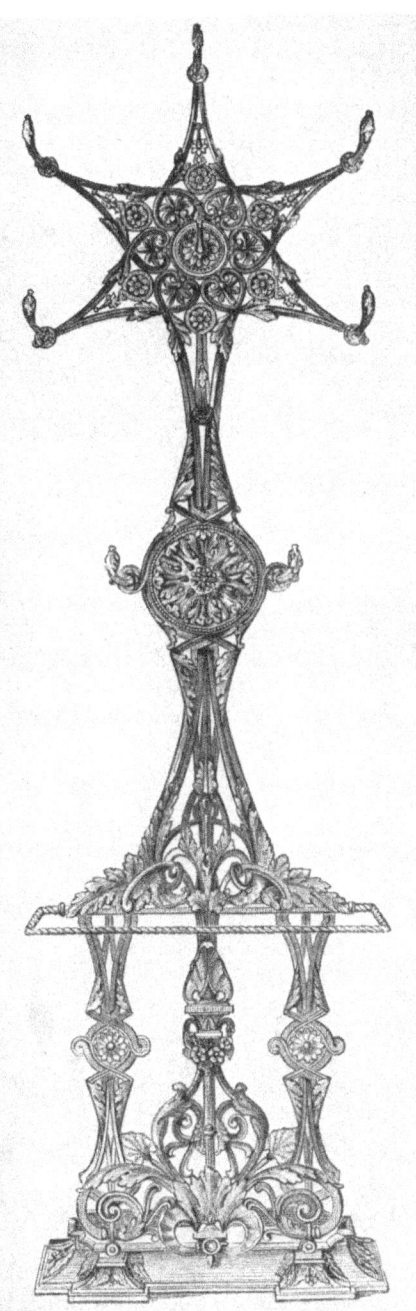

Fig. 384 (*Above*). A hall stand in cast iron. The maximum of complexity has been achieved in satisfying the simplest needs.

Fig. 387 (*Above*). Octagonal garden frame in cast iron.

Fig. 388 (*Above*). Square garden frame in cast iron.

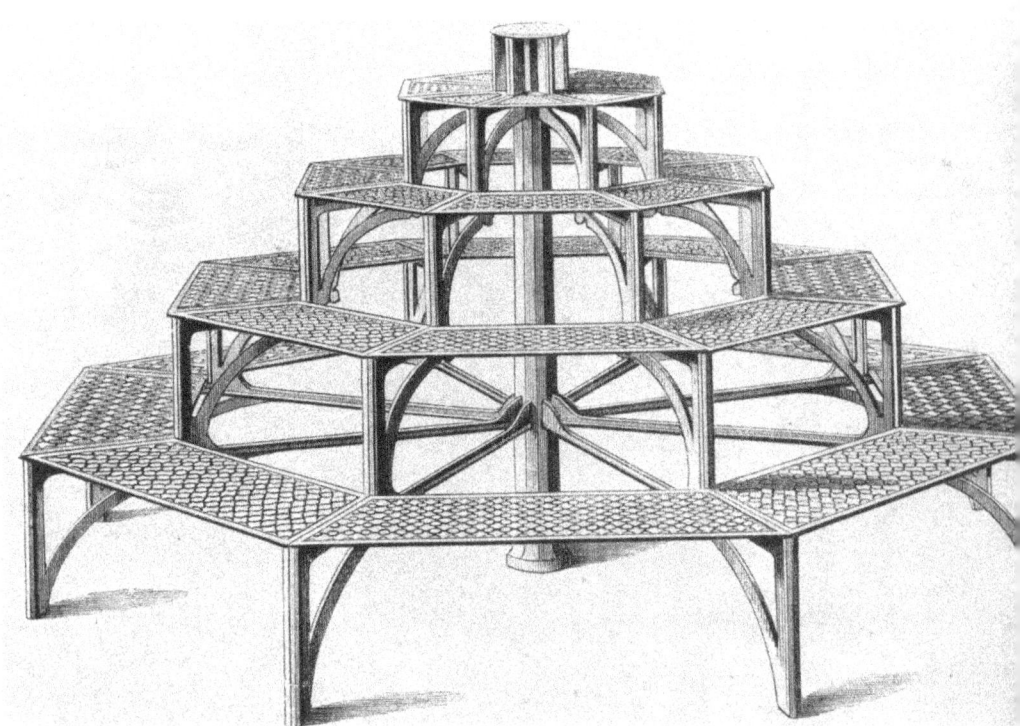

Fig. 389 (*Above*). A cast iron plant stand, making use of forms particularly suitable for the material.

Fig. 390 (*Below*). Two mud scrapers in cast iron bearing the impress of the Gothic Revival; the sort of scrapers that might be used in the house guarded by the gates shown in Fig. 336 on page 276.
The illustrations on this page and opposite are from the catalogue of the Coalbrookdale Company, published in 1875.

Fig. 391 (*Left*). A cast iron water fountain that appears to combine the functions of an umbrella and a foot-bath for the sculptured figure. The dolphins at the base, and the two basins have some connection with an earlier period of design: the figure itself may conceivably have been inspired by a caryatid, though it lacks the solemn dignity of the Greek prototype.

Fig. 392 (*Right*). A cast iron fountain that retains some respect for good proportion, though the moulded detail is coarse.

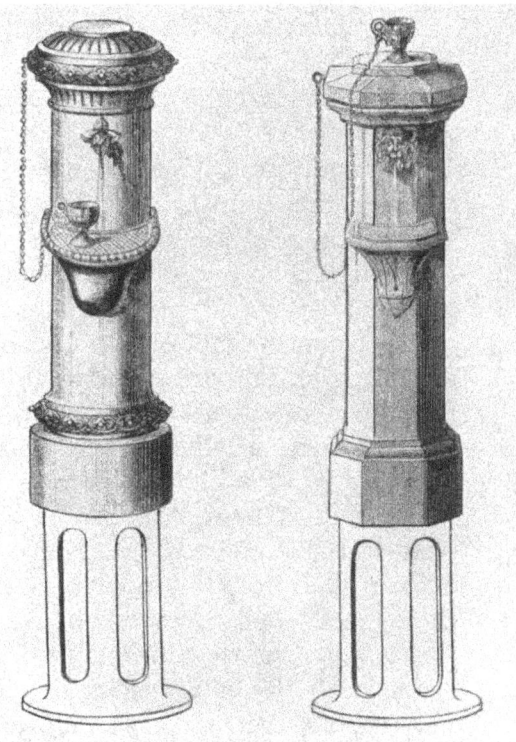

Fig. 393 (*Right*). Two street drinking fountains, sparingly embellished and reasonably well proportioned.

305

Fig. 394 (*Above*). A semi-circular headed cast iron window frame with glazing bars of the same material. The curved lintel is also of cast iron.

Fig. 395 (*Right*). Another example of a cast iron window frame.

The Industry.

The continuous expansion of industrial Britain established new foundries all over the country, and to-day there is an iron foundry in almost any sizable town in the country. Some of these grew into large and important concerns which exercised a great and beneficent influence on architecture; but many were almost backyard foundries with little chance of adequate management or metallurgical control, and their work was largely responsible for much of the prejudice against cast iron, for their products were cheap but unreliable.

Some idea of the growth of the large and influential foundries can be obtained from William Nimmo's *History of Stirlingshire*, published in 1880.[11] "The Falkirk Iron Works," he writes, "now the second largest in Scotland, were started some sixty years ago by a number of enterprising workmen from Carron, and only fell into the hands of the present proprietors in 1848. From the outset, their progress has been steady, but especially during the last thirty years the development of the foundry in its various branches has been remarkable. The buildings now cover eight acres of ground, and the employees, numbering 900 (men and boys) turn out over 300 tons of castings

FIG. 396 (*Above*). Cast iron windows and fanlights from Ewing Matheson's *Works in Iron* (1877). He describes how, at that date, most windows were made in cast iron, varying in price from 1/- to 1/6 per square foot. These are typical church and warehouse windows of the period.

FIG. 397 (*Above right*). Cast iron window frame with sill and lintel of the same material.
This example, and those shown opposite, in Figs. 394 and 395, are still in existence at a foundry in Coalbrookdale.

FIG. 398 (*Right*). A cast iron door lintel in a Coalbrookdale foundry.

307

per week. Here an extensive trade is done in the ornamental or artistic class of goods; but, during the Crimean War, 16,000 tons of shot and shell were manufactured by the Messrs. Kennaird; while their orders for guns of all sizes, for mercantile ships, are considerable. The firm, with Mr. George Binnie as manager, have also executed several foreign contracts of importance. Amongst these were castings for some of the principal iron bridges in India, Italy and Spain. But fountains for the Calcutta Water Company, and tubular telegraph posts for South America have likewise been supplied. The weightiest portions of work recently made, however, were the columns for the Solway viaduct. These were cast in 10 and 20 feet length, to be bolted together as the complete column. No establishment in Britain can cope with the Falkirk Foundry in its elegant and varied stock of patterns for such goods as the following: register stoves, hat and umbrella stands, garden-seats, verandahs, iron stairs, statuary groups, mirror-frames, inkstands, etc. A small figure of a stag, browsing, was shown by the Messrs. Kennaird, at the Exhibition of 1862, along with a variety of other castings, as illustrating the capabilities of the sand-moulding process. In order to have the stag cast in one piece, the mould had to be made in upwards of a hundred parts, each part being simply a clod of moist sand, held together by compression. Sugar pans for the West Indies;

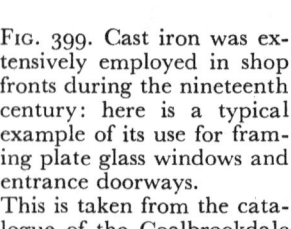

Fig. 399. Cast iron was extensively employed in shop fronts during the nineteenth century: here is a typical example of its use for framing plate glass windows and entrance doorways.
This is taken from the catalogue of the Coalbrookdale Company, published in 1875.

grates, pots and pans for the million, are only a further sample of the great variety of iron goods manufactured here. Few foundries, in fact, have risen so rapidly into fame and importance; and it may safely be affirmed that none show greater promise of being able to 'hold their own' in the vast competitive field of iron manufactures.

"A short distance west the canal bank are the works of the Burnbank, the Gowanbank, the Grahamston, the Parkhouse, and the Camelon iron companies; while at Lock 16 we have the Union Foundry, with the Port-Downie and the Forth and Clyde iron works. In addition to these eight establishments, there are three of recent date to the eastward of the Falkirk iron works. These are the Abbot's, the Gael, and the Etna foundries—the last mentioned being a branch of the Etna works in Glasgow. There are likewise two new foundries situated close to the branch of the North British Railway at Grahamston, the one being called the Callendar, and the other the Vulcan iron works. Here is also the extensive engineering establishment of the Messrs. Blackadder. The reason of so many foundries having been thus recently started to the north of Falkirk is not far to seek. Middlesbro' 'pigs', which are now chiefly used in the manufacture of castings, are brought by steamer to Grangemouth, and thence conveyed per Forth and Clyde Canal to Glasgow (Port-Dundas); while the manufactured goods are also forwarded along the latter route to Grangemouth for shipment to London. Hence the great and double saving in carriage to the Falkirk ironfounders—the distance to Glasgow from Grangemouth being nearly seven times the distance to Grahamston. The following table will show the rise and progress of iron-working throughout the foundry district of the county:

	Started	Hands employed in 1880
"Carron Iron Works	1760	2,500
Falkirk Iron Works	1819	900
Union Foundry	1854	100
Abbot's Foundry	1856	120
Burnbank Foundry	1860	140
Carron Bank Foundry (Denny)	1860	30
Bonnybridge Columbian Stove Works	1860	250
Bonnybridge Foundry	1860	400
Gowanbank Iron Works	1864	300
Grahamston Iron Works	1868	350
Denny Iron Works	1870	90
Larbert Foundry	1870	150
Camelon Iron Company	1872	180
Parkhouse Iron Company	1875	100
Gael Foundry	1875	40
Port Downie Iron Works	1875	100
Forth and Clyde Iron Works	1876	80
Springfield Iron Works	1876	20
Etna Foundry	1877	120
Callendar Iron Company	1877	80
Bonnybridge Malleable Iron Works	1877	8
Total Iron Workers		6,058"

This contemporary record of the growth and extension of Scottish enterprise in iron founding is of exceptional interest, for it shows an active expansion of the industry, an ability to explore and follow up new markets to compensate for changes in the structural needs of building, and a desire to accommodate fresh developments. Other great foundries had been established in the North, notably Macfarlane's at Glasgow, which was started in 1875; also the Lion Foundry at Kirkintilloch, near the same city, which began work in 1880. From all these extensions of the industry a new, inventive capacity was derived, which had a particular significance for the use of cast iron in architecture. Apart from the increased use of the material for fittings and equipment, new large scale uses were tried out. Windows, panels and apron pieces, shop fronts, and indeed the whole façades of buildings, were successfully cast, and were used in Britain, and exported to all parts of the world. Inevitably such early work bore the mark of Victorian taste; but a new technique for cast iron in building had arisen; and it is from these late nineteenth-century experiments, that present-day designers and iron founders have, alike, drawn the most instructive examples for the guidance and invigoration of their work.

SOURCES OF REFERENCE IN SECTION FOUR

[1] *Works in Iron*, by Ewing Matheson, M.Inst.C.E. Second edition, published by E. & F. N. Spon, 48 Charing Cross, London, 1877. Pp. 1 and 2.
[2] *Ibid*, pp. 6 and 7.
[3] *Ibid*, pp. 16, 17, 19.
[4] *Ibid*, p. 52.
[5] *Architecture Arising*, by Howard Robertson, F.R.I.B.A. Faber and Faber, 1944. Section II, p. 28.
[6] *Cast Iron in Building*, by Richard Sheppard, F.R.I.B.A. George Allen & Unwin Ltd., 1945. Section V, p. 63.
[7] *Works in Iron*, by Ewing Matheson, M.Inst.C.E. Second edition, published by E. & F. N. Spon, 48 Charing Cross, London, 1877. p. 227.
[8] *The Missing Technician in Industrial Production*, by John Gloag. George Allen & Unwin Ltd. Second edition, 1945, Chapter IX, p. 96.
[9] *Cast Iron in Building*, by Richard Sheppard, F.R.I.B.A. Section V, p. 78.
[10] *Reports of Artisans selected by the Mansion House Committee to visit the Paris Universal Exhibition*, 1889. Published by C. F. Rowarth, 5-11 Great New Street, Fetter Lane, E.C., 1889.
[11] *The History of Stirlingshire*, by William Nimmo. Hamilton, Adams & Co., London, and Thomas D. Morison, Glasgow, 1880.

SECTION FIVE

SECTION FIVE

TWENTIETH CENTURY DEVELOPMENTS:
1900 - 1945

1. GENERAL BUILDING USES.

IT was recorded in the previous Section that, after the work of Henry Bessemer in 1865 and the subsequent improvements in the manufacture of steel by Thomas & Gilchrist in 1877, there was a considerable time-lag before steel came into general use as a new structural material. In these early steel manufacturing processes 100 per cent pig-iron was used, but in 1861 C. W. and F. Siemens, two German engineers, built a furnace, not unlike the puddling furnace, which produced sufficient heat to melt, economically, not only the pig-iron, but also steel scrap. This was known as the "open-hearth" process, and further experiments in 1884 resulted in the "basic open-hearth process" in which phosphoric pig-iron could be used instead of the more expensive hematite, and also large quantities of cheap steel scrap, cast-iron scrap and wrought-iron scrap.

This progressive cheapening of steel was one factor that encouraged the more general use of the material in building from 1890 onwards. E. Piwowarsky, referring to the diminishing use of cast iron as a constructional material, suggested that this was "the logical consequence of the development of the classical theory of the strength of materials based upon their elastic properties. Modern times have, however, witnessed a partial abandonment of former conceptions of strength and the gradual development of a new theory of design. It is this trend which, combined with considerable improvements in the strength properties of cast iron, has given a new impetus to the employment of this material".[1]

To-day cast iron beams are rarely, if ever, used. But it is still a useful material for stanchions or columns of every description, and large numbers are produced annually for railway work, factory work and particularly for the warehouses and factories in the Dominions and Colonies.

Richard Sheppard, F.R.I.B.A., writing in the *Official Architect* on the "Structural Uses of Cast Iron", has said that "steel is more ductile and can withstand considerable deformation before the elastic limit is reached and it is equally strong in tension and compression. These characteristics, which have made the use of steel universal in the construction of framed buildings, are not possessed by cast iron and it has in consequence been largely displaced for this purpose. Nevertheless, there are certain functions for which cast iron is as suitable as steel and if the skill is available it is probably no more difficult to erect.

"It is possible that the use of steel has become so much a matter of course, a thing taken for granted, that any alternatives for structural purposes are seldom considered. Exactitude in the selection of the appropriate materials is both an economic and æsthetic factor in the design of buildings. The fulfilment of these factors is one of the aims of good design, and opportunities have sometimes been wasted by this myopic attitude. The engineers of the early part of the nineteenth century with their arched forms of construction had shown the way in which structural forms based on the compressive strength of iron could be developed in a way which combined a long life with economy in maintenance and low initial cost. Whether cast iron will be widely used again will not depend upon its treatment at the hands of a few gifted designers who know how to exploit its possibilities. Primarily its use will depend, as always, upon cost. If cast iron can once again show itself cheaper and more convenient than other materials for those uses to which it is particularly suited, then it can recover a measure of its former popularity. The designer has to work in terms of strength/cost ratio and this ratio must be considered in terms of the individual structure . . ."

He also pointed out that "the factory and the warehouse with a row of cast iron columns in the centre supporting longitudinal or transverse beams represents a most economical building type and facilitates good lighting conditions. It might well be considered as a standard for buildings of this

Fig. 400. Cast iron used in a Liverpool shop for the complete façade, and a bridge to another building.

Fig. 401. After the ornate developments of the latter part of the nineteenth century, cast iron architecture retained a few classical affinities, although it carried a less oppressive weight of ornament. Above is a typical example of seaside architecture in the material.

type in the future, as we shall need large numbers of them, and in the shortest possible time, and we must standardise the structural frames as well as the dimensions that govern our building materials if we are to get the maximum production. Cast iron is particularly suitable for standardisation; a mould automatically standardises an article. If we decide that a certain type of building—a class room, factory, a warehouse or storage unit—can be adapted to a standard set of dimensions then we are in a position to standardise the framing components of which it is made".[2]

The rolling process in the manufacture of steel members automatically limits the sectional outline of the members, and here cast iron has a distinct advantage which, as yet, has not been used in modern building. Once the mould has been made any number of stanchions can be easily produced, designed to suit the particular strains or stresses to be taken; and, in addition, any kind of ribbing, fluting or patterning can be incorporated to accentuate a particular note in the design. Nash used this method in his vertical supports

315

in the Pavilion at Brighton, and the same technique is apparent, often brilliantly executed, in many early nineteenth-century lamp-posts and bollards: it was frequently marred by the addition of stylistic detail, in the railway architecture later in the century. The extensive use of cast iron for piers and seaside architecture during the last fifty years, has already been mentioned. Its use was commended partly because of its high resistance to sea air and water, and partly because the material facilitates the cheap reproduction of complicated detail.

Structural cast iron work was used with considerable success in many underground railway tunnels and in such engineering works as the Mersey road tunnel, for which Mr. H. J. Rowse, F.R.I.B.A., was the architect, and Sir Basil Mott, C.B., F.R.S., the consulting engineer. For the work, great iron segments were carefully cast and all bolted in position before the lining of the tunnel was completed. For the road surface, cast iron road setts with a specially hardened surface were used in order to reduce repairs and thus obviate interruption of traffic. (Figs. 431 to 434, pages 336 to 338.)

In the last thirty years, Scottish foundries have frequently influenced the use of cast iron in architectural design. Many of them can show a considerable number of examples of the material used structurally as panels, in-filling, windows, pavilions, shelters, and great gates and doors, apart from the usual uses in building and domestic equipment. These structural uses represent productive co-operation between the foundry, the architect and the sculptor. The designs of the sculptor may appositely suggest the "poured" quality of the material. All too often in the past has cast iron been made to imitate another material in designs produced without regard for its inherent qualities and properties. On account of its resistance to wear and tear, cast iron has for a century and a half, been used almost exclusively for street furniture. In the early nineteenth century the bollards, signs and lamp-posts had elegance. Few of the lamp-posts, bollards, signs, street name panels, shelters and kiosks, of the last fifty years can qualify for that description. W. R. Lethaby, with his intense and informed interest in materials, never lost his regard for cast iron, and always held that it was undeservedly neglected by architects. In a series of articles, published in the *Builder* during October, November and

FIG. 402. A shelter in cast iron designed for 'bus passengers. There is still some superfluous detail; but the general effect is restrained, and it represents a good use of the material.

Fig. 403. Compare this shelter with Fig. 326 on page 269, in the nineteenth century examples of Section 4; evidence of stylistic treatment still remains, but the slender columns are used appropriately, and there is no extraneous ornamentation.

December, 1926, he wrote:—"Any attempt to reform and perfect the use of cast iron in modern building would have to be done with great reticence and modesty in a straightforward and functional way, as elegant engineering rather than ornamental designs in sham styles."

Lethaby realised what very few people do realise, that an immense amount of the everyday things that we accept as our background, are made of cast iron. In drawing attention to what has been called "street furniture" he said: "Our pillar post-boxes are efficient and inoffensive, we have become accustomed to them, and they are without much 'design' nonsense; their colour certainly helps them. Done in due course and without any pretence, they are one of the best triumphs of nineteenth-century art: many of us have quite an affection for them. There have been rumours from time to time that they were to be 'improved' by the imposition of style design, but fortunately it has not yet happened."

Sir Giles Gilbert Scott's telephone kiosks, designed in cast iron for the G.P.O., were not then in general use; for they certainly fulfil Lethaby's desire for "elegant engineering". Richard Sheppard makes an instructive comment on these kiosks. He says: "In point of mere numbers, the G.P.O. telephone kiosk must be the most successful piece of pre-fabrication ever designed. The latest pattern consists of four panels and a roof unit which are rebated and interlocking and fixed by bolting. The panels themselves vary as sides and door are made for glazing, while the rear unit is solid. To overcome the weakness of castings incorporating such a large opening, the door

x

Fig. 404. Cast iron window surrounds, panels and mullions in Unilever House, London. The elaborate broken pediment and the other classical details, have direct continuity with the discovery made by John Nash one hundred years earlier, that the classical orders of architecture and all their mouldings and ornamental details, could be admirably reproduced in cast iron. (Architects: Sir John Burnet, Tait and Lorne.)

FIG. 405. Window mullions and apron panels in the Scottish Legal Building, Glasgow. Although the stonework façade is classical in character, the design of the cast iron panels has vigorous independence of character, though completely in harmony with the façade of the building.

Fig. 406 (*Above*). Cast iron window breast panels in Lothian House, Edinburgh. Architect: Stewart Kaye, A.R.I.B.A. Sculptor: Pilkington Jackson. A bold and able design, exemplifying an appropriate use of the material.

Fig. 407 (*Right*). Cast iron window bays in the new Adelphi Building, London. The external finish on these bays is sprayed aluminium paint. Architects: Stanley Hamp, F.R.I.B.A. (Colcutt & Hamp).

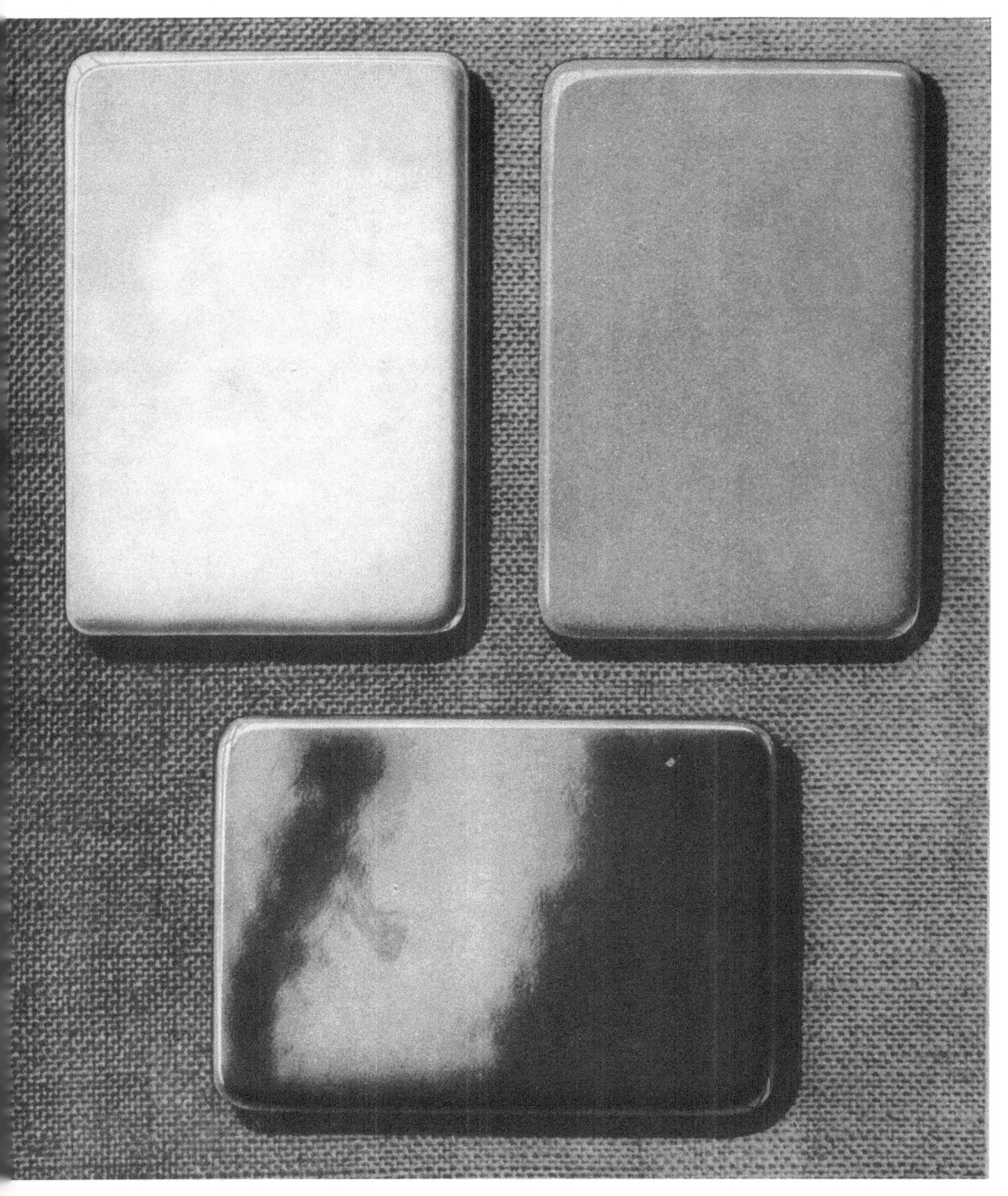

PLATE V. Three types of surface finish for cast iron
(*Above left*) Porcelain white enamel.
(*Above right*) Organic paint finish in colour.
(*Below*) Processed vitreous enamel finish.

Fig. 408. The G.P.O. telephone kiosk in cast iron, designed by Sir Giles Scott, O.M., R.A., PP.R.I.B.A. This is an admirable example of the use of cast iron for small prefabricated structures, and was designed for large scale mass production. The vertical reeding used in the corner sections, has a structural function.

Photograph: E. R. Jarrett.

and window units are formed with curved projections which raise them in front of the panel. For the same reason vertical reeding is used in the corner sections. The design of small structures of this type calls for a very close study of their individual requirements. The G.P.O. kiosk is an example of the close approximation of means to ends, and if iron is applied to a great range, a corresponding study of detailed requirements must be made. If the number of structural components can be reduced to a minimum, and the panels themselves standardised so that they are interchangeable, then it is possible to manufacture a range of such units which can be adapted for different uses. It would enable standard panels to be produced, with a number of different finishes, and would allow definite scope for individuality in the treatment of each type. The dimensional standards of these units—electric sub-stations, transport shelters, kiosks, police boxes, fire alarms and ambulance boxes—as well as their assembly, would form a productive study in technique".[3] The standard G.P.O. pillar boxes, which Lethaby described as "one of the best triumphs of nineteenth-century art", have outgrown their early ornamental complexities, and so have many lamp standards and bollards, particularly those designed before 1939 by H.M. Office of Works.

The ironfounding industry is taking a large part in the programme of the British Standards Institution in standardising some of the governing dimensions of cast iron building equipment, which have hitherto shown extraordinary variations.

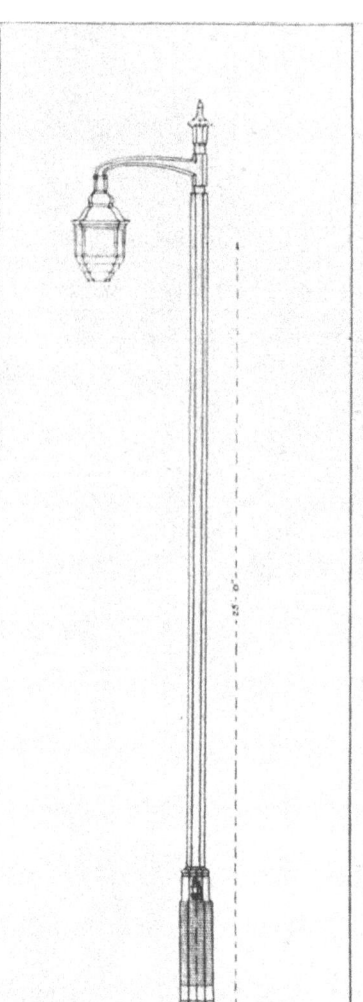
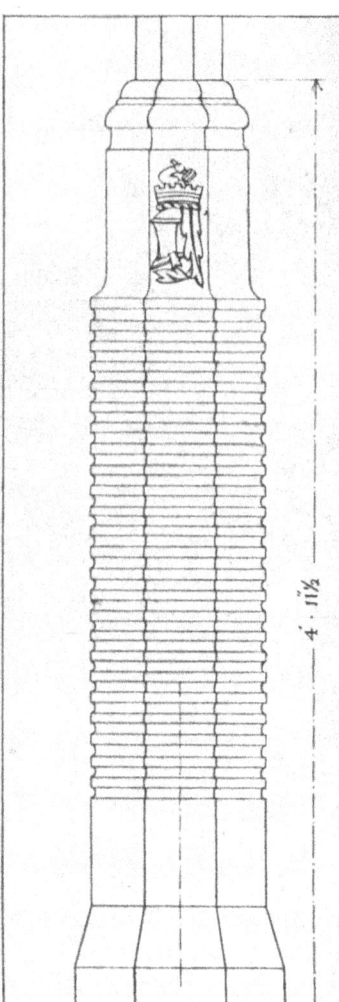

Fig. 409 (*Right*). Lamp standards designed for Birmingham by the Technical Committee of the Birmingham Civic Society, working in collaboration with the City Lighting Department. These standards were erected in 1938. The heraldic device on the base was modelled by a student of the Birmingham City Art School. This is an excellent contemporary use of the material. Compare the treatment of the base of this lamp standard with the bollard shown in Fig. 419, designed by the former Office of Works.

Reproduced by courtesy of the Birmingham Civic Society.

Fig. 410 (*Left*). A cast iron street lamp post at Hong Kong.

In building equipment, the familiar items are cookers, stoves, baths and basins, though the extent to which these everyday articles depend on cast iron is seldom appreciated. The greatest improvement in equipment in the last fifty years has been in kitchen ranges, heat-storage cookers, fires, and gas and electric stoves. To-day there are innumerable cookers, all varying slightly in character and performance, but all bright, easily cleaned, free from unnecessary knobs and enrichments, and all made of cast iron. During the 1939-45 war, considerable research was undertaken into the various methods of reducing fuel consumption and smoke production, and securing the full benefit of heat from fuel, so that cooking, space heating and water heating could be accomplished by the same unit.

Although a comparatively small amount of cast iron has been used structurally in the present century, the annual tonnage of the material employed in building and civil engineering is enormous. In 1945, there were nearly 1,800 foundries in Great Britain with a total annual output of approximately $2\frac{1}{2}$ million tons of cast iron. Castings for building and domestic uses represent about one-sixth of this figure, and pipes and castings for civil engineering works another sixth; so that architectural and building requirements total about one-third of the annual output.

322

A full list of building uses would be enormous, but the principal applications in building and street architecture, are as follows: Rainwater pipes, gutters, and rainwater heads; hot water and steam pipes and fittings; gas, water, sewage pipes and fittings; baths, basins, sinks, tanks and lavatory cisterns; gas, electric and solid fuel and heat-storage cookers, grates and stoves; refrigerators, boilers, furnace pans, coppers, radiators and fittings; door furniture, locks, keys and hinges; manhole covers, gratings, soot doors and air bricks; gates, fences, parapets and railings; signs, nameplates, lettering, milestones, bollards, curbs and road setts; gas and electric lamp standards, telephone boxes, pillar boxes, police and first-aid boxes, and shelters; staircases and tunnel segments.

FIGS. 411 and 412 (*Below*). Fig. 411 shows a Victorian cast iron pillar box; one of the designs that demonstrated how cast iron could be used to produce strong, satisfactory shapes, appropriately embellished: none of the ornamentation was overdone. Compare this with Fig. 412: in this modern version there is a little moulded detail, and ornamentation is confined to some fluting below the overhanging top.

Reproduced by courtesy of the Post Master General.

FIG. 413 (*Above*). A Victorian posting box, almost purely functional in character: every piece of moulded detail on its surface has a purpose and there are no extraneous trimmings.

Reproduced by courtesy of the Post Master General.

FIG. 414 (*Above*).

FIG. 416 (*Below*).

FIGS. 414 to 417 inclusive. Four examples of road traffic signs which show how a straightforward standard design may be reproduced with satisfactory results.

FIG. 415 (*Above*).

FIG. 417 (*Below*).

Fig. 418. Railings to the Embankment Gardens, London. These are the direct descendants of the railings designed by James Gibbs for the Senate House at Cambridge (see Figs. 42 and 43 in Section I on pages 46 and 47), and they also have something of Sir Robert Taylor's simple adaptation of a classical order in the railings of Ely House, Dover Street, London, shown in Fig. 120, Section II, on page 119. Victorian fussiness has been shed completely and these railings reflect the classical revival in architectural design which characterised the opening decades of the present century, inspired by the work of such architects as Norman Shaw.

Fig. 419. Bollard designed by the former Office of Works for refuge islands in Hyde Park, London. (These bollards are shown in position in Fig. 420, opposite.) The horizontal serrations provide a decorative treatment suitable for casting, which adds structural strength as well as lightening the appearance of the surface. The devotees of functional austerity might assume erroneously, that this decorative treatment lacked purpose: but in the technique of casting, it has structural significance. (See Richard Sheppard's remarks quoted on pages 317 and 321, relating to the reeding of the G.P.O. telephone kiosk shown in Fig. 408.)

Fig. 420. Lamp standard and bollards on a refuge island in Hyde Park, London, designed by the former Office of Works. Both lamp standard and bollards have the same horizontal decorative treatment, deliberately designed to give additional strength. (See Fig. 419 opposite.) The lodge gates and posts in the background are also of cast iron.

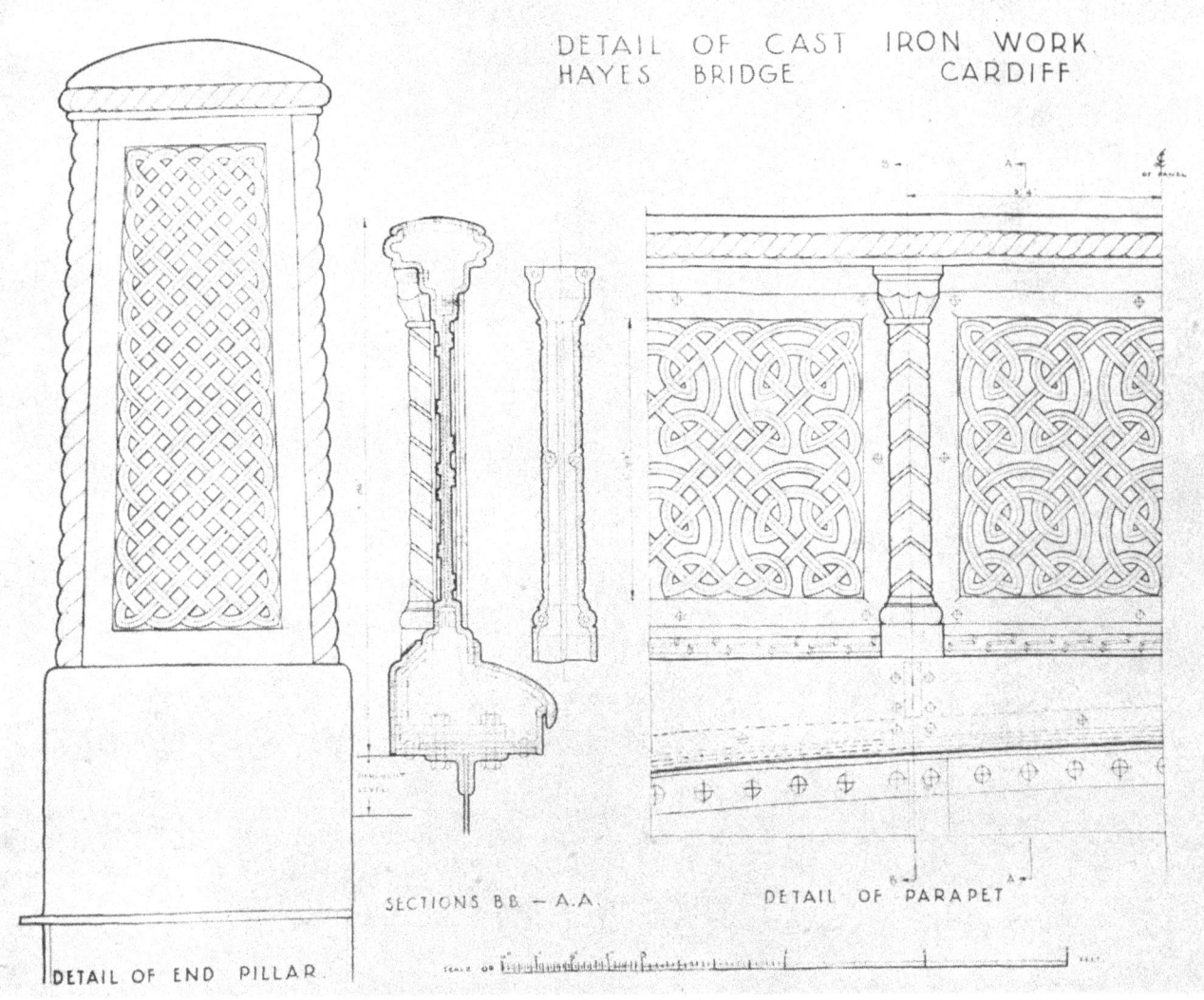

FIG. 421. Hayes Bridge at Cardiff. Above is a detailed drawing of the decorative work on the end pillar and the panels. A general view of the parapet appears on the opposite page, in Fig. 422.

If the panels on the end pillar and the parapet were isolated, shown, as it were, out of their context, they might be mistaken for some Northumbrian mason's work, in that brief but vivid civilisation that existed in North Eastern Britain in the eighth and ninth centuries. These interlocking designs, which may have begun as a conscious attempt at "Celtic revivalism" in ornament, provide an apt ornamental treatment for cast iron surfaces.

These details are drawn by E. G. Membery from drawings lent to the authors by the City Surveyor of Cardiff.

Fig. 422. This view of the end pillars and parapet of Hayes Bridge, Cardiff, shows the admirable continuity of effect gained by the decorative treatment. The ornamental forms have been selected with great judgment, and associated with a masterly sense of proportion. The forms may be archaic in character; but this is not "copybook" work; it is superior to the mere "lifting" of classical and Gothic forms of ornament which the scissors-and-paste men of the nineteenth century practised with such infelicitous zeal. This work gives new life to ancient traditions of form, and has given new life to cast iron as a material.

Fig. 423. Croxdale Bridge, Sunderland, Durham. Designed by A. E. Brookes, 1924 - 1926. The total length is 405 ft., each span being 48 ft. 2 ins. long. There is a 30 ft. carriageway, and 7 ft. 6 ins. footways. The concrete piers are faced with masonry, and the main spans are of steel plate girders, with cast iron parapets.

Reproduced by courtesy of the Public Works, Roads and Transport Congress.

Fig. 424 (*Above*) and Fig. 425 (*Below*). Cast iron bridge parapet in Cardiff. The same free and vigorous treatment, apparent in the Hayes Bridge shown in Figs. 421 and 422, is in evidence here. The fact that the parapet is pierced instead of being composed of cast panels, gives the design greater fluidity of character; but even when a more rigid treatment has been adopted, as in Fig. 425, vigour has not been sacrificed. This second example of a bridge parapet at Cardiff draws some of its inspiration from mediæval sources; but it is no slavish copy of an antique model—it is nourished and not paralysed by the past.

Fig. 426. Cast iron window at Sassoon House, Shanghai.
Architects: Palmer and Turner.

FIG. 427. Cast iron doors, Sassoon House, Shanghai. *Architects:* Palmer and Turner.

Fig. 428 (*Left*). Cast iron flower box designed by Frederick Gibberd, F.R.I.B.A., for the British Iron and Steel Federation house at Northolt, Middlesex. The designer has used reeding for the surface, a simple and most effective treatment.

Fig. 429 (*Right*). Cast iron staircase, designed by Professor Sir Charles Reilly, F.R.I.B.A., for the Students Union at Liverpool University. This design, with its classical elegance, picks up the threads of tradition where they were broken in the early nineteenth century. There is an intricate delicacy about the balustrade which suggests the work of the brothers Adam, although it is an original treatment and not based on any traditional pattern.

Fig. 430. Cast iron gates to Clive Buildings (now known as Gillander House), Calcutta. Designed in 1910 by H. S. Goodhart-Rendel, pp.r.i.b.a. These gates, particularly the large circular panels, show the great delicacy that may be achieved in cast iron.

Fig. 431 (*Above*). The Mersey Tunnel connecting Liverpool with Birkenhead. A view of the cast iron segments, bolted together.

Fig. 432 (*Right*). Bolting the cast iron segments. See Fig. 433 opposite, and Fig. 434 on page 338.

PLATE VI. Surface finishes for cast iron.
(*Above left*) Nickel plated surface.
(*Above right*) Surface sprayed with aluminium.
(*Below*) "Parkerised" finish.

FIG. 433. The Mersey Tunnel after completion. See Figs. 431 and 432 opposite, and Fig. 434 on the following page.

Fig. 434. Laying the special cast iron setts on the roadway of the Mersey Tunnel. These have a backing of tough cast iron, with a special hardened "white iron" surface to resist wear.

Fig. 435.

Fig. 436.

Fig. 437.

Figs. 435, 436 and 437. Figs. 435 and 437 show non-slip finish patterns in cast iron floor tiles, and Fig. 436 shows the underside of the tile.
Reproduced by courtesy of the Butterley Co. Ltd.

Fig. 438 (*Above*). Cast iron paving tiles used on the floor of a warehouse, designed to resist heavy wear and tear.
Reproduced by courtesy of the Butterley Co. Ltd.

Fig. 439 (*Left*). Cast iron road setts used where there is excessive wear caused by the stopping and starting of traffic.

Fig. 440 (*Above*). A 200,000 gallon external water supply tank, made from standard castings bolted together.

Fig. 441 (*Below*). A 5,000 gallon water storage tank made from standard castings bolted together.

Fig. 442 Fig. 443

Fig. 444.

Fig. 442 (*Above left*). A Victorian kitchen range, with many projections that cause extra work.

Fig. 443 (*Above right*). The same stove as it was re-designed in the period between the wars. It had an enamel finish and many of the ornamental features had been simplified to facilitate such a finish.

Fig. 444 (*Immediately above*). Here is a contemporary version of the same design with far fewer parts, thus reducing labour costs for assembling. Insulation has been applied to the doors, sides and back. No hinges are visible. The general form is smooth and the surface is easy to keep clean.

Fig. 445 (*Above*). A solid fuel fire and cooker in cast iron with a vitreous enamelled finish.

Fig. 446 (*Below*). This shows the neat finish of the flue surround, door backs and interior when the appliance is open.

Fig. 447 (*Above*). A solid fuel cooker in cast iron with a vitreous enamelled finish of biscuit colour and black.

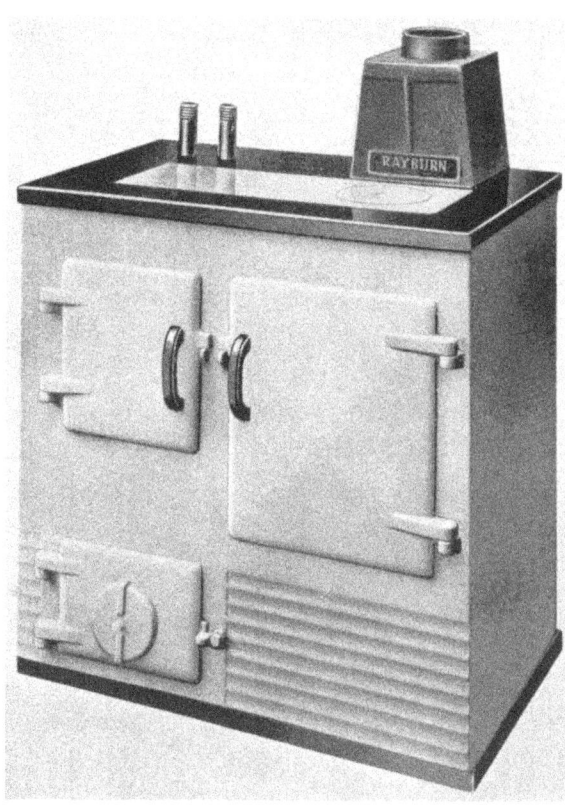

Fig. 448 (*Above*). Another solid fuel cooker with similar finish to that shown in Fig. 447.

Fig. 449 (*Above*). A heat storage insulated cooker and water heater in vitreous enamelled cast iron.

Fig. 450 (*Above*). A continuous burning insulated cooker and water heater in black and cream vitreous enamelled cast iron.

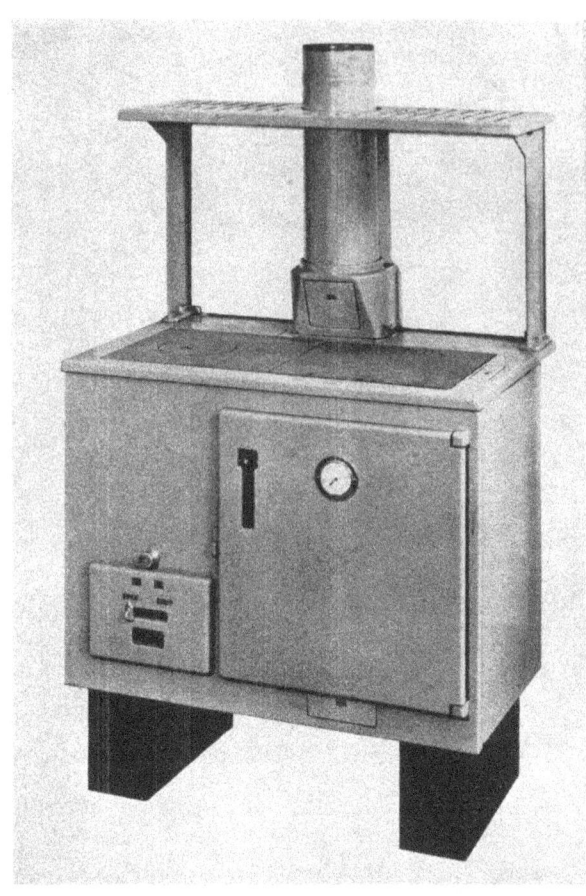

Fig. 451 (*Above left*). A slow burning solid fuel cooker in black and white vitreous enamelled cast iron.

Fig. 452 (*Above right*). A solid fuel cooker with flue and plate rack, in vitreous enamelled cast iron.

Fig. 453 (*Right*). An insulated solid fuel cooker with enclosed flue and plate rack, in vitreous enamelled cast iron.

FIG. 454 (*Above*). A domestic hot water heater vitreous enamelled cast iron.

FIG. 455 (*Below*). A domestic hot water boiler black and white vitreous enamelled cast iron.

FIG. 456 (*Above*). A solid fuel cooker, wall panels and table top in vitreous enamelled cast iron.

345

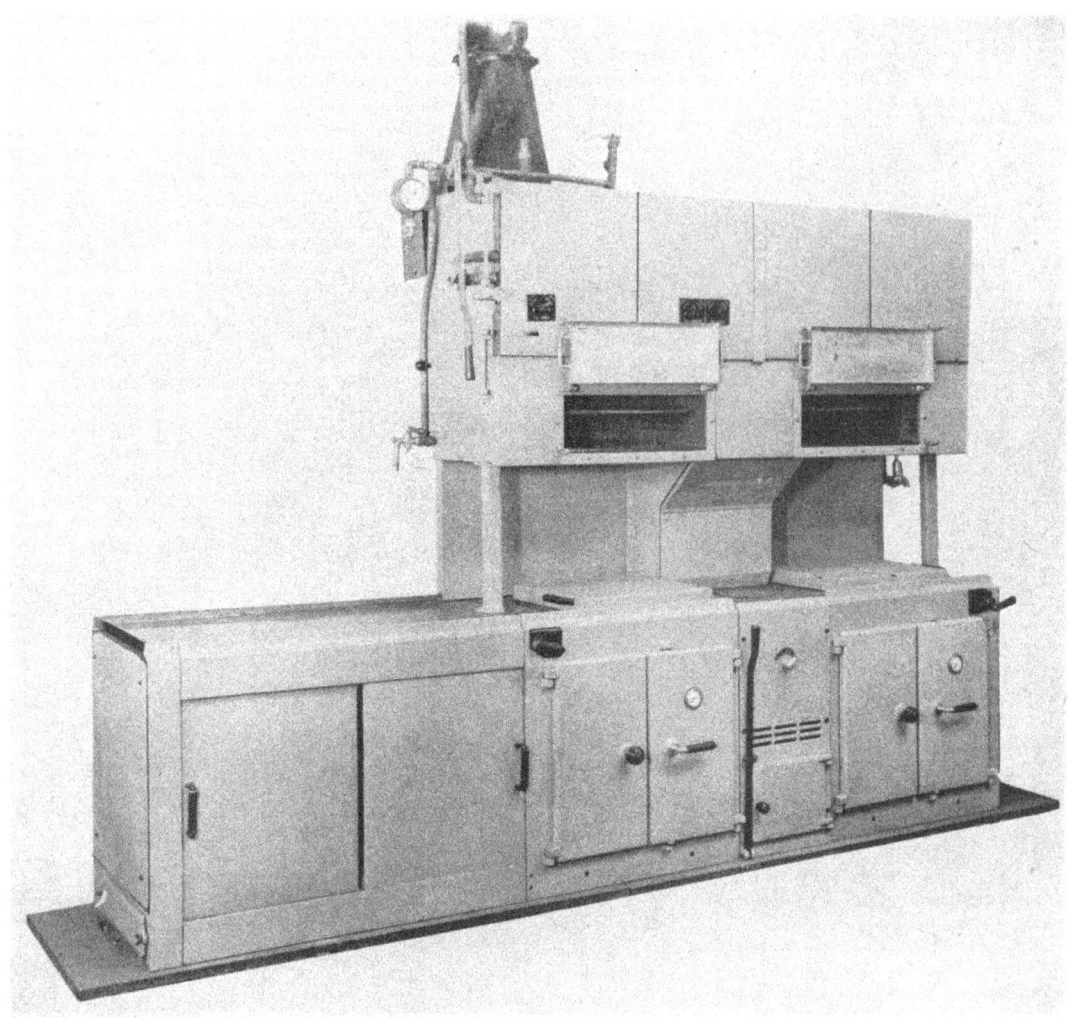

Fig. 457 (*Above*). Solid fuel cooker and hot closets, designed for the kitchen quarters of a railway restaurant car, in vitreous enamelled cast iron.

Fig. 458 (*Below*). Cookers, boilers and steamers for institutional cooking. The finish on these cast iron appliances is vitreous enamel.

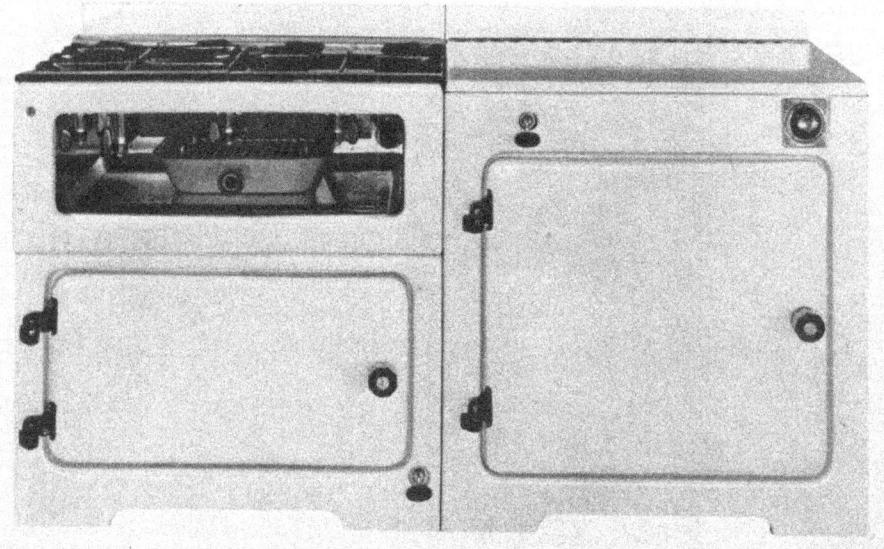

Fig. 459 (*Above*). Gas cooker in vitreous enamelled cast iron.

Fig. 460 (*Right*). Gas cooker in sheet metal and vitreous enamelled cast iron.

Fig. 461 (*Left*). Cast iron and sheet steel are used in this cabinet gas fire, with oven, gas rings and cover top.

Fig. 462 (*Right*). Gas cooker in vitreous enamelled cast iron. The enamelling of the parts in direct contact with the gas rings and burners is specially resistant to heat.

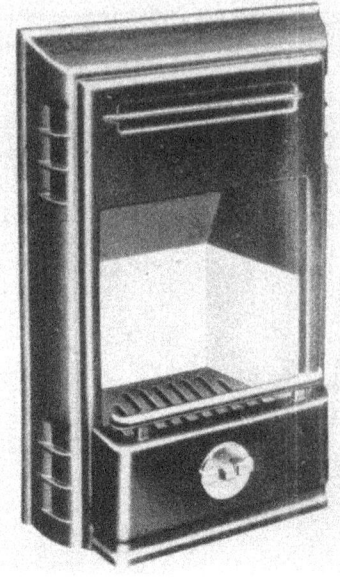

Fig. 464 (*Below*). A slow burning solid fuel space heater, designed to fit into an opening and to stand almost flush with the wall surface. The finish is vitreous enamel.

Fig. 463 (*Above*). A nineteenth century design for an independent stove, compared with a modern projecting fire: both are in cast iron. The contrast in the designer's approach to the use of the material reflects the Victorian preoccupation with ornament and the contemporary respect for functional frankness.

Fig. 465 (*Below*). Domestic hot water boiler finished in vitreous enamel.

Fig. 466 (*Left*). Slow burning heating stove in vitreous enamelled cast iron.

Fig. 467 (*Above*). Solid fuel slow combustion heating stove finished in vitreous enamel.

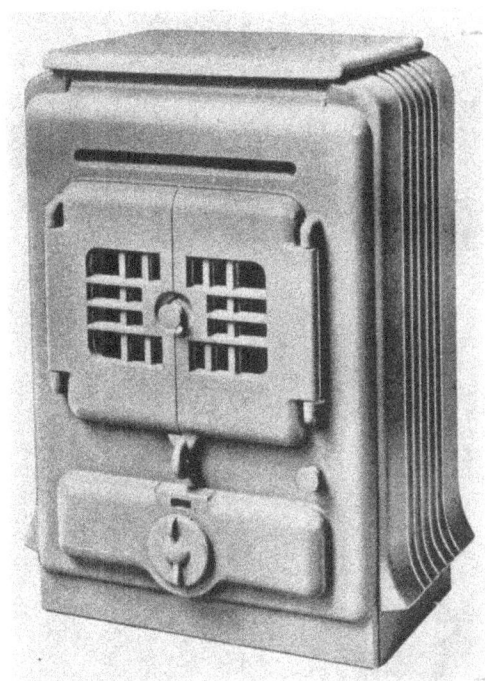

Fig. 468 (*Above*). Slow burning solid fuel space heater designed to stand free. Finished in vitreous enamel.

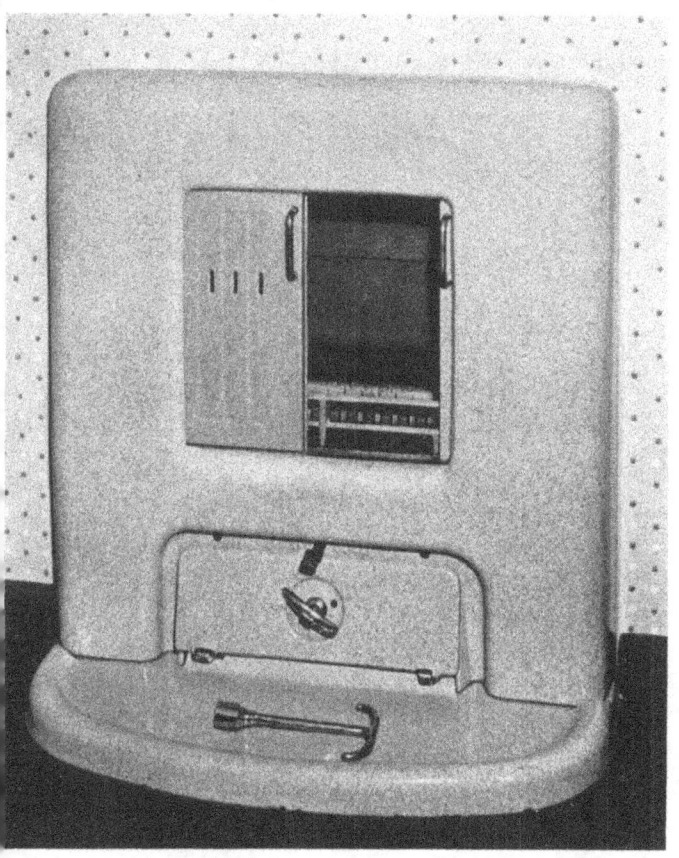

Fig. 469. A slow burning stove that can be closed up completely or left open; finished in vitreous enamel.

Fig. 470. A slow burning stove for space heating, finished in vitreous enamel.

Fig. 471 (*Above*). A gas fire in cast iron, finished in stove enamel.

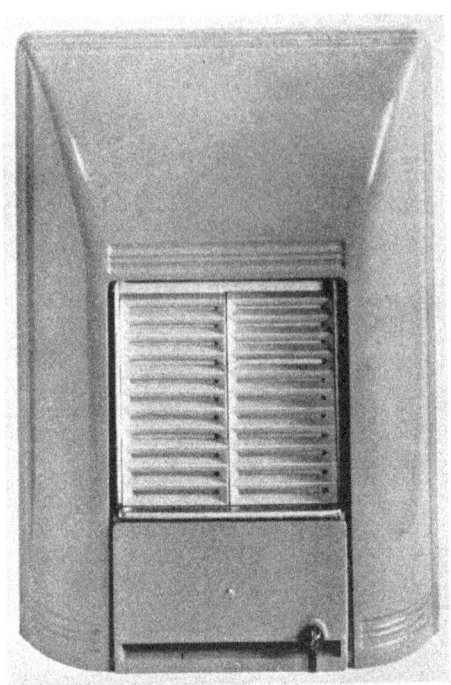

Fig. 472 (*Above*). A gas fire finished in vitreous enamelled cast iron.

Fig. 473 (*Above*). A gas fire with a new type of burner. This design allows for convected hot air heating through the grilles.

The designs on this page show the possibility of producing the whole surround for a gas fire in a simple form with a clean and pleasant finish.

Fig. 474 (*Above*) and Fig. 475 (*Right*). Two gas fires in cast iron, finished in stove enamel.

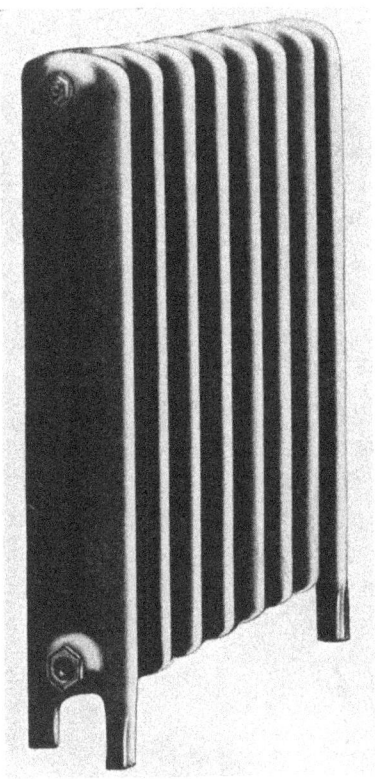

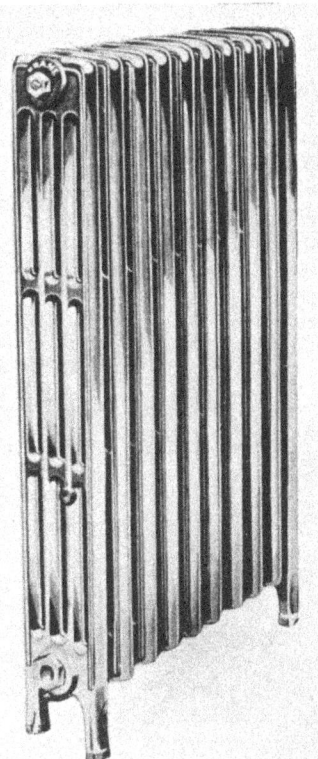

Fig. 476.   Fig. 477.   Fig. 478.

Figs. 476, 477 and 478 (*Above*). These are typical examples of contemporary design in cast iron radiators.

Fig. 479 (*Right*). A gas heater in cast iron finished in bronze stove enamel.

Fig. 480 (*Left*). An electric heater in cast iron with a smooth enamelled finish.

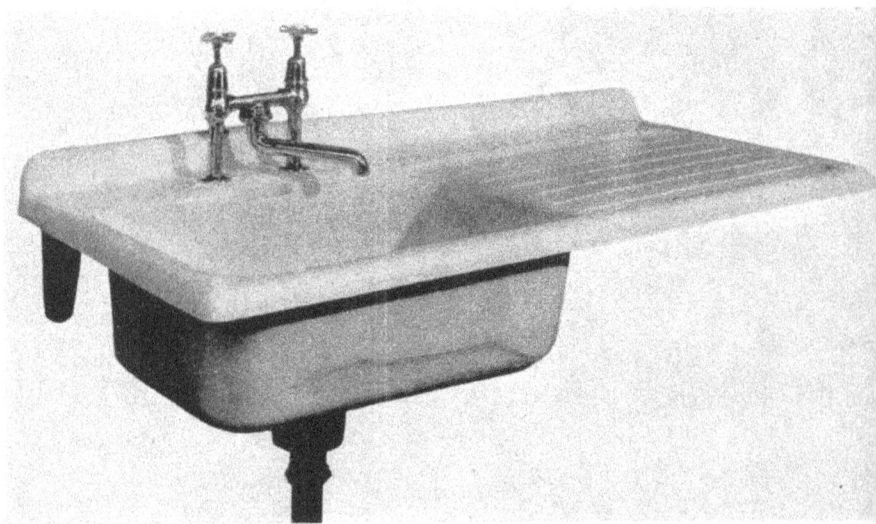

Fig. 481 (*Left*). A range typical cast iron rainwat heads exemplifying an a and restrained use of t material.

Fig. 482 (*Right*). A cast iron sink with draining board, finished in acid-resisting vitreous enamel.

FIG. 483 (*Below*). A wash basin of vitreous enamelled cast iron with built-in supports of the same material and finish.

FIG. 484 (*Above*). A kitchen installation with a sink, draining boards and splashback tiles of enamelled cast iron.

FIG. 485 (*Below*). A cast iron shower bath tray in vitreous enamel finish.

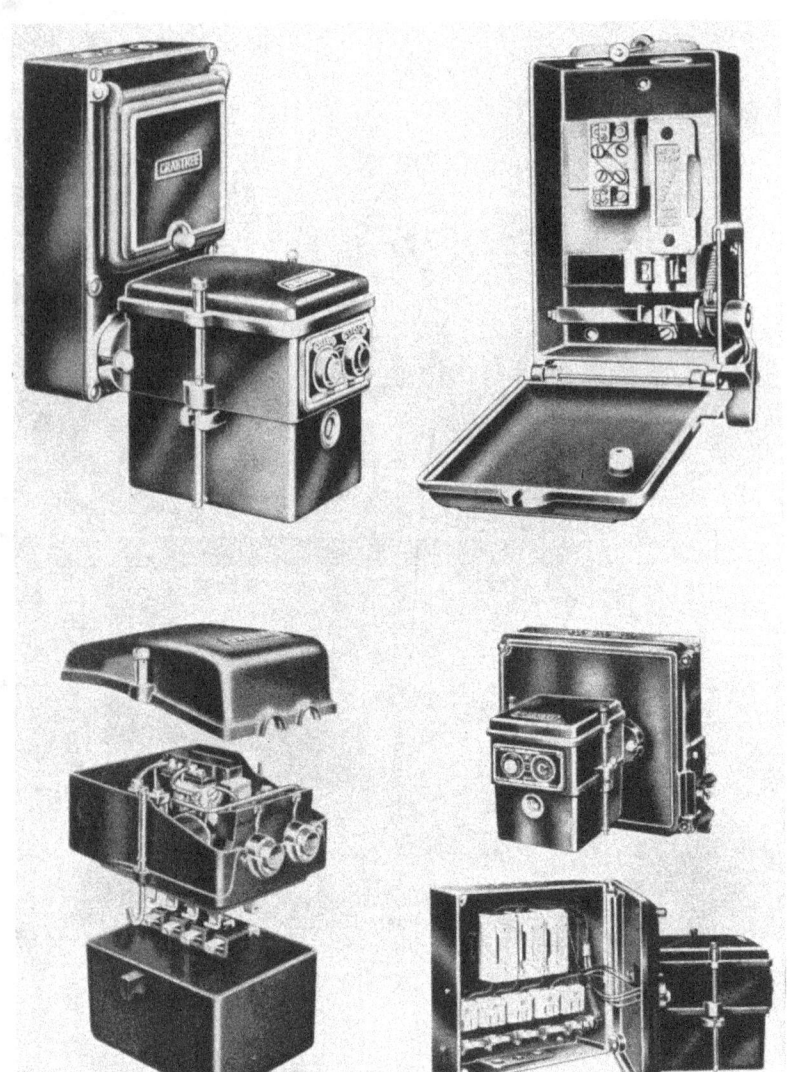

Fig. 486 (*Left*). Fuse boxes and switch gear for electric circuits, in cast iron. The material is used extensively in this form of electrical equipment.

Fig. 487 (*Below*). Further examples of fuse boxes and switch gear in cast iron.

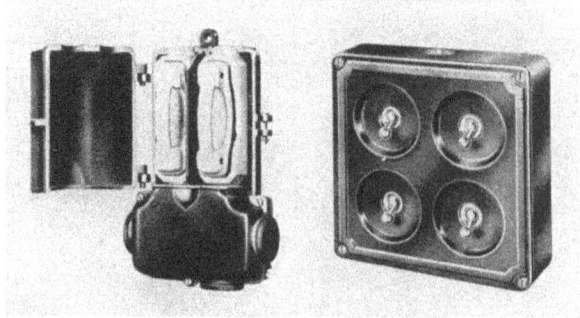

2. INDUSTRIAL AND SCIENTIFIC DEVELOPMENTS.

Since the end of the 1914-18 war, three fundamental developments have affected the technique, organisation and progress of the industry. Examination of these developments must be prefaced by a brief account of the principal operations in the production of a simple casting in a foundry. Foundry work may be described under these five headings: Metal; Pattern; Moulding; Core-making; and Assembly.

Metal. The Metal used to produce castings is obtained by re-melting pig iron and scrap in a furnace known as a cupola. The composition of the metals to be charged into the cupola is generally decided by the foundry manager or metallurgist and is of great importance, for on this depends whether the resulting metal is suitable for thin castings, heavy castings of thick section, castings to resist wear, or castings to resist pressure.

Pattern. Before any mould can be made from which to cast the finished article, it is necessary to have a pattern. This is generally made in wood for ease of working; but where many castings are to be made, the pattern is often formed in metal or other materials to resist wear and tear. Patterns are made slightly larger than the size of the finished casting to allow for the shrinkage which takes place on cooling, the allowance being one-eighth inch in a foot. Pattern making is highly skilled work, for not only must allowance be made for shrinkage, but often it is necessary, because the casting has adjacent areas of varying thicknesses, to allow a correction for warping as the casting cools.

Moulding. The moulder has to make a cavity in sand corresponding exactly in shape and size with the pattern. All sands are not suitable for moulding purposes, and the sand used must contain the correct proportions of clay and of water to make the mixture plastic and workable. Research has shown how moulding sand can be treated so that the right mixture for a required operation is secured; and most foundries have facilities for testing strength, the degree to which the sand is permeable, the moisture content and other properties.

The pattern is put in a four-sided metal frame called a "moulding box" and the sand rammed round the pattern. The box is made in two halves, the lower being known as the "drag" and the upper as the "cope", with pins and lugs on each to hold the two tightly together in the correct position. The pattern is so placed in the double box that, when all the sand has been rammed, the two halves can be disengaged and the pattern withdrawn. When the two halves are again connected a cavity is thus left in the sand—the actual mould—which, when filled with molten metal, produces a casting of the required shape. In order to allow the metal to be poured into this mould suitable channels and passages are cut into the sand, connecting with an opening at the top of the sand mould called a "gate". Another channel allows the exit of surplus molten metal.

Core-making. In a hollow casting something must obviously fill the hollow temporarily to avoid the casting being entirely solid. This filling is termed

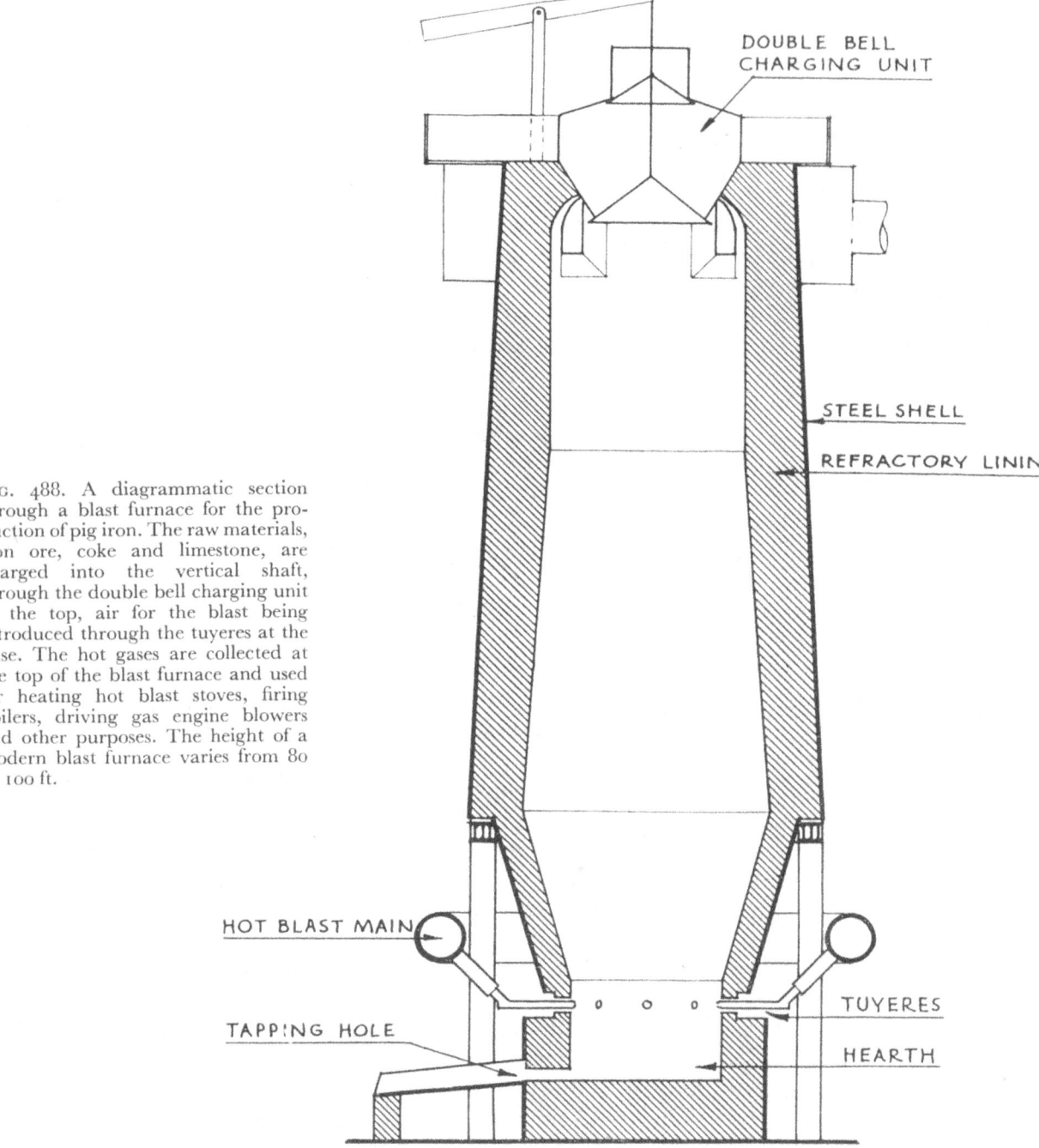

Fig. 488. A diagrammatic section through a blast furnace for the production of pig iron. The raw materials, iron ore, coke and limestone, are charged into the vertical shaft, through the double bell charging unit at the top, air for the blast being introduced through the tuyeres at the base. The hot gases are collected at the top of the blast furnace and used for heating hot blast stoves, firing boilers, driving gas engine blowers and other purposes. The height of a modern blast furnace varies from 80 to 100 ft.

a "core". It is the pattern maker's responsibility to produce a "core box" which is a mould, with a cavity corresponding exactly to the final cavity in the finished casting. This core box is filled and rammed by the "core maker" with special sand mixtures and then dried or baked to form the hard core, which is inserted and kept in position in the cavity in the mould.

Assembly. In assembling the mould and cores, the latter are first placed in position in the lower half of the mould and fixed in position, and the upper half of the mould is then carefully placed on top and the two parts clamped together ready for pouring. After pouring the molten metal into the mould it is allowed to cool; the box is then opened and the sand roughly cleared from the casting, which is removed to the dressing shop where it is finally cleaned with wire brushes, grinding wheels or other machines to

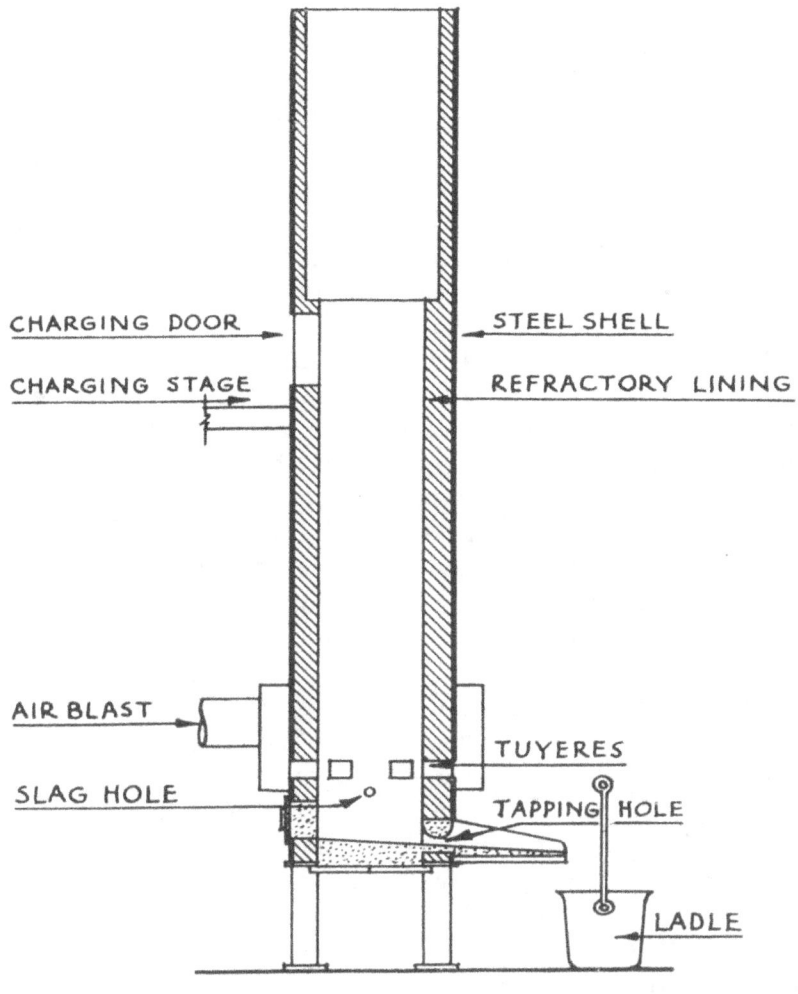

Fig. 489. Diagrammatic section through a cupola for melting pig iron before casting. The raw materials, pig iron, scrap iron and coke, are charged into a vertical shaft after a fire has been started at the base. Air for the full combustion of the fuel enters through the tuyeres. The molten metal is tapped out at the base. The height of a modern cupola from floor to charging stage varies from 15 to 20 ft.

remove any fins of metal where the two halves of the mould joined, or pieces of metal adhering where the channels in the mould adjoined the casting. A finish on the casting is often obtained by sand-blasting in a special chamber.

Though this description of making a simple casting by hand only deals with one aspect—albeit a basic aspect—of the subject; it reveals the extent of the skill and the exacting training and experience required to produce a good pattern-maker, moulder or core-maker, and shows the considerable amount of physical work necessary for the handling of the various materials.

The three fundamental developments concerning the technique, organisation and progress of the industry may be grouped under three headings:

(a) Mechanisation, which affects technique.
(b) Associations, which affect the organisation of the industry, and
(c) Research, which affects the technique, organisation, character and future of the industry.

(a) *Mechanisation.*

From the description that has been given of foundry methods, dependent as they appear to be almost wholly upon the skill of trained men, it would seem difficult to apply methods of mass production to the industry. Indeed such methods are not always applicable to the foundry; and in the past there has been much prejudice against mechanisation, often emanating from the workmen themselves, for the spirit of the Luddites still smoulders, and is often fanned into flame by improvements in industrial technique. Mechanisation can revolutionise the output of a foundry where large quantities of castings are to be made from a few standard patterns, particularly with such small castings as those needed for gas, electric or solid fuel cookers, stoves and other heating appliances, electric meters, switch boxes and so forth. John Dearden, the author of *Iron and Steel Today*, describes a typical modern foundry where "each moulding machine has a hopper immediately above it, from which the sand can be delivered straight into the moulding box without any effort. The finished moulds are taken from the machines and placed on a continuous conveyor, usually travelling at 6-12 ft. per minute. This carries them to a point near the cupola where they are poured as they pass. After a further 10-20 minutes' journey, during which time the metal cools and solidifies, the moulds arrive opposite a jolting grid, on to which they are pushed or lifted. The jolting action knocks out the sand, which falls through the grid, to be conveyed away to a reconditioning plant for further use. The red-hot castings are set aside to cool, and the empty boxes returned to the moulding machines".[4]

A machine known as a "sand slinger" is used where large moulds are to be filled. It is capable of propelling a stream of sand at sufficient speed for the sand to be automatically rammed by impact and at a rate of 5-10 cubic feet per minute. Core blowing machines blow sand into the core box by means of compressed air, and moulding machines have been developed to the point when "anything from 30 to 60 complete moulds can be produced in an hour by a pair of machines provided with sand-hoppers, mould conveyor, and the other necessary services".[5]

An instance of an old method of casting being replaced by a semi-mechanised method, is afforded by the centrifugal casting of straight iron pipes for gas, water, sewage and other technical purposes. The traditional method is still used, but in 1914, a French scientist, D. S. de Lavaud, introduced a method of centrifugal casting for cast iron pressure pipes, and in 1917 commercial production with this method began in Canada. Subsequently, large plants were erected in Britain for the production of centrifugally cast pipes in metal or sand-lined cylindrical metal moulds. No core is needed to form the inside of the pipe except at the socket end. The correct amount of molten metal is poured into a channel running down the centre of the cylindrical mould, which is revolved on its long axis at high speed. The channel is slowly withdrawn as it pours molten metal into the mould at a controlled rate, and centrifugal action forces the metal against the side of the mould. The whole operation is completed in about a minute. The force with which the metal is projected tends to consolidate the casting, and in some instances has reduced the thickness of pipes by as much as 25 per cent less than the normal thickness giving a lighter pipe of undiminished strength.

(*b*) *Associations.*

During the interval between the two world wars there was a tendency in the industry to form groups whereby several foundries became associated

FIG. 490 (*Right*). The illustrations on this page and the two that follow show the different stages in the casting of the front of a slow combustion stove. On the right the pattern and the finished article are shown.

FIG. 491 (*Right*). The bottom half of the box rammed and turned over, with the pattern still in position. The top half of the box is being lifted on.

FIG. 492 (*Left*). Ramming the top half box. The line of the pattern may be seen between the two boxes.

359

Fig. 493 (*Left*). Ramming completed; lifting the top half of the box.

Fig. 494 (*Right*). The top half of the box lifted; removing the pattern.

Fig. 495 (*Right*). The pattern removed; mould being blown free of dust with bellows.

Fig. 496 (*Right*). The completed mould being assembled, top half of box locked in position.

Fig. 497 (*Left*). Pouring the molten metal.

in order to overcome some of the drawbacks inherent in small units. This tendency has led to greater measures of standardisation in the products of the industry, and has thus anticipated some of the conditions created by the second world war. The British Standards Institution, which is continually scrutinising, creating and revising standard terms, definitions, codes of practice and specifications for materials and fabricated articles, has received constructive help from associations in the foundry industry. The industry has collaborated in the Institution's task of clarification, and is represented on the B.S.I. Committees which are dealing with the appropriate and essential standardisation of building equipment.

Some of the ironfounders' trade associations are based on the manufacture of a common product, such as builders' castings; others are regional in character, grouping founders in a specific industrial area. During the second world war, in order to simplify contact with Government Departments, the *Council of Ironfoundry Associations* was created, which linked existing sectional and regional associations, and formed one body.

The principal raw material of the foundry industry, pig iron, has to be of a special grade suited for ironfounding, differing from the type of pig iron used in steel manufacture, although both the iron and the steel trades use the low phosphorous pig iron known as hematite. The producers of foundry pig iron, who were also represented by several trade associations, formed the *Council of Iron Producers*. The common interests both of pig iron producers and ironfounders—producers and consumers of the major raw material, pig iron—were recognised in 1945 by the creation of the *Joint Iron Council*, and this body undertook long term planning for the industry, in such matters as education and training, research and re-equipment. Individual membership of a trade association is voluntary.

(c) *Research.*

Extension of research work has followed general recognition throughout the industry of its immense practical significance.

It was recorded in Section Three how Fairbairn in the mid-nineteenth century had advocated a thorough investigation into the properties of cast iron, and had envisaged some of the results achieved by subsequent research; but this research work was desultory and unsystematic. Only in the present century has scientific research been given its proper place in industry; and gradually important foundries organised special departments which dealt with the testing and control of all raw materials. In 1924 the British Cast Iron Research Association was formed. It was one of the Government-aided research associations designed to redress the technical deficiencies revealed in British industry during the first world war. Membership on the part of the producers was voluntary; and the association now includes over 500 individual firms, representing more than half of the output of the industry. The Association was first housed in Birmingham, but the headquarters were moved in 1942 to Alvechurch, some ten miles from that city. The Association works closely with the industry and is divided into four main sections; administration and offices; Library and Information Bureau; Research Department and Development Department. In 1944 a Building Uses Department was established to provide architects, engineers, surveyors, industrial designers and builders with comprehensive and impartial information on the building applications of cast iron. Similar services are available for engineering applications of the material.

3. SPECIAL CAST IRONS.

Research work conducted during the last twenty-five years has introduced cast irons to suit many special purposes. To-day metallurgists regard cast iron not as one material but as a series of materials with slightly varying properties. This metallurgical revolution has followed the application of science to the improvement of the materials and the methods of the industry.

Ordinary cast iron has a tensile strength of only 10-15 tons per square inch as compared with the 26-30 tons per square inch of mild steel. Now, by the addition of alloys, thermal treatment and special foundry processing, special

Fig. 498. A mechanised foundry. This shows the mechanised sand plant, sand hopper delivering sand to the moulders, casting runway with the ladle being used to pour the molten metal, and the mould conveyor.

Fig. 499 (*Above*). Moulding and conditioning plant.

Fig. 500 (*Above opposite*). Another view of the moulding and conditioning plant.

Fig. 501 (*Below opposite*). The moulding machines, showing the orderly arrangement. Each machine has two attendants, and each machine has an output of one mould per 40 seconds of an eight-hour day.

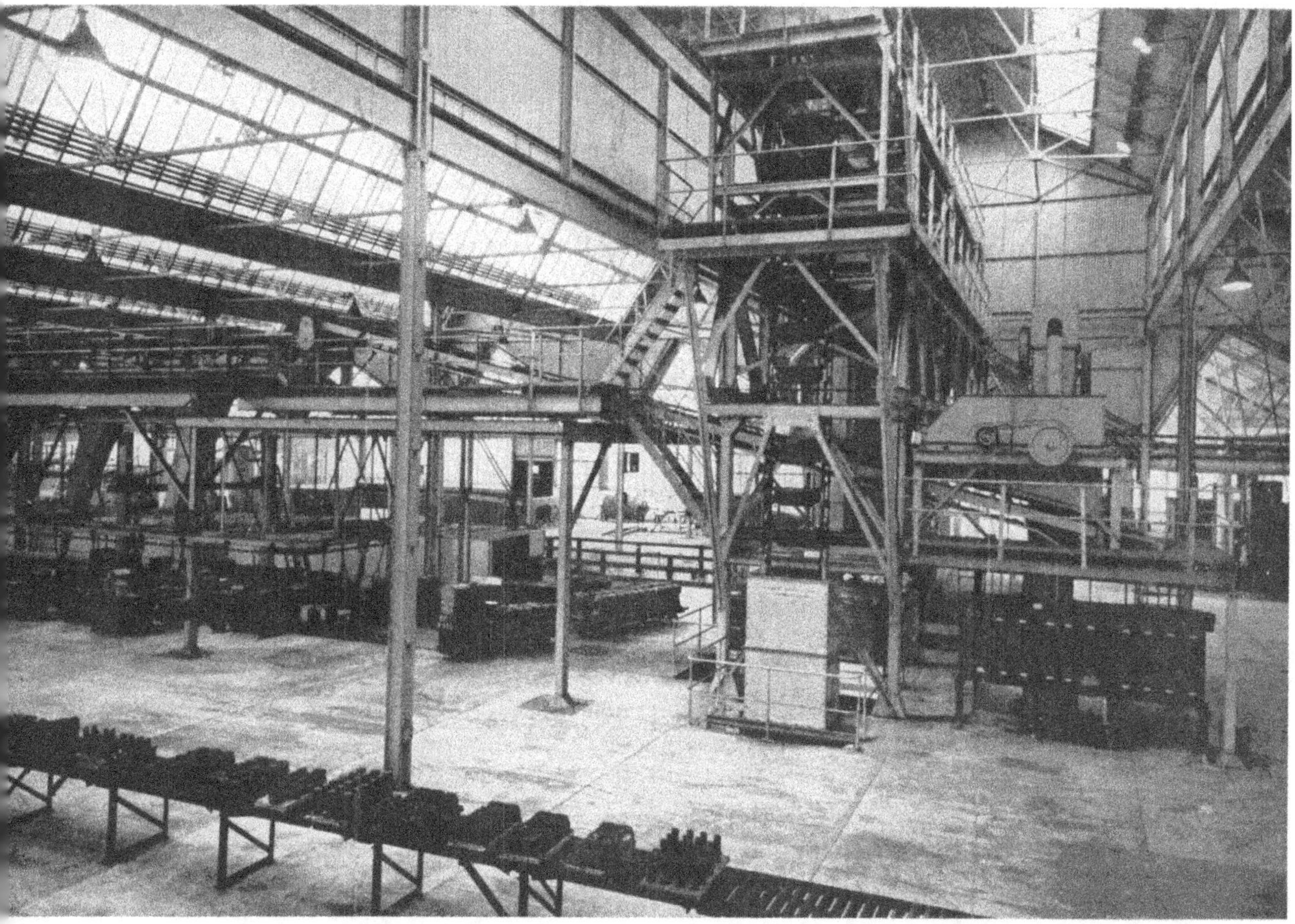

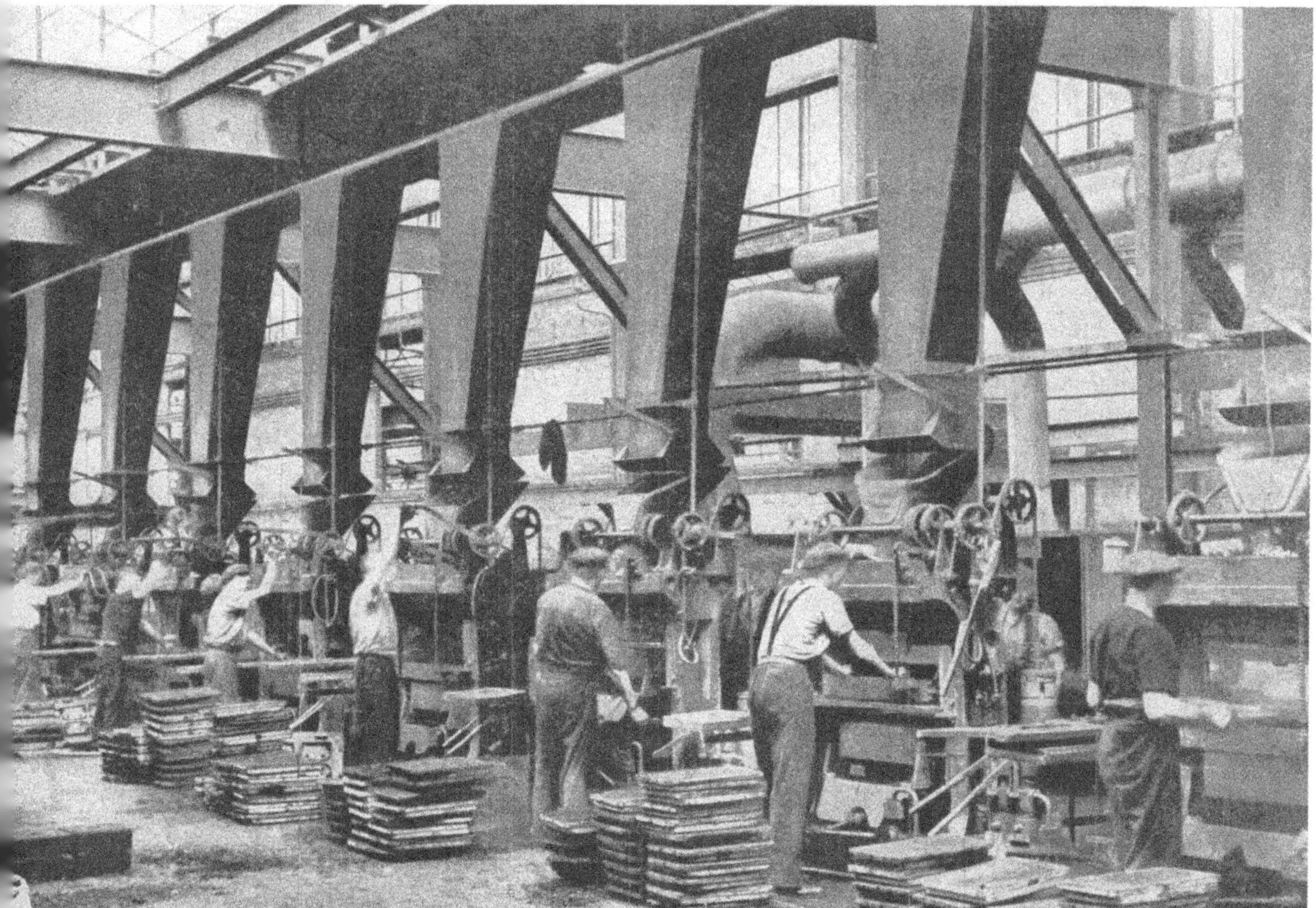

cast irons are available, both with and without brand names, with a tensile strength of 20-30 tons per square inch, and austenitic irons such as "Nomag", "Ni-resist" and "Nicrosilal", which give high resistance to corrosion and heat, and are soft, non-magnetic, ductile, resistant to shock, and can take a high polish. Martensitic irons are very hard and resistant to abrasion, erosion and wear. So-called acicular irons have been developed, which attain greater strengths than have been previously available.

Nodular cast iron. Late in 1946, on the occasion of the twenty-fifth anniversary of the British Cast Iron Research Association, the President of that body, Dr. Harold Hartley, announced the invention of a new type of cast iron. The results of this British invention are, as *Iron and Steel* pointed out: "almost immediately destined to revolutionize the foundry industry". This iron is produced by a simple treatment, which begins "with a straight cast iron and ends with a straight cast iron of vastly improved properties. That has a great significance in foundry economics".[6] The invention allows the use of unalloyed cast iron for many of the purposes for which alloy irons have hitherto been essential; with a consequent lowering of cost for such castings. The significance of its application to industry was thus summarised by the *Foundry Trade Journal*: "Whilst the engineer will appreciate the higher levels of shock resistance, tensile and transverse strength associated with ease of machining, the foundryman will welcome the ability to manufacture this new iron without taking all those precautions so necessary for the production of the high-duty alloyed or inoculated material."[7] Sir William Fairbairn's views, expressed during the eighteen fifties, on the possible improvements in the properties of the material to be derived from research and experiment have been abundantly justified as the result of twenty-five years of consistent research work. The properties of nodular cast iron may conceivably lead to the reinstatement of the material in structural work in building, as the improvement and perfection of a multiplicity of new finishes have extended its employment for all kinds of domestic equipment and ancillary uses in contemporary building.

Reinforced cast iron. This has been successfully used in Germany for the manufacture of certain types of machinery and equipment. In 1936 the Russian Academy of Architecture made a careful examination of the watchtower built in 1725 at the Newjanks Iron Works, in the Urals. This unexpectedly revealed that the cast iron beams serving as roof members were reinforced by steel to give greater resistance to tensile stresses, much as concrete is reinforced with steel for the same purpose to-day. In a paper on the subject Dr. Saller has said:

"The Russian term for this to-day is, literally, steel cast iron. The Russian engineer, Lewanow, gave a lecture before the Technical Council of Science of the Commissariat for Heavy Industry on the possible uses of cast iron, when reinforced in this way, for building and in engineering constructions. The advantage of this building material is that the steel reinforcement considerably increases the resistance to tensile stresses, while the steel part takes the major portion of the tensile stresses, and the mechanical properties of the cast iron

Fig. 502. View from the cupola charging platform of a modern mechanised foundry. There are four cupolas, working alternately; each pair is mechanically charged. Each cupola produces 45 tons of molten metal daily. Before the second world war, the floors were swept with a vacuum plant, but during the war period this was discontinued owing to shortage of labour.

base are improved, due to the diffusing of some carbon into the steel, and the rate of cooling being altered. This carburisation of the steel at the same time changes its properties to those of a high-carbon (high-duty) steel. Technologically the process, which can be used in every type of foundry, differs only in so far as, during the casting procedure, the reinforcement is inserted before pouring. Further advantages claimed for these reinforced cast iron structures are that their weight is 10 per cent less than that of steel and that any desired shape can be obtained with the casting process. Their absolute weight is four to five times less than that of ferro-concrete, their resistance to dynamic stresses and shock, etc., is said to be very good whilst their corrosion resistance is high. The Soviet Union intends to spend large sums for further research, and successful experiments have in fact already been made with slabs, beams and tunnel sections, etc., of reinforced cast iron. Tests have apparently shown that the use of about three to five per cent by weight of steel reinforcements increases the resistance to bending from 2.5 to 3.5 per cent. Furthermore, calculations appear to indicate that reinforced cast iron structures are from one and a half to two times as cheap as reinforced concrete, and 35 to 45 per

Fig. 503 (*Above*) Sand casting pipes vertically in a pipe pit.

cent less than steel. It is intended to use reinforced cast iron for tunnel sections in the Moscow Underground, for electric lamp-posts, for small railway bridges, for tunnels, and also for railway cuttings. It is possible that foundrymen here may question whether the process has any practical advantage, but in any case the whole problem is certainly worth investigating".[8]

Malleable cast iron. The graphite in ordinary grey cast iron is in flake form and because these flakes act as potential cracks, grey cast iron is brittle and will break before it will bend. Réaumur, in 1722, and Samuel Lucas, in 1804, attempted to overcome this by packing castings in boxes surrounded by red hematite iron ore and heating them for a week or more at 950°-1,000° C. The heat diffuses the carbon to the skin of the castings where it is removed by oxidation promoted by the iron ore. As a result, soft malleable castings are produced with a tensile strength of 20-25 tons per square inch and ductility, measured as "elongation per cent", of 10 per cent. On breaking such castings show a steely fracture which causes the material to be known as "white heart". In the finished product graphite exists in a nodular form usually known as temper carbon.

In 1826, Boyden, in America, developed a similar process but used sand or crushed slag to surround the castings at a temperature of 850° - 900°C, the object being not to remove the carbon but to precipitate it in the form of

Fig. 504 (*Below*). Centrifugal casting of pipes, showing the tilting ladle pouring the molten metal along the trough into the revolving cylinder.

similar round nodules in the iron, as these did not have the fracturing effect of the carbon in flake form. Castings made on this principle have a tensile strength of 20-25 tons per square inch and ductility in the region of 15 per cent as compared with 1 per cent in ordinary grey iron. On breaking, such malleable cast iron shows a sooty black fracture which causes the material to be known as "black heart".

Both methods are used considerably to-day, particularly in the small components of building, and about 90,000 tons of malleable cast iron is made annually in Great Britain.

4. FINISHES.

The oxidation of iron, which forms iron oxide, occurs whenever iron is exposed to air and moisture. Cast iron, of all the ordinary ferrous metals, is the least affected by oxidation; and when the oxide is formed it provides, to some extent, a protective coating. Lethaby once said that the best protection which cast iron could have from the weather was its own rust. Perhaps this is true when the metal is used in places where it will not be subjected to wear that would disturb the coating; but to-day, cast iron in building or building equipment is almost invariably given a surface treatment in one of many possible ways, in order to prevent chemical action or corrosion and to provide an attractive appearance and a surface that can be easily cleaned and maintained, and will resist wear or erosion.

The number of these different and highly specialised industrial finishes has become so large that the old "rule of thumb" method of control has been supplanted by what is virtually a new and distinct branch of engineering science. The laboratory study of the properties and problems of finishes is comparable to the research work undertaken for many other of the raw materials of industry before they can be effectively and economically used. Finishes for cast iron may be divided into three principal groups, though occasionally the treatments are combined and the groups overlap.

Group 1. In this group, a modification of the chemical composition of the surface of the iron gives it greater resistance to wear and corrosion than the basis metal. The three principal methods are known as Parkerising, Bonderising and Bower-Barffing.

(*a*) *Parkerising*. This phosphate treatment was originated by Thomas Watts Coslett, of Birmingham, in 1906, and is still sometimes referred to as Coslettising; but it has been highly developed in the U.S.A., under the name of Parkerising. The parts to be treated are boiled in a solution of iron and manganese phosphates; the surface of the basis metal becoming primarily iron and manganese phosphates in a heavy crystalline form. This protective coat should again be protected by oil or paint for which it forms an excellent base. Because of its crystalline form, the surface is somewhat rough and cannot be used where a lustrous smooth paint finish is required. It is used for the cast iron parts of mangles, wringers and other domestic equipment.

(*b*) *Bonderising*. This treatment is a variant of Parkerising but produces a finish of iron and copper phosphates on the basis metal which is even more resistant to corrosion than a Parkerised surface. It also forms an excellent base for paint. It is used for much the same purposes as Parkerising, but is also frequently specified as an undercoat for many types of iron castings which are normally finished with paint. Large and practical opportunities for the use of this treatment are afforded by gutters, rainwater pipes, street furniture, and any castings where the maintenance of external paintwork is considerable.

(c) *Bower-Barffing*. This treatment is applied by heating an article in the presence of steam, which converts the surface of the basis metal to the black magnetic oxide of iron. This protective coat will resist much wear and tear under varying temperatures, and is used extensively on boilers and coppers.

Group 2. This embraces permanent, or almost permanent, coatings of a different material, some of which are mere skins while others form an iron alloy, deposited on the basis metal. The chief methods are vitreous enamelling, hot-dipping, sherardising and calorising, metal plating and metal spraying.

(a) *Vitreous enamelling*. By this method an article is enclosed by a coating of glass or porcelain so that the basis metal is covered. This is done either by covering the article with a fine powder—a mixture of glass, opacifiers and colour oxides—or spraying it with similar materials in water suspension, and finally heating to a high temperature in a furnace when the coating melts and runs over the whole surface. Almost any colour can be obtained, and as a result of twenty-five years of research work, it is now possible to produce vitreous enamels which have great resistance to wear, heat, acids and alkalis. The most familiar use of this finish is for baths, basins, sinks, cooking stoves, heating stoves, fireplaces, refrigerators, and a large number of domestic articles. Well designed cooking and table ware has been produced in a thin cast iron, coated inside and out with vitreous enamel.

(b) *Hot dipping*. By dipping a cast iron article into hot molten zinc, a certain amount of alloying takes place, producing a bond with the basis metal and forming a protective skin; a process that is generally known as galvanising. This process is commonly used and is satisfactory, for it is possible and usual to give relatively thick coats which afford good resistance to exposure and weathering. It is used on many of the smaller castings employed in building, such as tanks, manhole covers, gratings, air bricks, soot doors and equipment which does not require a high finish.

The same method is used in applying a coating of tin to cast iron. It is used chiefly for articles that are likely to come into continual contact with food.

(c) *Sherardising and Calorising*. An alloying effect can be produced without dipping. A coating metal is used in the form of fine powder and is heated in revolving drums with the basis metal. This method is extensively used for forming a protective coat on screws, nuts, bolts, metal windows, and small cast iron fittings. When zinc is used for the coating, it is called Sherardising; when aluminium is used, it is called Calorising. The latter method is chiefly useful for conferring heat resistance.

Fig. 505. Cooking table ware in cast iron, finished in pastel shades of vitreous enamel.

(d) *Metal plating*. Electro-plated castings are largely used in building, the chief plating metals being chromium, brass, copper, nickel, cadmium, tin and silver. These form excellent thin protective coatings for the basis metal, but are generally used for their decorative value, particularly for shop fittings, fittings for grates, heating and cooking stoves, sanitary ware and kitchen fittings. The coating has protective value against corrosion, and can also be used for building up worn parts.

(e) *Metal spraying*. It is possible by means of spraying metal through a special pistol, to coat any solid object with a layer of any of the common metals and alloys. Lead, zinc, tin, aluminium, copper, iron, nickel, brass or bronze, can thus be deposited on a casting and to any desired thickness, from about 0.002 inches upwards. The original casting must be slightly rough to form adhesion, as no alloying action takes place. This system has been used to deposit aluminium on fire grate bars and other fire grate parts to resist heat and corrosion. Zinc is used as a deposit on cast iron air bricks, ventilators and gratings.

The more decorative spray finishes can be used effectively on window aprons, sculptured panels, doors and such large areas. One of the chief assets of the system is that modern spraying pistols are light and highly efficient, and the whole plant can be taken to a building and the metal work sprayed *in situ*, when erection or assembly has been completed.

Group 3. A relatively temporary thin coating of paint, varnish or lacquer is often applied as a protection to the metal beneath or to provide a coloured finish. Many of these finishes are heated after application, and the uses of Berlin black, mottled paints and varnish finishes are familiar on such castings as the cheaper cooking and heating equipment, gas fires and radiators.

There are many purely utilitarian uses of the protecting systems described. A bituminous finish is always applied to the inside and sometimes to the outside of soil pipes, and a similar treatment is applied to water supply pipes. In special cases, a concrete lining is sometimes used. Glass-lined cast iron pipes and containers are common in breweries, milk depots and chemical works, and so is vitreous enamelled ware.

The contemporary architect or industrial designer can and does choose from a very large range of finishes in order to obtain not only a non-corrosive surface, but one of almost any desired texture or colour. Victorian taste for excessive enrichment, excrescences in the shape of knobs and other bits and pieces, has nearly passed away, and the achievements of modern research have emphasised the strange Victorian inability to produce any finish to cast iron other than a surface which demanded interminable black-leading. Inventiveness in the matter of finishes may have been dormant because there was a vast supply of underpaid, unconsidered domestic labour in the nineteenth century; for the servants in the cavernous basements of the mid-Victorian house, black-leading was just another bit of the hard and totally unnecessary work that arose from bad industrial and architectural design.

FIG. 506. Cast iron letters finished in bright coloured vitreous enamel.

FIG. 507. One of the iron panels cast in the Nafvequarn Foundry, Sweden, and used on the Oratory Central Schools, Chelsea, London. A standard permanent oxydised finish is used. Architect: Christian Barman, F.R.I.B.A. Sculptor: Laurence Bardshaw.

The curse of black-leading has been a potent influence in preserving a prejudice against cast iron; and so lasting is the prejudice, that it is with astonishment, almost with incredulity, that people learn that their clean, gleaming heat-storage cooker or gas stove in some pastel shade, their cooking and table ware and the bright metallic panels on many buildings, are made of a material they have unjustly despised and disliked. The material in the mid-twentieth century is metallurgically far in advance of the cast iron that was available fifty years ago; and the new surface finishes allow architects to use it in contemporary design with the knowledge that they are using a contemporary material with unique properties.

5. CONCLUSIONS.

The sections of this book have shown how cast iron, and the art, craft and science of iron founding, have influenced the character of the first industrial revolution and the architecture that arose from and served that revolution. For three quarters of a century progress in the technique of producing cast iron has been fairly continuous; but there has been no corresponding progress in design. The lack of interest in design is revealed by the paucity of literature on the subject. Before the publication in 1945 of Richard Sheppard's *Cast Iron in Building*, the last outstanding works were William

Fairbairn's *The Application of Cast and Wrought Iron to Building Purposes*, which reached its third edition in 1864, and Ewing Matheson's *Works in Iron*, the second edition of which was published in 1877. Apart from the detailed accounts of early uses to be found in *The Lives of the Engineers*, by Samuel Smiles, there are only passing references to the architectural uses and significance of the material in works dealing with various aspects of the industrial history of the nineteenth century. Occasionally some scholarly critic like Lethaby would write about the material; and some of his discerning conclusions about the uses of cast iron have been quoted earlier in this section. The validity of his views are unimpaired: what he wrote in 1926 is still applicable, and his summing up of the possible future of cast iron in architecture cannot be bettered. Writing in the *Builder* he said:

"My primary purpose is to show that casting iron has been a large and characteristic English industry, and to suggest that an earnest intelligent effort should be made to re-establish the art on sound common-sense principles as a national craft... If we could get the idea of a suitable treatment of material for reasonable ends applied in the case of cast iron, the notion might slowly spread to other building materials, and 'architecture' might in the end find itself reformed.

"The business of iron founding is a vast and important industry, and anyone might be proud to be engaged in it. I wish I could incite an ambition to refound it as a sound and reasonable 'art of use'. Art is universal, all materials are alike, if not equally vehicles for its expression, and each one can give us something special for itself. The easy contempt we show for iron is only the result of our own treatment of it. It is not some special material—bronze or white marble—that is 'artistic', but art comes of the mind and heart embodied in any material."

It has been our primary purpose in this book to show (in Lethaby's words) "that casting iron has been a large and characteristic English industry"—though for English we would substitute British, for Scottish influence on the industry has been potent and Scotland's contribution to it prodigious—and to record its architectural achievements and indicate present and future potentialities. It is to be hoped that during the second half of the twentieth century, the full services of a needlessly neglected material may be recovered for the benefit of architectural design.

SOURCES OF REFERENCE IN SECTION FIVE

[1] *Die Giesserei*, article by E. Piwowarsky. Vol. 30, No. 12/13. June, 1943.
[2] "Structural Uses of Cast Iron", by Richard Sheppard, F.R.I.B.A. *The Official Architect*, Vol. VIII, No. 5, May, 1945, pp. 240 and 242.
[3] *Cast Iron in Building*, by Richard Sheppard, F.R.I.B.A. Section V, pp. 77, 78.
[4] *Iron and Steel To-day*, by John Dearden, B.Sc., A.M.I.Mech.E. Oxford University Press, 1943, p. 61.
[5] *Ibid*, p. 62.
[6] *Iron and Steel*, January, 1947.
[7] *Foundry Trade Journal*, December 19, 1946.
[8] "Reinforced Cast Iron", by Dr. Ing. Saller. *Die Giesserei*, July 29, 1938. Vol. 25, p. 379.

APPENDIX I

Letter written about 1775, *by Abiah Darby, wife of the second Abraham Darby.*

Esteemed Friend,

Thy very acceptable favour of the 9th ulto., claim'd my earliest acknowledgments, which I should immediately have made had not thy kind condescension in taking notice of my late honour'd Husband; and requesting to be inform'd of any circumstance which may be interesting relating him, caused my delay—to recollect what might occur concerning his transactions or improvements in the Manifactory of Iron so beneficial to this Nation, but before I proceed further, I cannot help lammenting with thee in thy just observation, "that it has been universally observed, that the Destroyers of Mankind are recorded and remembered, while the Benefactors are unnoticed and forgotten": This seems owing to the depravity of the mind, which centres in reaping the present advantages and suffering obscurity to vail the original causes of such benefits; and even the very names of those to whom we are indebted for the important discoveries, to sink into oblivion, whereas if they were handed down to posterity, gratitude would naturally arise in the commemoration of their ingenuity, and the great advantages injoyed from their indefatigable labours. I now make free to communicate what I have heard my Husband say, and what arises from my own knowledge: also what I am informed from a person now living, whose father came here as a workman at the first beginning of these Pit Coal Works.

Then to begin at the original: It was my Husband's Father, whose name he bore (Abraham Darby, and who was the first that set on foot the Brass Works at or near Bristol) that attempted to mould and cast Iron Pots &c. in sand instead of Loam (as they were want to do, which made it a tedious and more expensive process) in which he succeeded, this first attempt was tryed at an Air Furnace in Bristol. About the year 1709 he came into Shropshire to Coalbrookdale, and with other partners took a lease of the Works, which only consisted of an Old Blast Furnace and some Forges. He here cast Iron Goods in sand out of the Blast Furnace that blow'd with wood charcoal; for it was not yet thought of to blow with Pit Coal. Sometime after he suggested the thought, that it might be practable to smelt the Iron from the ore in the Blast Furnace with Pit Coal. Upon this he first try'd with raw coal as it came out of the Mines, but it did not answer. He not discouraged, had the Coal Coak'd into Cynder, as is done for drying Malt, and it then succeeded to his satisfaction. But he found that only one sort of pit coal would suit best for the purpose of making good iron. These were beneficial discoveries, for the Moulding and casting in Sand instead of Loam was of great service, both in respect to expence and expedition, and if we may compare little things with great—as the invention of printing was to writing, so was the moulding and casting in Sand to that of Loam. He then errected another Blast Furnace, and enlarged the Works. This discovery soon got abroad and became of great utility.

This place and its environs was very barren, little money stirring amongst the Inhabitants, so that I have heard they were obliged to exchange their small produce one to another instead of money, until he came and got the Works to bear, and made money circulate amongst the different parties who were employed by him. Yet notwithstanding the service he was of to the Country, he had Opposers and ill wishers—and a remarkable circumstance of awful memory occurs; of a person who endeavour'd to hinder the horses which carried the Iron Stone and coal to the Furnaces, from coming through a road that he pretended had a right to oppose; and one time when he saw the horses going alone, he, in his passion, wished he might never speak more if they should ever come that way again. And instantly his speech was stop'd, and altho' he lived several years after, yet he never spoke more!

My Husband's Father died early in life; a religious good man, and an Eminent Minister amongst the people call'd Quakers.

My Husband Abraham Darby was but six years old when his Father died—but he inherited his genius—enlarg'd upon his plan, and made many improvements. One of consequence to the prosperity of these Works was: as they very short of water—that in the Summer or dry seasons they were obliged to blow very slow, and generally blow out the furnaces once a year, which was attended with great loss. But my Husband proposed the errecting a Fire Engine to draw up the water from the lower Works and convey it back into the upper pools, that by continual rotation of the water the furnaces might be plentifully supplied; which answered exceeding well to these Works, and others have followed the example.

But all this time the making of Barr Iron at Forges from Pit Coal pigs was not thought of. About 26 years ago my Husband concieved this happy thought—that it might be possible to make bar from pit coal pigs. Upon this he sent some of our pigs to be tryed at the Forges, and that no prejudice might arise against them, he did not discover from whence they came, or of what quality they were. And a good account being given of their working, he errected Blast Furnaces for pig iron for Forges. Edward Knight, Esq., a Capitol Iron Master urged my Husband to get a patent, that he might reap the benefit for years of this happy discovery; but he said he would not deprive the public of such an acquisition which he was satisfyed it would be; and so it has proved, for it soon spread and many Furnaces both in this neighbourhood and several other places have been errected for this purpose.

Had not these discoveries been made, the Iron Trade of our own produce would have dwindled away, for wood for charcoal became very scarce, and landed gentlemen rose the prices of cord wood exceeding high—indeed it would not have been to be got. But from pit coal being introduced in its stead, the demand for wood charcoal is much lessen'd and in a few years I apprehend will set the use of that article aside. Many other improvements he was the author of; one of service to these Works here, they used to carry all their mine and coal upon horses backs, but he got roads made and laid with sleepers and rails, as they have them in the North of England for carrying them to the Rivers, and bring them to the Furnaces in waggons. And one waggon with three horses will bring as much as twenty horses used to bring on horses backs. But this laying the roads with wood begot a scarcity and rose the price of it, so that of late years the laying of rails of Cast Iron was substituted, which, although expensive, answers well for ware and duration. We have in the different Works near twenty miles of this road, which cost upwards of eight hundred pounds a mile. That of Iron Wheels and axle trees for these wagons was I believe my Husband invention. He kept himself confined to the Iron Trade and the necessary appendages annex'd thereto. He was just in his dealings of universal benevolence and charity, living strictly to the Rectitude of the Divine and Moral Law, held forth by his great Lord and Saviour; had an extraordinary command over his own Spirit, which thro' the assistance of Divine Grace enabled to bear up with fortitude above all opposition, for it may seem very strange so valuable a man should have antagonists, yet he had those called Gentlemen filled with an envious spirit, could not bear to see him prosper; and others covetious, strove to make every advantage, by raising their Rents of their Collieries, and lands in which he wanted to make roads, and endeavoured to stop the Works—but he surmounted all; and died in peace, beloved and lamented by many.

INDEX

A
Abbeys, owning ironworks, 9
Aberdare, 37
Abergavenny, Lord, 17
Abermule Bridge, *Fig. 102*
Absolon, Professor K., 6
Acid process, 239
Acts of Parliament, relating to the industry, 18, 78, 92
Adam, the brothers, 54, 70, 117
   ,, James, 70
   ,, John, 70
   ,, Robert, 70, 116
Adzes, early iron, 6
Aire Bridge, Haddlesey, *Fig. 107*
Albert Hall, *Fig. 275*
Albert Suspension Bridge, London, 249, *Fig. 298*
Albion Flour Mills, 106
Aldridge Road Bridge, Birmingham, *Figs. 206, 207*
Alhambra Theatre, London, 196
Alvechurch, Birmingham, 362
America, 198, 252
American Institute of Mining and Metallurgical Engineers, 3
American War of Independence, 67, 73
Amsterdam, railway station, 251, *Figs. 313-316*
Analysis of Ornament, 208, 250
Anderson, James, *quoted, 66*
Andirons, cast iron, 11, 16, 53
Anglesey, 179
Anthracite, 162
Anvils, 11
Aqueducts, cast iron, 54, 104, 105, 106
Arcades, cast iron, 250, *Fig. 304*
Architecture Arising, 252
Area guards, cast iron, *Fig. 289*
Armaments, 16th century, 18
Aristotle, 5
Ashton, T. S., 42, 43, 56, *quoted 9, 18, 19, 33, 39*
Assyria, early evidence of iron, 2, 3

B
Babraham Bridge, Cambs., *Fig. 218*
Backbarrow, Lancs., 44, 59, 79
Bacon, Anthony, 78
Baker, Sir John, 18
   ,, Sir Richard, 18
Balconies, cast iron, 115, *Figs. 122-149, 153-155, 283-286, 341-347, 349, 353*
Balusters, cast iron, *Figs. 167, 169, 280, 290, 339, 340*
Balustrades, cast iron, *Fig. 286*
Bandstands, cast iron, 250, 254, *Fig. 326*
"Bank", 81
Banks & Barry, 250
Baptist Mills, 41
Bar iron, 39
Barges, cast iron, 64
Basic open-hearth process, 313
Basiliscus, 17
Basins, cast iron, 322, *Fig. 483*
Bath, New Iron Bridge, *Fig. 100*
Bath tray, cast iron, *Fig. 485*

379

Baths, cast iron, 278, 322
Baude, Peter, 17
Beads, early iron, 2
Beams, cast iron, 54, 82, 179, 192, 198, 239, 250, 313
Beck, L, 42, *quoted 63*
Beighton, Henry, 49
Berlin black, 372
Bersham, 59, 60, 63, 104
Bessemer, Henry, 239, 250, 313
   ,,   process, 239, 240
Bibliothèque Nationale, Paris, 195
Bill hook, early iron, 6
Bilston, Staffs, 64
Birkenshaw, 167
Birmingham, 175
Birmingham-Derby railway line, 167
Birmingham Market Hall, *Fig. 236*
Bishops, of Durham, 9
Bituminous finish, 372
"Black heart", 369
Black Yajurveda, 3
Blackfriars Bridge, London, *Fig. 293*
      ,,   Railway Bridge, London, *Figs. 294, 295*
Blast furnaces, 162, *Fig. 488*
Blisworth Bridge, *Figs. 186-188*
Bloom, of iron, 8, 10, 11
Bloomery, 16
Bloomfield Colliery, Tipton, 60
Boilers, cast iron, 48
Bollards, cast iron, 316, 321, *Figs. 166, 201, 419, 420*
Bonderising, 369
Boodles Club, London, 117
Boring machine, 60, 66, 73
Boston Town Bridge, Lincs., 94, *Fig. 98*
Boulton, Matthew, 53
Boulton & Watt, 60, 62, 63, 78, 106, 113, 239
Bower-Barffing, 369, 370
Boyden, S. 367
Brackets, cast iron, 169, *Figs. 327, 346*
Bradley, 60, 63
Brass, 116
Bremer Bridge, Australia, 249, *Fig. 299*
Bridges, cast iron, 54, 62, 73, 76, 82, 86, 90, 100, 120, 167, 178, 179, 186, 247, 249, *Figs. 80-107, 114, 180-193, 206-218, 220-235, 293-295, 299-302, 421-425*
Bridgewater, Duke of, 104
Bridgewater House, London, *Fig. 237*
Bridgnorth, 11, 56
Brighton, the Pavilion, 120
Brindley, James, 49, 53, 104
Bristol, 41
Britain, mediæval cast iron, 10
    ,,   ,,   ironworks, 9
    ,,   pre-Roman iron, 6
    ,,   Roman iron, 8, 9
Britannia Bridge, North Wales, 186, *Fig. 226*
British Cast Iron Research Association, 362, 366
British Museum, *Figs. 269, 270*
British School of Archæology, 2
British Standards Institution, 321, 361
Bronze Age, 2
Brunel, I. K., 167, 172, 189
Buckhurst, Lord, 17
Buckingham Palace, 120
The Builder, 316, 374
The Builders' Magazine, 116

Buildwas Bridge, 86, 100, 104, *Fig. 87*
Burton, Decimus, 202, 250
Burwash, Sussex, 11
Buxted, Sussex, 17
Býči Skála cave, 6
Byng, the Hon. John, 57, *quoted 86*

C

Cabrey, 167
Cadell, William, 65, 78
Cæsar, Julius, 6
Caissons, cast iron, 247
Calorising, 370
Cambridge, Senate House, 116, *Figs. 42, 43*
Camden Town depôt, North Western Railway, *Fig. 195*
Canals, 53, 104, 105
Canal banks, cast iron, 54
Candlesticks, cast iron, *Fig. 282*
Cannon, cast iron, 11, 17, 53, 59, 60
  ,,   lathe, 60
  ,,   patents for casting, 17, 18
Cantilevers, cast iron, 169, 251
Capitals, cast iron, *Fig. 354*
Capitol, Washington, 252
Cardiff, 8
Carlton House Terrace, London, 120
Carron Company, 66, 67, 70, 71
Carron Works, 48, 64, 65, 66, 78, 106, 116
Cast iron, design in relation to, 11, 15, 204, 206, 209, 250, 278, 285, 316, 373
  ,, ,, early, 5, 6
  ,, ,, foreign influence, 15, 17
  ,, ,, high duty, 366
Cast Iron in Building, 252, 282, 373
Cast Iron Pipe, Its Life and Service, 79
Cast iron, malleable, 78, 368, 369
  ,, ,, mediæval British, 10, 11
  ,, ,, nodular, 366
Casting, art of, 39, 206
  ,, boxes, 42
  ,, hollow ware, 41
  ,, process, 15, *Figs. 490-497*
Castings, Chinese, 3, 5
  ,, origin of, 3
  ,, softening, 78
Central America, discoveries in, 2
Centrifugal casting of pipes, 81, *Fig. 504*
Ceriog, 105
Cesera, Aroanus de, 17
Chairs, cast iron, 278 (See also Garden Furniture)
Chambers, Thomas, 76
Chandos House, London, 117
Charcoal, 10, 18, 37, 42, 44, 53, 56, 59
Charlesworth, M. P., *quoted 8*
Chat Moss, 163
Chatsworth, 200-202
Chelsea Suspension Bridge, London, *Figs. 231, 232*
Chepstow Bridge, 100, 167, 189, *Fig. 97*
Chester and Birkenhead railway, 167
  ,, ,, Crewe railway, 167
Chester-Holyhead railway, 179
Chester railway station, 172
Chettoe, C. S., 179
China, early evidence of iron, 3
Chirk, 105, *Figs. 44, 109*
Churches, use of cast iron, 194, 199

City Temple, London, *Figs. 239, 240*
Civil War, 33, 37
Clapham, Dr., 42
Clark, Edwin, 186
Cleveland Bridge, Bath, *Fig. 99*
Clocks, cast iron, 278
Clyde Ironworks, 78, 162
Coal, patents in relation to, 18, 56
Coalbrookdale, 42, 44, 48, 54, 56, 58, 60, 86, 104, 162, *Figs. 38-41*
Coalbrookdale Company, 57
Cog wheels, cast iron, 106
Coke, smelting, 42, 44, 53, 54, 56, 63, 76, 162
Colins, Joan, 11, *Fig. 8*
Collingwood, R. G., 9
Columns, cast iron, 54, 80, 105, 106, 120, 169, 171, 172, 175, 196, 198, 199, 201, 202, 239, 247, 250, 251, 278, 313, *Figs. 170, 171, 173-177, 202, 310-312*
Conishead, 9
Conservatories, cast iron, 199, 200, 250, 251, *Fig. 303*
Continent, British influence, 62
Conway Bridge, North Wales, 90, 99
   ,, river, 186
Cookers, cast iron, 322, *Figs. 445-453, 456-458*. (See also Cooking Appliances, Electric Stoves, Gas Stoves, Stoves, Ranges)
Cooking appliances, cast iron, 278
   ,, utensils, cast iron, 11, *Fig. 505*
Corbridge, 8
Core-making, 355
Cores, moulding, 5
Corstopitum, 8
Cort, Henry, 56
Coslett, Thomas Watts, 369
Coslettising, 369
Cottingham, L. N., 116
Council of Ironfoundry Associations, 361
   ,, ,, Iron Producers, 362
Cradley, forges, 18
Cranage, George, 56
   ,, Thomas, 53, 56
Craigellachie Bridge, 90, *Figs. 88, 89*
Crawshaw, John, 73
Crawshay, Richard, 78
Crowhurst, Surrey, 11
Crucible process, steel, 73
Croxdale Bridge, Sunderland, *Fig. 423*
Crystal Palace, 201, 202, *Figs. 245-252, Plates I-IV*
Culpepper, Alexander, 18
Cunningham, W., 42
Cupola, adoption of, 48
   ,, foundry, 63, *Figs. 50, 489*
Currency bars, early iron, 6
Cyfarthfa Ironworks, 56, 78
Cylinders, cast iron, 59, 60, 66, 76, 247, *Fig. 52*
Czechoslovakia, 6

D

Dacre, Lord, 17
Daggers, early iron, 6
"The Dale", 54, 57, 58
Dale Company, 54, 56, 66
Dance, George, 54, 116
Darby, Abiah, 42, 54
   ,, Abraham, the elder, 40-42, 44, 48, 53, 54, 56, 58, 162
   ,, ,, ,, younger, 42, 48, 49, 53, 54, 56, 60
   ,, Alfred, 58
Davey, N., 179

Dean, Forest of, 8, 9, 33, 43
Dearden, John, *quoted 10*
Dearman, Richard, 60
Dee, river, 105
Delhi, 3
Derby, Earl of, 17
Derbyshire, 79
Design, related to cast iron, 11, 115, 204, 206, 209, 250, 278, 285, 316, 373
Devonshire, Duke of, 200
Diary of a Tour through the Midlands, 90
Dickens, Charles, *quoted 201*
Dixon, J., 167
Dodds, Isaac, 90
Dogs, cast iron, *Figs. 334, 335*
Doors, cast iron, 64, 313, *Fig. 427*
Door furniture, cast iron, 76, 78
Door lintels, cast iron, *Fig. 398*
Dowlais Ironworks, 78
Drainpipes, cast iron, 81
Dudley, Dud, 18, 19, 42
Durham, Bishops of, 9
Dutch fireback, *Fig. 31*
Dutch-Rhenish Railway, 251
Dutch stoveplate, *Fig. 31*
   „ workmen, 41

E
Eagle Foundry, Birmingham, 60
Eastleigh Bridge, Hants., *Fig. 225*
Edington, Thomas, 78
Egypt, early evidence of iron, 2
El Gerseh, 2
Electric stoves, cast iron, 322
Elizabeth, Queen of England, 17
Ellesmere Canal System, 105
Ely House, London, 116, *Fig. 120*
Enamelling, on cast iron, 81, 82. (See also Vitreous Enamelling)
Engine parts, cast iron, 56, 67
Euston railway station, London, 171, 172, *Figs. 200-204*
Europe, early evidence of iron, 5
Excavations, of iron objects, 2, 3, 5, 6, 8
Exhibition buildings, use of cast iron, 199, 201, 202. (See also Crystal Palace)
Exports, of cast iron, 67, 248, 249, 250, 310
Extraction, of iron ore, 1, 5, 6, 8, 9, 10, 11, 18, 42

F
Factories, use of cast iron, 117, 194, 239, 313
Fairbairn, Sir William, 113, 115, 179, 186, 239, 240, 241, 362, 366, 374, *quoted 192*
Falkirk Ironworks, 78
Fanlights, *Figs. 281, 396*
Fenton, Vivian, 49
Ferguson, James, *quoted 202*
Fernhurst, Sussex, 37
Field, Joshua, *quoted 90*
Files, early iron, 6
Finials, cast iron, 116, *Figs. 156, 157*
Finishes, on cast iron, 82, 369, 372, *Plates V, VI*
Fire engine, invention of, 48
   „ „ use by Abraham Darby the younger, 49, 54
   „ „ parts, 59
Firebacks, cast iron, 11, 15, 16, 44, 48, 53, *Figs. 13-22, 27, 28, 31, 32, 35-37*
Firedogs, cast iron, *Figs. 33, 34*
Fireplaces, cast iron, 70, 78, *Figs. 75, 261, 262*. (See also Grates)
Flaxley, Abbey of, 9
Flemish stoveplates, *Figs. 23, 25*

Floor tiles, cast iron, *Figs. 435-438*
Folkestone, Kent, 8
Footbridges, cast iron, 63, *Fig. 180*
Ford, Richard, 48, 54
Foreign influence, 15, 17
  ,, workers, 41
Forest of Dean, 8, 9, 33, 43
Forged steel, 78
Forges, mediæval, 16
  ,, 18th century, 43, 44
  ,, 19th century, 162
Forks, cast iron, 78
Forster, Anne, 11, *Fig. 9*
Foundries, mediæval, 11, 16
  ,, Sussex, 17, 37
  ,, Mr. Stringer's, 42
  ,, 19th century, 82, 162, 306
  ,, 20th century, 322, *Figs. 498-502*
Foundry cupola, 63
Foundry processes, 16, 48, 53, 81, 162, 192, 240, 241, 355
Foundry Trade Journal, 366
Fountains Abbey, 9
Fountains, cast iron, 163, 278, *Figs. 391-393*
Fowke, Captain, 251
Fox, Sir Charles, 249
France, 62, 196, 202
French firebacks, *Figs. 35, 36*
  ,, firedogs, *Fig. 34*
  ,, Revolution, 86
Friend, J. Newton, *quoted 6*
Fuel, supplies of, 33, 37, 39, 42, 43, 44
Furnace, reverberatory, 48, 56
Furnaces, early, 1, 2, 8, *Figs. 5-7*
  ,, German, 11
  ,, mediæval, 10, 11, 16
  ,, restrictions on, 43
  ,, Thos. Tilley's air, 48
  ,, 17th century, 42, *Fig. 38*
  ,, 18th century, 43, 44, 56
  ,, 19th century, 63, 78, 80, 162
Fuseboxes, cast iron, *Figs. 486, 487*

G
Galton Bridge, 90, *Figs. 90, 91*
Galvanizing, 370
Garbett, Samuel, 65
Garden furniture, cast iron, *Figs. 380-382*
  ,, rollers, cast iron, 48
Gardner, Starkie, 11
Gas fires, cast iron, *Figs. 471-475*
  ,, stoves, cast iron, 322, *Figs. 459-462*
Gasworks, plant for, 78
Gate posts, cast iron, *Figs. 163, 201, 203, 204*
Gates, cast iron, 70, 82, 172, 175, 278, 316, *Figs. 158, 162, 201, 203, 204, 270, 272-274, 276-278, 281, 283, 336, 350-352, 430*
German stoveplates, *Figs. 24, 26*
Germany, 11
Gibbs, James, 54, 116
Giedion, Siegfried, *quoted 198*
Gilchrist, Percy, 240, 313
Gilpin, Gilbert, 64
Gingerbread Hall Bridge, Gt. Baddow, *Figs. 214-216*
Girders, cast iron for bridges, 178, 179, 192
Gizeh, 2
Glamorganshire, 56

Glastonbury, 6
Gloucestershire, 8, 9
Goldney, Thomas, 48, 54
Gooch, T. L., 167
Gorton depôt, Sheffield & Manchester Railway, *Fig. 194*
Gospel End Ironworks, 99
Gothic Revival, 199, 206
Gouges, early iron, 6
Gowland, William, *quoted 1, 2*
G.P.O. pillar boxes, 321, *Figs. 411-413*
  ,, telephone kiosk, 317, *Fig. 408*
Grammar of Ornament, 251
Grand Junction Canal Bridge, Blisworth, *Figs. 186-188*
Grates, cast iron, 43, 67, 70, 78, 82, *Figs. 74, 75, 356-370*
Gratings, cast iron, *Figs. 376, 378, 379*
Grave slabs, cast iron, 11, 44, 53, *Figs. 8-12*
Great Exhibition, 175, 201, *Figs. 256-262, 274*. (See also Crystal Palace and Exhibition Buildings)
  ,, Western Railway, 167
Grecian Iron Age, 5
Greeks, use of iron by, 5
Greenhouses, use of cast iron, 199
Grenoside, 173
Gresham, Sir Thomas, 18
Grilles, cast iron, 278, *Figs. 275, 287-289*
Gt. Barr Street Bridge, Birmingham, *Fig. 220*
Gt. Britain. (See Britain)
Guest, John, 78
Guns, cast iron, 59, 66

H
Haarlem railway, 192
Hadrian, 8
Haldane, J. B. S., *quoted 159*
Hallstatt, Austria, 6
Ham Hill, Somerset, 6
Hammersmith Suspension Bridge, London, *Fig. 230*
Hampstead Bridge, London, *Figs. 184, 185*
Handyside & Company, 249, 251
Harbours, use of cast iron, 247
Hardwicke, Philip, 172
Hartley, Harold, 366
Harvo, John, *Fig. 22*
Hawkins, Robert, 66
Haworth, Henry, 70, 71
    ,, Samuel, 71
    ,, William, 70, 71
Hayes Bridge, Cardiff, *Figs. 421, 422*
Hazzledine, John, 100
Hearths, primitive, 8, *Fig. 6*
Heathfield, Sussex, 37
Heating appliances, cast iron, 278, *Figs. 463, 464, 466, 479, 480*. (See also Gas Fires, Grates, Hob Grates)
Henry VIII, 17
Hickling, Samuel, 82
High-duty cast iron, 366
Hinges, cast iron, 76, 78
Hinkley Foundry, San Francisco, 252
Hirth, Dr. Friedrick, 3
History of Stirlingshire, 306
History of the Modern Styles of Architecture, 202
Hob grates, cast iron, *Figs. 59-63, 74*
Hodgkinson, E., 115, 179, 186, 192, 239
Hogge, Ralph, 17
Holinshed, 17
Holland, 41
Hollow ware, cast iron, 41, 43, 76

Holmes Works, 76
Homer, 5
Hooks, early iron, 6
Hoole, H. G. & Company, 78
Horizontal casting of pipes, 80, 81
Horsehay, 44, 56, 57
Horseley Bridge Company, 90
  ,,   ,, & Engineering Company Limited, 90
  ,, Furnace, 90
Horticultural equipment, cast iron, *Figs. 377, 387-389, 428.* (See also Garden Furniture)
Hot blast, 162
  ,, dipping, 370
Household Words, 201
Howard, John G., 116
Humfray, the brothers, 78
Hunsbury Camp, Northants., 6
Huntsman, Benjamin, 73
Hyde Park, *Figs. 272, 273*

I
Ickneild Street Bridge, *Fig. 94*
Imports, 8, 39, 41, 239
India, early evidence of iron, 3
Industrial Revolution, 48, 159, 373
Industry, mechanisation of the, 357
  ,, organisation of the, 357
International Exhibition, Paris, 285
Ireland, 17
Iron Age, 2, 5
Iron and Steel, 366
Iron and Steel in the Industrial Revolution, 43, 56
Iron, bar, 39
  ,, castings, origin of, 3
  ,,   ,, early Chinese, 3, 5, *Figs. 1-4*
  ,, early cast, 5, 6
  ,,   ,, evidences of, 2
  ,, excavations, 2, 3, 5, 6, 8
  ,, malleable, 10, 16, 78, 368, 369
  ,, mediæval, 10, 11, 15
  ,, meteoric origin, 1
  ,, ore, 1, 5, 6, 8, 9, 10, 11, 162
  ,, patents for casting, 17, 18, 33, 41, 42, 56, 59, 78, 82, 104, 162
  ,, pig, 11, 16, 44, 47, 54, 56, 59, 63, 78, 162
  ,, production, 18, 33, 43, 44, 48, 54, 56, 162
  ,, tax on, 3
  ,, wrought, 3, 10, 11, 16, 56, 70, 115, 116, 179, 186, 239, 247, 249, 250
Irons, smoothing, 43
Ironbridge, Shropshire, 62, 82, 86, 100, *Figs. 80-82*
Ironmasters, 16, 17, 18, 37, 54, 67
Ironstone, 162
Ironworks, 8, 9, 33, 37
Islington Row Bridge, Birmingham, *Fig. 221*
Izons & Company, 76

J
Jardin d'Hiver, Paris, 202
Jars, Monsieur, 48
Jenkins, Rhys, 40, 41, 47, *quoted 48*, 57
Joint Iron Council, 362
Jones, Owen, 251
Journal of Abraham Darby, 42, 43

K
Kent, 9, 11
Kentish firebacks, *Fig. 30*

Ketley, Shropshire, 44, 57, 100, 105
Kettles, cast iron, 43
Kew, Surrey, 202, 206, *Figs. 253, 254*. (See also Palmhouses)
Keys, early iron, 6
Kings Cross railway station, 175
Kiosks, cast iron, 250, 316, *Fig. 306*
Knives, cast iron, 78
   ,,   early iron, 6
Kwan Yin, *Figs. 2, 3*

L
L'Art d'Adoucir le Fer Fondu, 63, 78
La Tène, Switzerland, 6
Labrouste, Henri, 196
Lamberhurst, 37, 115
Lambeth Suspension Bridge, London, *Figs. 228, 229*
Lamp posts, cast iron, 163, 316, *Figs. 329, 330, 410*
Lamp standards, cast iron, 80, 321, *Figs. 150, 265, 282, 331, 332, 409, 420*
Lamps, cast iron, *Figs. 279, 281, 290, 327, 328*
Lansdowne House, Berkeley Square, 117, *Fig. 118*
Lapsley, G. T., 9
Le Creusot, 62, 63
Leeds Infirmary, 250, *Fig. 309*
Lenard, Richard, 16, *Figs. 32, 33*
Lethaby, W. R., 279, 316, 321, 369, *quoted 15, 115, 317, 374*
Lewes Castle, 115
Lightmoor, 44
Lillie, 192
Lindale, 64
Links Foundry, 80
Lintels, cast iron, 54, 59
Lion, cast iron, *Figs. 268, 271*
Lion Foundry, 310
Little Clifton Furnace, 58
Lives of the Engineers, 66
Liverpool and Manchester Railway, 159, 163, 167
Locke, J., 167
Locks, cast iron, 54, 78, 105, *Figs. 113, 114*
London, 17
   ,,   & Birmingham Railway, 167, 179
   ,,   Bridge, 90, *Fig. 86*
   ,,   Bridge Works, 48, 106
   ,,   & North Western & Great Western Joint Railways, 251
   ,,   Tower of, 17
   ,,   water supply to, 48
"Lorn" brand cast iron, 59
Lucas, Samuel, 78, 368

M
Macfarlanes, 310
Machinery, cast iron, 48, 67, 78, 106
Mackintosh, Charles Rennie, 279
Macpherson, D., *quoted 66*
Madeleine, Paris, 198
Madeley Wood, 44
Malleable iron, 10, 16, 78, 368, 369
Malom, James, 78
Malt Mills, 40
Manchester, 113
   ,,   & Birmingham Railway, 167
   ,,   & Leeds Railway, 167
Manganese, 239
Manhole covers, cast iron, 278
Mantelpieces, cast iron, 278
Mantoux, 42

Maquis Viaduct, Chile, 374, *Figs. 296, 297*
Marseilles, 79
Masborough, 73
Master of Ordnance, 18
Masterson, S., 286, 288
Matheson, Ewing, 374, *quoted 240-245, 247, 248, 278, 279*
Matthews, Edward, 18
Maurier, du, 286
Mayan civilisation, 2
Meade, R., 42
Mechanisation of the industry, 357, *Figs. 498, 502*
Mediæval casting, 15
    ,, foundry processes, 16
    ,, ironworks, 9
Meikle, Andrew, 53, 92
    ,, John, 92
Menai Straits, 90, 94, 179
Mersey river, 105
    ,, Tunnel, 316, *Figs. 431-434*
Merthyr Tydfil, 37
Metal spraying, 371
    ,, plating, 371
Metallum Martis, 18
Metallurgical knowledge, 6, 11, 16, 240, 241, 362, 366
Metals, early use of, 1
    ,, ,, production of, 1
Meteorites, 1
Metropolitan Tabernacle, London, *Figs. 241, 242*
Midland Railway, 167
Millington Hall Bridge, Retford, *Fig. 223*
Mills, use of cast iron, 162, 194
    ,, cotton, 113, 115, 117, 179, *Figs. 115-117*
    ,, flour, 106
Ministry of Transport, 178
    ,, ,, Works, 86
Mitchell, G. R., 179
Monograms, cast iron, *Fig. 51*
Montague, Lord, 17
Montelius, 5
Morris, Peter, 48
Mortars, cast iron, 11, 43, 48
Mott, Sir Basil, 316
Moulding, 5, 76, 78, 206, 355, *Figs. 40, 499-501*
Moulds, 15, 42
Mullions, cast iron, *Figs. 404, 405*
Murdoch, W., 53
    ,, John, 106
    ,, William, 106
Mushet, David, 43, 54, 162, 239
Myres, J. N. L., 9
Mythe Bridge, *Fig. 95*

N
Nameplates, cast iron, *Fig. 165*
Napoleonic Wars, 73
Nash, John, 54, 100, 104, 120, 200, 315
Nash Mill Bridge, *Fig. 193*
National Gallery, London, 199
Natural History, Pliny's, 5, 6
Natural History of Staffordshire, 37
Nene river, bridge for, *Fig. 192*
Neogi, Panchaman, 3
Neolithic Age, 1, 2
New Bersham Company, 60
New Willey Company, 60

New Iron Bridge, Bath, *Fig. 100*
New York Crystal Palace, *Fig. 255*
Newcastle, 48, 179
Newcomen engine, 48, 54
Newels, cast iron, *Fig. 290*
Newjanks Ironworks, 366
Newland Iron Company, 79
Newton, George, 78
"Nicrosilal", 366
Nielson, J. B., 162
Nimmo, William, *quoted 306, 308, 309*
Nimrod, Assyria, 3
Nine Elms Goods Depot, Southern Railway, *Figs. 196, 197*
"Ni-resist", 366
Nodular cast iron, 366
"Nomag", 366
Norman ironworks, 9
Norris, W. G., 42
North Parade Bridge, Bath, *Fig. 101*
North Western Railway, *Fig. 195*
Northumberland, Earl of, 17
Norway, 44
Nottinghamshire, 79

O
Office of Works, 321
Official Architect, 313
Open-hearth process, 313
Ordish, R. M., 249-251
Ordnance, casting of, 18, 60, 73, 82
Ores, extraction of, 1, 5, 6, 8, 9, 10, 11, 18, 42
    ,, imports, 8, 39, 41, 239
Organisation of the industry, 357
Osmund Furnace, *Fig. 7*
Otis Elevator Company, New Jersey, 252
Our Iron Roads, 172, 188

P
Paddington railway station, 172, *Figs. 198, 199*
Paine, Tom, 53, 73, 86, 104
Paint finishes, 372
Palmhouses, use of cast iron, 199, 250, *Figs. 253, 254*
Panels, cast iron, 70, 310, 316, *Figs. 64-73, 355, 404-406, 507*
Pantiles, cast iron, 252
Paris, 62, 202
Park Street Bridge, *Figs. 189, 190*
Parkerising, 369, 370
Patents, for using cast iron:
    box iron, 59
    bridges, 104
    cannon, 17, 18
    enamelling, 82
    guns, 59
    hot blast, 162
    pig iron with coal, 56
    pots, 41
    softening castings, 78
    for using coal, 18, 33, 42
Pattern-making, 78, 355
Pavilion, Brighton, 120, 200, 316
Pavilions, cast iron, 250, 254, 316
Paxton, Joseph, 200-202, 250
Pearce, Thomas, 64
Peel, Sir Robert, 206
Pelham, Sir John, 18

Pensnett, 18
Pennsylvania, 163
Percy, Dr., 42
Perrier, 62
Pestles, cast iron, 43
Petrie, Sir William Flinders, *quoted 2*
Phantassie, 92
Phoenix Foundry, 78
Phillips, J., *quoted 105*
Phillips & Lee, 113, 179
Piers, cast iron, 169, 249, 254, *Figs. 279, 350*
  ,, seaside, 247, 252, *Fig. 307*
Pig iron, 11, 16, 44, 47, 54, 56, 59, 63, 78, 162, 313, 361
Pigs, iron, 43
Pilasters, cast iron, 200
Pillarboxes, 321, *Figs. 411-413*
Pillars, cast iron, 120
Pinel, Maurice L., 3
Pipes, cast iron, 48, 59, 60, 62, 78, 80, 278
  ,, centrifugal casting of, 81, *Fig. 504*
  ,, horizontal casting of, 80, 81
  ,, vertical casting of, 80, 81, *Fig. 503*
Piwowarsky, E., *quoted 313*
Place, Victor, 2
Plaques, cast iron, *Figs. 263, 264, 266*
Plastics, 53
Pliny, 5, 6
Plot, Robert, *quoted 37-39*
Poland, 202
Pont-Cysylltau, 105, 106, *Figs. 110-112*
Portman Square, London, 117
The Postman, 42, 47
"Pressure" pipes, 80
Prestonpans, 65
Price control, of cast iron, 60
Pritchard, Thomas Farnolls, 82, *Fig. 45*
Privy Council, 18
Prussia, 63
Puddling process, 56
Pugin, A. Welby, 208, 209, *quoted 206*
Pulpit, cast iron, 64
Pyrenees, 8

Q
Quadrant. (See Regent Street)

R
Radiators, cast iron, *Figs. 476-478*
Radyr, 18
Railings, cast iron, 37, 54, 70, 82, 115, 116, 163, 171, 172, 278, *Figs. 42, 43, 121, 145-148, 150-153, 156, 157, 159-161, 163, 164, 201, 267, 269, 273, 274, 337, 338, 341, 343, 345, 348, 351, 418*
Railway lines, cast iron, 54, 62, 159, 163
Railways, use of cast iron, 163, 169, 313, 316
Rainwater equipment, cast iron, 278, *Fig. 481*
Ranges, cast iron, 322, *Figs. 57, 374, 375, 442-444*
Ransom, 53
Rastrick, John, 54, 100
Razors, cast iron, 78
Read, Thomas T., 3, 5
Reaping hooks, early iron, 6
Reáumur, Réne Ferchault de, 63, 78, 368
Records of Cardiff, 39
Refining, of iron, 16, 18, 37, 42
Regents Canal Bridge, *Figs. 181, 182*
Regents Park, London, 120, 202

Regent Street, London, the Quadrant, 120
Reinforced cast iron, 366, 367
    ,,   concrete, 53
Rennie, John, 53, 92, 94, 99, 106, 120, 159, 186
Repton, Humphry, 200
Research, 3, 5, 357, 362
Reverberatory furnace, 48, 56
Reynolds, Sir Joshua, 70
    ,,   Richard, 57, 82
    ,,   William, 100
Richborough Castle, 8
Rickman, Thomas, 199
Rievaulx, 9
Rights of Man, 86
Rivet, early iron, 6
Road and Rail Traffic Act, 1933, 178
Road setts, cast iron, *Figs. 434, 439*
Robertson, Howard, *quoted 252*
Rockingham Forest, 9
Roebuck, John, 53, 64, 65
Rolling, of steel, 315
Roman Britain and the English Settlements, 9
Roman Britain, iron in, 8, 9
Romans, metallurgical knowledge, 5, 6
Roofing plates, cast iron, 252
Rosettes, cast iron, *Fig. 291*
Ross-on-Wye, 8
Rotherham Ironworks, 86
Rotterdam, 39
Rovenzon, John, 18, 42
Rowse, H. J., 316
Royal Coat of Arms, *Fig. 333*
  ,,   Institute of British Architects, 120
  ,,   Lodge, Windsor, 200
  ,,   Horticultural Society winter garden, 251, *Fig. 305*
Russia, 44, 202
Russian Academy of Architecture, 366

S
St. Etienne, France, 79
Saint-Fond, Faujas de, 161
St. George's, Birmingham, 199, *Fig. 244*
St. Leonard's, Shoreditch, 116
St. Louis, U.S.A., *Fig. 243*
St. Marie's, Oscott, 206
St. Martin's-in-the-Fields, London, 116
St. Pancras railway station, London, *Fig. 205*
St. Paul's Cathedral, London, 37, 115, 116
Saller, Dr., *quoted 366*
Saltash Bridge, 167
Sand, moulds, 15
  ,,   use of, 5, 59
Santiago & Valparaiso Railway, Chile, *Figs. 296, 297*
Sargon, King of Assyria, 2
Savery, Captain, 48
Saws, early iron, 6
Schubert, Dr. H., 17
Schuylkill river, 86
Scissors, cast iron, 78
Scotland, 78, 79, 162, 288, 310, 316, 374
Scott, Sir George Gilbert, 250
  ,,   Sir Giles Gilbert, 317
Scrapers, cast iron, *Fig. 390*
Scrivenor, Harry, 42, *quoted 9*
Seaside architecture, 315, 316. (See also Piers, seaside)

Sedgeley, 40
Senate House, Cambridge, 116, *Figs. 42, 43*
Service pipes, cast iron, 79, 80, 82. (See also Drainpipes, Pipes and Water Mains)
Severn, river, 105
Seymour, of Sudley, Lord, 17
Shafts, cast iron, 106
Sheffield, Sussex, 17
   ,,   & Manchester Railway, *Fig. 194*
Sheffield-Rotherham Railway, 167
Shelters, cast iron, 250, 254, 316, *Figs. 402, 403*
Sheppard, Richard, 373, *quoted 252, 282, 313, 317*
Sheradising, 370
Shop fronts, cast iron, 310, *Figs. 399, 400*
Shot, cast iron, 17
Shrewsbury, 105
Shropshire, 42, 43, 54, 56, 82, 86, 90, 100, 288
Sickles, early iron, 6
Siemens, C. W. & F., 313
Signal cabins, cast iron, 163
Signs, cast iron, 163, *Figs. 414-417*
Sills, cast iron, 54
Sinks, cast iron, *Figs. 482, 484*
Sion House, Isleworth, 116
Slag, 6, 8
Smeaton, John, 53, 66, 106
Smelting, of iron, 16, 18, 37, 42, 43, 44, 53, 66
Smethwick Foundry, 63
Smiles, Samuel, 48, 66, 374, *quoted 64, 82, 106, 163*
Smirke, Edward, 120
   ,,   Sir Robert, 54, 120, 121
   ,,   Sydney, 120, *quoted 121*
Smith and Founders Director, 116, *Figs. 276-292*
Smith, Messrs, H. B., New Jersey, 252
   ,,   R. A., *quoted 8*
Smoothing irons, 43, 59
Soane, Sir John, 54
Society of Arts, 285
Soho Works, Birmingham, 60, 62
Sorocold, 48
Southern Railway, *Figs. 196, 197*
Southwark Bridge, London, 99
Spa Road Bridge, *Fig. 191*
Spain, 8, 11, 44, 239
Spearheads, 6, *Figs. 280, 281, 337*
Spey Bridge, Fochabers, *Figs. 210-213*
Stanchions, cast iron, 54, 82, 313, 315
Stafford Bridge, Oakley, *Fig. 217*
Staffordshire, 56
Staircases, cast iron, 278, *Figs. 172, 429*
Stanford Bridge, 104
Stanton-by-Dale, 80
Stanton Ironworks Company, 79
Stations, railway, use of cast iron, 163, 169, 171, 175, 251
Staveley, 79
   ,,   Coal & Iron Company Limited, 80
Steam engine, 53, 62, 66, 73, 76
Steel, forged, 78
   ,,   production, 73, 239, 313, 315
   ,,   Roman, 5, 6, 8
   ,,   used with cast iron, 70, 116
Stephenson, George, 53, 90, 163, 167, *Fig. 178*
   ,,   Robert, 90, 163, 167, 169, 179, 186, *Figs. 179, 201*
Stevens, Alfred, 78
Stewart, D. Y., 80
Stokesay Bridge, *Fig. 96*

Stone Age, 1, 2
Stone Buildings, Lincoln's Inn, 116, *Fig. 119*
Storrie, J., 8
Stoves, cast iron, 5, 67, 76, 78, 322, *Figs. 53-56, 78, 82, 371-373, 467-470.* (See also Cookers and Ranges)
Stoveplates, *Figs. 23-26*
Stringer's Iron Foundry, 42, 47
Stuckofen Furnace, 11
Sturtevant, Simon, 18 42
Sugar rolls, cast iron, 59
Summerson, John, 117, 118, 120, 198, *quoted 104*
Surrey, 11
  ,,   Earl of, 17
  ,,   Iron Railway, 159
Sussex, 8, 9, 11, 17, 18, 37, 43, *Figs. 13-22, 29, 32, 33*
  ,,   Archæological Society, 115
Sussex Weekly Advertiser, 120
Swanwick, F., 167
Sweden, 44
Switchgear, cast iron, *Figs. 486, 487*
Sydneye, Sir Henry, 18
Sykes, Godfrey, 78
Symart, 17

T
Tables, cast iron, 278
Tai Shan, 252
Tanks, cast iron, *Figs. 440, 441*
Tarbotton, M. O., 249
Taylor, Sir Robert, 54, 116
Telephone kiosks, cast iron, 317, *Fig. 408*
Telford, Thomas, 53, 86, 90, 92, 99, 104-106, 120, 159, 186, *quoted 90, Fig. 46*
Temple, London, 117, 118, 198
Tern, 105
The Application of Cast and Wrought Iron to Building Purposes, 374
The Complete Body of Architecture, 116
The Lives of the Engineers, 374
Thetford Town Bridge, Norfolk, *Figs. 208, 209*
Thomas, Hartland, *quoted 209*
  ,,   John, 41
  ,,   Sydney, 240, 313
Thorncliffe Ironworks, 78, 99
Thos. Tilley's air furnace, 48
Thrashing machine, invention of, 92
Timber, shortage of, 18, 44
  ,,   supplies, 41, 42, 43
Tindal Bridge, Birmingham, *Fig. 222*
Tirwett, Sir Robert, 18
Tomlinson, Elizabeth, 18
Tools, cast iron, 53
Toronto, 116
Torrington, Fifth Viscount, 57
Tower of London, 17
Toynbee, A., 42
Trade associations, 357, 358, 361, 362
Trade Routes and Commerce of the Roman Empire, 8
Traffic signs, cast iron, *Figs. 414-417*
Tramways, cast iron, 54, 56, 57, 82, 161, 162
Tredgold, T., 115, 179, 192, 239
Trent Bridge, *Fig. 227*
  ,,   ,,   Nottingham, 249, *Figs. 300-302*
Trevithick, Richard, 53, 100
Trilby, 286
Troughs, cast iron, 105, 106
Trusses, cast iron, 169, 172, 175
Ts'angchou, *Fig. 4*

Turner, Richard, 202

U
Umbrella stands, cast iron, 278, *Figs. 383-386*
United States Pipe & Foundry Company, New Jersey, 252
University College, London, 198
Urns, cast iron, 70
Uses, of cast iron, modern, 322
Utensils, cast iron, 53, 76, *Fig. 505*
,, cooking, 11

V
Van Cullen, 17
Varnish finishes, 372
Vases, cast iron, 70, 278
Vauxhall Bridge, London, *Fig. 105*
Verandahs, cast iron, 70, 278, *Figs. 122-126, 280, 349*
Versailles, 79
Vertical casting of pipes, 80, 81
Viaducts, cast iron, *Figs. 296, 297*
Victoria railway station, London, *Figs. 320-325*
Vignolles, C., 167
Vintners Hall, London, *Fig. 278*
Virgil, 6
Viroconium, 8
Vitreous enamelling, 370, 372
Voysey, C. F. A., 279

W
Wadhurst, 11, *Figs. 10-12*
Wales, 8, 44, 78, 162, *Fig. 37*
Walker, Aaron, 73
Walker colliery, 49
Walker, Jonathan, 73
,, Samuel, 73
Walton, George, 279
Ware, Isaac, *quoted 116*
Warehouses, use of cast iron, 117, 194, 239, 313
Warrington Road Bridge, Lancs., *Fig. 224*
Washington, the Capitol, 198
Wasterkirk, 86
Water heaters, cast iron, *Figs. 454, 455, 465*
,, mains, cast iron, *Figs. 76-79*
,, power, 44, 49, 66, 73
,, wheel, *Fig. 39*
Waterloo Bridge, Bettws-y-Coed, 90, *Figs. 92, 93*
Watt, James, 53, 60, 66, 73, 94, 106
Weald, Sussex, 18, 33, 37
Wear Bridge, Sunderland, 73, 86, 99, 104, *Figs. 83-85*
Weardale, 9
Welbeck Abbey, *Fig. 238*
Wellington, Shropshire, 11
Wenlock, 9
Westminster Bridge, London, *Figs. 233-235*
Wheels, cast iron, 78
Whitehaven, 162
Wilkins, William, 54, 198
Wilkinson, Isaac, 44, 58, 59
,, John, 53, 59, 60, 62-64, 82, 105, *Figs. 47-49*
,, William, 48, 53, 59, 60, 62, 63
Willey, 44
Williams, Frederick S., *quoted 172, 188, 189*
Wilson House, Lindale, 59, 64
Wilson, T., 86
Winbolt, S. E., 8

Windmill shafts, cast iron, 106
Window bays, cast iron, *Fig. 407*
   ,,   frames, cast iron, 64, *Figs. 394, 395, 397, 404*
   ,,   guards, *Figs. 284, 285*
Windows, cast iron, 54, 278, 310, 316, *Figs. 396, 426*
Winnington, Sir Edward, 100, 104
Winter gardens, cast iron, 251
   ,,   Palace, Dublin, *Fig. 308*
Woodside railway station, Birkenhead, 251, 252, *Figs. 317-319*
Wookey Hole, 6
Works in Iron, 374
The World, 54
Wornum, Grey, *quoted 71*
   ,,   Ralph N., 209, 250, *quoted 208*
Worth, Sussex, 17
Wreaths, cast iron, *Fig. 291*
Wright, Thomas A., 3
Wrought iron, 3, 10, 11, 16, 56, 70, 115, 116, 179, 186, 239, 247, 249, 250
Wurttenberg, 81
Wyatt, W. D., 172

Y

Yates, William, 99
York & North Midland Railway, 167

Z

Zimmer, G. F., 1